AF462640

The Engine of Scientific Discovery

The Engine of Scientific Discovery

How New Methods and Tools Spark Major Breakthroughs

ALEXANDER KRAUSS

OXFORD
UNIVERSITY PRESS

Oxford University Press is a department of the University of Oxford.
It furthers the University's objective of excellence in research, scholarship,
and education by publishing worldwide. Oxford is a registered trade mark of
Oxford University Press in the UK and in certain other countries.

Published in the United States of America by Oxford University Press
198 Madison Avenue, New York, NY 10016, United States of America.

CIP data is on file at the Library of Congress.

ISBN 9780197829790

DOI: 10.1093/oso/9780197829790.001.0001

Printed by Marquis Book Printing, Canada

The manufacturer's authorized representative in the EU for product safety is
Oxford University Press España S.A. of Parque Empresarial San Fernando de Henares,
Avenida de Castilla, 2 – 28830 Madrid (www.oup.es/en or product.safety@oup.com).
OUP España S.A. also acts as importer into Spain of products made by the manufacturer.

Dedicated to those who trigger new discoveries—and those who want to understand how to trigger them

How do we spark new scientific discoveries? Why do some breakthroughs seem even accidental? And most importantly, how can we accelerate them and push the boundaries of science? These are some of the biggest unsolved questions in science. Many believe breakthroughs arise by chance or serendipity. But here we show that it is powerful new tools and methods that enable discovering what we often did not even know existed: improved microscopes uncovered the hidden world of microorganisms and viruses, x-ray methods unlocked the structure of our DNA, particle accelerators detected subatomic particles that make up our world, and advanced telescopes revealed galaxies we never imagined. The Engine of Scientific Discovery explores, for the first time, science's biggest discoveries—spanning all nobel-prize and major non-nobel discoveries throughout history. What we find is surprising: we have triggered science's over 750 major discoveries by first inventing a new method or tool that made the breakthroughs possible—often within just a few years. This is because new tools enable us to see, measure and understand the world in ways that are impossible without them. From discovering planets beyond our solar system that reshape how we understand the universe, to CRISPR gene editing that transforms how we fight disease and cancer. This consistent pattern reveals a powerful insight: our transformative new methods and tools are The Engine of Scientific Discovery—a fundamental principle of scientific progress that has been overlooked until now. By shifting our focus to inventing new discovery tools, we can spark a method revolution in science. Instead of waiting for breakthroughs, we can actively design and build new tools of discovery. What if the next great breakthroughs depend not just on asking better questions, but on developing better tools to ask and answer them—on entirely new ways of discovering? A new theory of discovery emerges here, offering a roadmap to accelerate the pace of new breakthroughs across science. This book is for anyone who wants to understand how we make discoveries—and what drives science.

Alexander Krauss is an interdisciplinary researcher focused on the science of science, metascience, innovation and economics of science. He is the author of *Science of Science* (Oxford University Press) and is a Research Associate at the London School of Economics and an Affiliated Professor at the Barcelona School of Economics and the Spanish National Research Council. After finishing his PhD, he worked with governments and the World Bank for five years—on topics from global health to energy and agriculture—and then taught at University College London.

Contents

PART III THE PRESENT LIMITS AND FUTURE OF SCIENCE AND DISCOVERY

Introduction

In the early 20th century, one of biology's biggest mysteries puzzled scientists: what is the structure of DNA? The great leap forward came with the development of x-ray crystallography in 1913—a method that makes exploring the structure of molecules possible. But it was not until the 1950s when Rosalind Franklin applied this powerful method to the DNA puzzle. At King's College London, she captured the first images that revealed the missing piece: DNA's double-helix structure.[1] With the key x-ray image in hand, Francis Crick and James Watson at Cambridge quickly published their famous one-page paper in *Nature* proposing the double-helix model in 1953. They shared the Nobel prize for the discovery together with Maurice Wilkins—a few years after Franklin passed away from cancer. It remains one of the greatest discoveries of the 20th century, and it was surprisingly missed for decades.

Also in the early 20th century, another major unsolved puzzle persisted but in astronomy: is our universe static or expanding? Many believed space was fixed and unchanging. But the mystery was uncovered in 1929 when Edwin Hubble—a law graduate who also studied astronomy—used a powerful new telescope, the world's largest at the time. At Mount Wilson Observatory in California, he discovered galaxies speeding away from us in every direction. The universe was expanding outward. This extraordinary breakthrough transformed how we understand the universe and our place in it.[2] These two discoveries—one in the hidden world of molecules, the other across the vast and hidden expanse of space—seem worlds apart. But both were driven by something fundamental: new ways of seeing that were impossible before. After Hubble's discovery, Einstein—who had introduced a cosmological constant to his equations to maintain a static universe—abandoned the constant and adopted Hubble's new understanding.

So how do we spark new discoveries in science? Why do some of our greatest breakthroughs seem to happen by chance or serendipity? And how can we accelerate them and break science's current limits? These are some of the greatest unsolved questions about science. While many credit flashes of serendipity, large teams or big funding, we uncover here a very different story: we trigger new discoveries by inventing new methods and tools that enable us to see, measure and imagine the world through entirely new lenses. From new electron microscopes uncovering viruses and nanoparticles, to statistical methods revealing hidden patterns in complex climate and genomic data, methods and tools discover what we often did not even know existed until we created them. To detect planets beyond our solar system, we first had to invent space telescopes beyond earth's atmosphere. To map the brain's activity, we required first creating high-resolution MRI scanners. In fact, these breakthroughs would be unimaginable without these advanced tools.

The Engine of Scientific Discovery. Alexander Krauss, Oxford University Press. © Oxford University Press (2026).
DOI: 10.1093/oso/9780197829790.003.0001

But taking a step back, how do scientific discoveries transform our lives in the first place? Many see our major scientific advances as humanity's greatest hallmark. Alone in the 20th century, our life expectancy almost doubled globally through vaccines and antibiotics, through better food production and improved hospitals. Breakthroughs shape the quality of our lives through electricity, the internet and smartphones that connect us to people around the world—and computers that shape how we think and work. Beyond our immediate lives, science has also vastly expanded how we view reality. We made remarkable discoveries that explain how the universe emerged about 14 billion years ago. Over the centuries, we revealed fundamental forces governing physical reality—gravity, electromagnetism, the strong force and the weak force. We uncovered the periodic table that captures the chemical elements that make up the world. We found that DNA carries genetic information that determines how living beings grow and function.

Most of this knowledge did not even exist a few hundred years ago. Our greatest scientific advances have happened in a very short span of human history since the 1700s—a period when disease, premature death, malnutrition and extreme poverty have reduced faster than ever before. Our breakthroughs have not just driven this progress but changed the way we think about life, time, space and our relationship to the world. Studying these breakthroughs and the methods behind them can also change the way we think about science and discovery itself. After all, our scientific innovations have even shaped our very sense of who we are and what is possible. Yet there is a paradox here: what brought us electricity and crop surpluses also partly contributed to global challenges like climate change, overpopulation and depleting resources. We once again turn to science to try and safeguard us—developing climate modelling tools and carbon capture systems to tackle these pressing challenges.

So if nothing shapes our lives more than science, why do we still understand relatively little about how scientific discovery works? We know a lot about how stars and planets form, how life evolves and how atoms behave. But we know relatively little about how new breakthroughs, new scientific tools and new scientific fields form, evolve and behave. And yet, they are what make those very discoveries possible. In fact, the nature of science and discovery is among science's biggest questions along with the nature of life, matter and the universe. But science and discovery—the force behind uncovering all of them—are arguably the least understood. So one of science's biggest puzzles is itself. Scientists are usually busy doing research and rarely have the time to step back and study these deeper questions. Yet this is the fundamental unsolved puzzle guiding much of this book: how do we spark new discoveries—and even entirely new scientific fields? This is also the central question in the field of *science of science*—and answering it would be the key advance in the field.[3–5] It would also enable us to make new breakthroughs faster. But there are two big challenges researchers face on this puzzle: first, it is difficult to study discoveries at scale and systematically—across scientific fields and history; second, it is also difficult to develop a general theory of discovery that explains what triggers breakthroughs, backed by such systematic evidence across science. This book takes on and aims to tackle these two challenges: with new evidence and a new theory.

So we return to the question, what actually triggers new discoveries? Most people have their own idea of how they emerge. Some researchers argue that breakthroughs arise from funding,[6–8] others highlight teamwork and collaboration.[9–11] Some emphasise more productive younger researchers,[12–14] others stress revolutionary theories.[15,16] Still others point to unpredictable moments of serendipity, insight or chance.[17–20] These different explanations are usually based on studying one individual piece of the puzzle—a particular sample of breakthroughs with a particular method and particular lens. So this leads to a deeper question: how can we better study and uncover what actually drives science's big breakthroughs? We argue that the best way to tackle the puzzle is by studying science's major discoveries as systematically and comprehensively as possible—all nobel-prize-winning discoveries in science and other major discoveries that were made before or did not receive a Nobel prize (as we unpack in the next chapter). These make up science's over 750 biggest discoveries that form the foundation of the physical, biological and social sciences. These breakthroughs define science: from quantum mechanics, relativity theory and Newtonian laws of motion, to the human genome, the structure of DNA, Mendelian genetics and the theory of evolution, to the periodic table of elements and the expanding universe.

Why is uncovering how we trigger discoveries so important? Understanding the engine of science and discovery is fascinating and crucial, not just because science and technology have revolutionised and redefined how we live and how long we live. But also because discoveries often seem unexplainable and random—even mysterious. In fact, most people want to understand how discoveries happen: citizens want to know how future breakthrough medicines and technologies can affect our lives. Researchers try to predict what questions we should focus on that could be the next big breakthrough—and impact science, society and policy. Funding agencies explore project proposals and academic journals assess submitted articles by trying to predict which ones may have the largest impact. University hiring committees try and predict which researchers are most likely to make the next big discoveries. Companies want to understand what breakthroughs are most likely to lead to innovative and profitable technologies and products. Governments try to forecast which research areas hold the greatest potential for new discoveries when determining where to spend public resources—and what challenges to prioritise: from fighting cancer and preventing future pandemics to slowing global warming.[4]

Major breakthroughs attract attention more than any other aspect of science. And there is tremendous demand across society to make discoveries more predictable and more frequent. Yet even with proposed explanations like serendipity, funding and teamwork, the paradox is that when zoomed in on individual scientists, breakthroughs often seem random and unpredictable. But what we uncover here challenges that belief: by zooming out and studying the aggregate factors behind science's major discoveries across fields and history, we reveal a common pattern and driving force we can actively shape.

In schools and universities, we are largely taught what science has discovered: the periodic table of elements, the theory of relativity, DNA's double-helix structure. We do not generally learn how we develop those discoveries. We celebrate the outcomes, not the process. Even leading scientists who study science focus on the end results, tracking article citations and publications.[3,4,14,21–24] And leading historians and philosophers of science explore final scientific theories.[16,17,25–28]

What have researchers said about scientific discovery so far? The most well-known and cited philosopher of science, the Austrian Karl Popper, published a highly influential book *The logic of scientific discovery*. Ironically, in that seminal book, Popper denies the very possibility of a logic of how scientific discovery emerges.[17,20] For him, 'there is no such thing as a logical method of having new ideas ... every discovery contains "an irrational element"'.[17] So he instead just focuses on how to justify scientific theories. As the nobel-prize-winning physicist Carlo Rubbia also said, 'Scientific discovery is an irrational act'. In fact, many researchers believe discoveries are blind and unpredictable—arising as a sudden flash of insight or a random event that we cannot study systematically.[17,18,20,29–32] This popular narrative catches the attention of scientists and the public—and they make for exciting stories.[19,33,34]

Other researchers who study scientists' careers also highlight that serendipity may drive breakthroughs.[30,35,36] Some argue that a scientist's most impactful work may emerge seemingly randomly across their career.[12,35,37] These findings have led scientists to the conclusion of 'the profound unpredictability that pervades many aspects of scientific discovery'.[4] Yet claiming that there is no logic or method of discovery, but that discovery is explained by serendipity or chance, is not actually an explanation at all. And more importantly: it is misleading and can waste researchers' time and resources. By systematically analysing science's over 750 major discoveries, we uncover a surprising and overlooked pattern here. Discoveries commonly labelled as serendipitous were actually sparked by using—for the first time—a new tool that enabled the unexpected observation we were not even looking for: *an improved microscope revealed cells, a discharge tube uncovered x-rays, gel diffusion revealed the Hepatitis B virus and a sensitive spectrograph detected the first exoplanet—a planet outside our solar system.* We find that discoveries that seem serendipitous at the time follow shortly after we create the new tool that sparks the discovery. What looks like chance is often just using a powerful new lens—as in each of these extraordinary discoveries. This is important and can change how we think about discovery: it is powerful new tools that help us strategically plan and predict new discoveries.

Some people, when thinking about discoveries, narrow the lens on big theories—like Einstein's relativity and Newton's laws. The most influential and cited explanation of scientific progress to date came from the American historian of science Thomas Kuhn in his landmark book: *The structure of scientific revolutions.* Kuhn was deeply interested in trying to explain how science evolved—and he focused on the evolution of scientific theories.[15] He argued that science does not progress cumulatively; instead, it is driven by radical paradigm shifts—fundamental changes in the theories of a field. Kuhn's bold hypothesis was based on a handful of theories largely in physics up to the early 20th century.[15] Since Kuhn, this central question has sparked heavy debate—and yet, no consensus or general theory yet exists. But by shifting the focus here to our powerful methods and tools that enable new breakthroughs—including theories—a very different picture emerges. No major methods and tools we use across fields have been entirely abandoned—from statistics and telescopes to centrifuges. No major scientific fields have been entirely replaced—from biology and chemistry to computer science. Instead, we keep extending and refining them. We uncover here that how science progresses is deeply cumulative.

The latest contribution to studying scientific progress comes from influential scientometricians and network scientists—using big data to study science. They track science through publication counts and citations (the number of times a paper is cited by other papers, as a rough measure of its impact). These scientists explore thought-provoking insights about the dynamics of science such as scientists' careers, research productivity and networks of scientists.[3,4,9,14,24,38,39] They also explore team collaboration—for example finding that individual researchers and small teams are more likely to generate breakthrough research than large teams.[10] But even these researchers recognise the limits of their approach in explaining science: 'this bias toward citations is reflective of the current landscape of the field, [and] it highlights the need to go beyond citations as the only "currency" of science' and high-impact research.[14,38,40] In fact, 'we may now be approaching the limit of what they can teach us about the scientific ecosystem and its production of discoveries'.[4]

What is this book's new and alternative approach? Until now, most researchers studying how science works focus on its outputs—exploring a sample of breakthroughs,[41–44] studying scientific theories[16,17,28] or tracking publications and how often they are cited.[10,12,22,39,45] But to tackle these fundamental questions, we have to take a completely different strategy—one that has been overlooked until now: here we study science's major discoveries, the methods and tools they used and the broad range of traits of the discoverers themselves—across science, at the same time.[38,40] Surprisingly, what drives science's greatest discoveries has not yet been studied in a systematic and comprehensive way across science and history and no studies have yet systematically examined the role of new methods and tools in sparking those discoveries.[3,4,14,15,17,22–24] This approach offers a powerful new lens and theory of discovery.

Sparking a *method revolution* in science

Our powerful new methods and tools trigger breakthroughs by enabling us to see, measure and imagine the world in ways that were impossible before them. Microscopes and telescopes opened hidden worlds too small or too distant for the naked eye—revealing unexpected parasites, bacteria and nanoparticles, and uncovering new distant star clusters and galaxies. Statistical methods stretch the limits of our mind, allowing us to process and analyse enormous amounts of data. They uncover new patterns we could not see otherwise—from tracking gene mutations to identifying climate trends. X-ray, MRI, ultrasound and CT devices make it possible to generate extraordinary images inside our body without a single incision. They have revolutionised medicine by letting us better diagnose, monitor and understand disease.

Spectrometers and radar devices enable capturing and measuring light, radio waves and signals that our senses cannot detect—from astronomy to cancer detection. Controlled experimental methods allow us to test treatments for countless diseases—from tuberculosis to scurvy—in rigorous ways that have saved millions of lives and were unthinkable before. Computer technology lets us simulate complex systems—from the earth's climate to neural networks—and even predict the spread of pandemics that would otherwise be too vast to grasp. AI and machine learning methods enable analysing immense datasets, can perform experiments and make predictions based on spotted patterns—with exceptional speed. New tools give us new questions to ask and new ways to answer them. Each powerful method reveals a new layer of reality that we often did not even know existed.

In fact, the efforts of millions of minds, eyes and hands studying the world are rarely as powerful as someone developing or getting their hands on a new tool. A single invention—whether an advanced telescope or x-ray method—gives us a new lens to perceive and probe further and deeper than humans can alone. It gives us a new way to trigger multiple discoveries with a single tool.

We use the microscope, throughout the book, as a prime example of a powerful discovery tool—transforming what we are able to uncover. The first microscopes radically amplified our limited vision to reveal a vast microscopic world we did not expect. This optical breakthrough was continually extended and refined over time. Eventually came the electron microscope: one of the most powerful instruments in the history of science.(46) A 26-year-old German graduate student Ernst Ruska invented the instrument in 1933 at the Technical University of Berlin. By using beams of electrons instead of light, the electron microscope depicts images of miniscule objects. And it vastly expanded our world of microorganisms, molecules and nanoparticles.(47,48) In fact, this tool sparked over a dozen major discoveries—from the cell structure in 1945 to ribosomes in 1955. What unites these discoveries is not a shared theory, research teams or more funding, it is a shared tool—the electron microscope—that made these breakthrough findings possible across different fields. It also launched new fields like modern cell biology that would have been impossible without it. Here we uncover how the story of science is largely driven by inventing and reinventing powerful methods and tools—from CRISPR gene editing to space telescopes—that trigger our big breakthroughs.

What are the top ten methods and tools most used to spark discoveries? Examining science's over 750 major discoveries, we identify the top ten—and highlight them in Figure 0.1. Interestingly, the top ten methods and tools are almost the same when looking at all nobel-prize discoveries only, except that particle accelerators and particle detectors make it into the top ten, while telescopes and thermometers do not. Take electrophoresis and chromatography methods—before inventing them, we faced many barriers to analysing chemical substances. But once developed, they enable us to separate and determine the composition of DNA, protein molecules and viruses—in ways impossible before. Electrophoresis was developed by the Swedish graduate student Arne Tiselius at Uppsala University—at just 28. It enabled making major discoveries and another powerful method: DNA sequencing.

Then there is partition chromatography—one of the most important methods in chemistry and biology. It was invented by the British graduate student Richard Synge at just 27, alongside his colleague Archer Martin at 31. By separating and analysing complex chemical mixtures, this extraordinary method enabled discoveries from the structure of insulin to how carbon dioxide is absorbed in plants. Today, we rely on it in many areas that affect our everyday lives—from pharmaceuticals to food safety and environmental testing. Surprisingly, Archer Martin, in his Nobel prize speech describing the method, said that: 'All of the ideas are simple and had peoples' minds been directed that way the method would have flourished perhaps a century earlier'.[49] What is remarkable is that we find many of these cases across science: the world-changing potential of tools that can sit unnoticed for decades simply because not enough researchers are dedicated yet to developing tools. Yet adopting tools much later comes with important opportunity costs—and delayed impact for science and society. It is also striking that these groundbreaking inventions—among the most important tools in science—were made by young graduate students doing hands-on experimental work as part of their PhDs. This tells us something crucial about how discovery works: progress is often not about greater funding, bigger teams or big theories. But about researchers refusing to accept the limits of existing methods and tools, and building new ones that unlock entirely new ways of seeing, testing and understanding—as we will explore throughout the book.

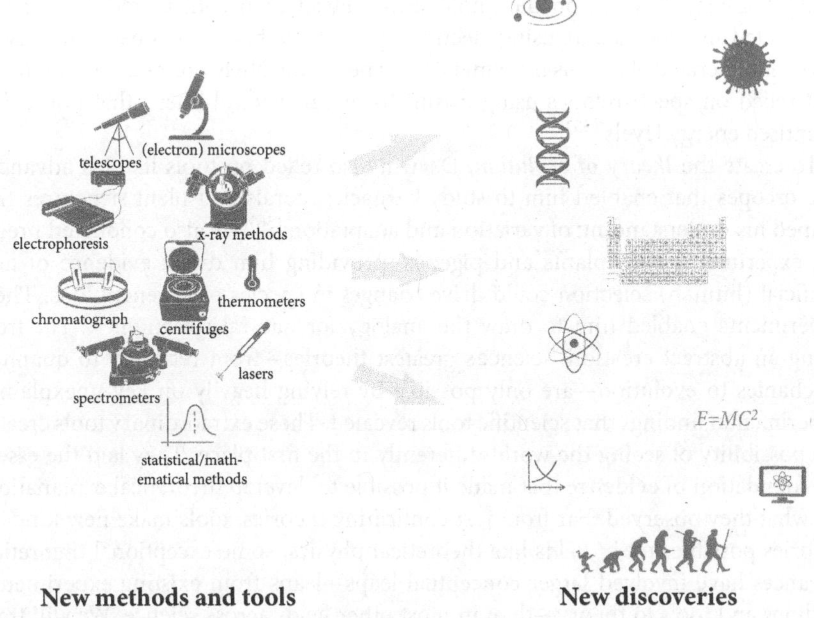

Figure 0.1 *New methods and tools trigger new discoveries*

Here we will be unpacking the fascinating stories of how and who developed our most powerful tools of discovery—and how they have revolutionised science by enabling completely new lenses to the world. Our ability to understand the world is limited by the tools we have developed so far. By the ways we can presently observe, measure and uncover—whether neural synapses, viruses, radio waves, the human genome or exoplanets that were unimaginable before our tools. For the tools we have created up to now shape the questions we can ask, the answers we can provide and the way we can explore and understand the world.

But are there major theoretical breakthroughs that are mostly just sparked by intuition or insight? What about some great theories of science—are relativity, quantum mechanics, evolution mainly a product of thought? Some believe they are. Here we will see how Einstein developed the *theory of relativity* by building on unexpected experimental findings, made possible by an innovative new instrument: the interferometer. Those puzzling findings showed that the speed of light remained constant regardless of the Earth's motion through space. This extraordinary evidence challenged the existing notions of absolute space and time.[50] It enabled developing Einstein's theoretical interpretation. In Einstein's words: 'the Michelson and Morley experiment had made it clear that phenomena obey the principle of relativity'.[51,52] Tensor calculus, a new mathematical method developed at the time, was also indispensable to capture the experimental findings and express gravitational fields in Einstein's equations—what we will unpack later. To develop *quantum theory*, Max Planck also did not just rely on thought, but took the first step by analysing unexplained findings from experiments on blackbody radiation, made possible with powerful tools—spectrometers and bolometers—that measured emitted light.[53] Einstein's work on the photoelectric effect then contributed further and was rooted in experiments using electroscopes and cathode ray tubes that showed how light liberated electrons from metal.[54] Then came Niels Bohr's quantum model that relied on spectroscopes using prisms to reveal spectral lines—that pointed to quantised energy levels.[55]

To create the *theory of evolution*, Darwin also relied on tools like his advanced microscopes that enabled him to study barnacles, corals and plant structures that shaped his understanding of variation and adaptation.[56] He also conducted breeding experiments with plants and pigeons, providing him direct evidence of how artificial (human) selection could drive changes in species over generations. These experiments enabled him to draw the analogy for natural selection.[57] Far from being an abstract creation, science's greatest theories—from relativity to quantum mechanics to evolution—are only possible by relying heavily on key unexplained experimental findings that scientific tools revealed. These extraordinary tools created the possibility of seeing the world differently in the first place. They laid the essential foundation of evidence that made it possible to develop theoretical explanations for what they observed. Far from just confirming theories, tools make new kinds of theories possible. Yet in fields like theoretical physics, some exceptional theoretical advances have involved larger conceptual leaps—leaps from existing experimental findings and tools to theory—than in most other fields across science. We will trace the origins of science's biggest breakthroughs in history as we work through the book.

Taking a step back, a striking insight emerges: each method and tool gives us a unique lens to seeing the world. The sum of these lenses largely determines how far we can currently see, explore and think—our present limits of science. To better understand the world and science, we have to better understand the tools we invent to study the world and do science. Because accelerating discovery depends on how well we understand our powerful methods and tools and how we develop them—as they lay the foundation that science is built on.

So how do we build strong evidence and develop a general theory? The National Research Council in the US confirms: 'No theory of scientific progress exists ... that allows prediction of the future development of new scientific ideas' but researchers instead often emphasise how uncertain discovery can be.[58] But there is a better approach to tackle the puzzle—one grounded in rich evidence. In general, there are two powerful ways to develop strong evidence and a general theory. The first is to explore the entire landscape—the total population as comprehensive as possible, not just a subset. That is what we do here: rather than typically studying a sample, we examine science's over 750 major discoveries across fields—all nobel-prize and major non-nobel discoveries. (We will unpack these discoveries in the next chapter.) By studying this much larger population, we get a clearer, more accurate picture of what actually drives discovery—by spotting the general patterns. In other words, what science's biggest discoveries have in common—and what they do not. And these two independent data sources enable us to test and compare the strength of the patterns and findings.

The second way is to approach the problem from multiple perspectives. Instead of relying on a single method, we use different methods and perspectives from diverse fields—from statistics and history of science to cognitive science, and beyond. This gives us multiple, independent ways to test and confirm the patterns we find—and the strength of the findings. Rather than just relying on one method that can be more narrow or have blind spots. This combination—using the most comprehensive data possible and diverse methods—lets us fill the gap: a general theory of how discovery emerges.

Here we develop the *new methods-driven discovery theory.* What the theory explains is straightforward: new methods and tools unlock new discoveries by enabling us to see, measure and understand the world in ways that are impossible without them. The ultimate test of a successful theory is: can it explain what drives science and discoveries—and can it advance and better predict discoveries? On this test, this theory can take us further than current explanations—as we will see. It can explain our past discoveries—and improves predicting future discoveries: the speed and direction of method innovations we make give us powerful signals that breakthroughs are soon to follow and where they can come from. The central thread connecting the different chapters of the book is exactly this expanding toolbox of ours.

Mapping discovery: the book's structure and key insights

This is the first book to systematically tackle the puzzle of what the engine of scientific discovery is. We analyse science's over 750 biggest discoveries—spanning across fields and history—and the broad range of traits of the discoverers, and provide a new lens on science's powerful methods and tools. This is an until now unexplored perspective to discoveries.[4,14,15,17,38,40] The book takes a big picture view: to better understand science, its methods, its limits and its future, we have to study them systematically, combining methods and evidence from across fields. So what exactly do we cover in the book?

Chapter 1: How do we spark new discoveries? To tackle the puzzle, we study science's biggest discoveries and link them to the methods and tools they used and to the discoverers themselves. We find a surprising pattern: across fields, science's major discoveries are sparked by a new method or tool that enables exploring the world in entirely new ways—from revealing unexpected bacteria to unknown particles. Often within just a few years. We uncover that powerful tools—from new x-ray devices and electron microscopes to statistical methods and spectrometers—have each unlocked dozens of major discoveries that were impossible without them. Inventing the laser for example triggered breakthroughs across fields that impact our lives from laser surgery to fibre optics—the backbone of high-speed internet and telecommunications we rely on. In Chapter 1, we lay out the core analysis that the other chapters build on—giving us a powerful new way to understand and accelerate discoveries. This leads us to explore how common expected and unexpected discoveries are in the next chapter.

Chapter 2: What role do serendipity and chance play in discovery? For many, this is the most puzzling and mysterious feature of discovery—because it seems unpredictable. By analysing science's major discoveries, we uncover a striking pattern behind the surprise: breakthroughs often seen as serendipitous are actually triggered by using a powerful new tool, for the first time, that makes the unexpected observation possible. From pulsars (neutron stars) using a new massive radio telescope, to superconductivity using a new liquefier. New tools systematically enable other researchers to arrive at the same findings applying these same discovery tools. Serendipitous discoveries generally reduce the role of theory as they are sparked by new tools before anyone knew what to look for or how to explain it. In short: new tools are what commonly unlock surprise. It is far less about stumbling upon serendipitous insights and far more about engineering discoveries by constructing better ways to see, sense and think—and accelerate unexpected findings. So what if much of the randomness and serendipity associated with discovery so far has not come from the scientific system, scientists and their minds, or even unpredictable timing—but from something else entirely: not yet focusing deliberately on our tools' blind spots? Understanding the role of surprise in discovery brings us to the next chapter, where we explore just how cumulative science actually is—and how our discoveries fit into the overall picture of scientific progress.

Chapter 3: Does science mainly go through revolutionary paradigm shifts—or is scientific progress mainly cumulative? Here we test this influential hypothesis of abrupt paradigm shifts—like the Copernican revolution. We uncover a remarkable pattern. Three key measures of scientific progress—science's major discoveries, methods and fields—each point to the same finding: the deeply cumulative evolution of science, not sudden ruptures that fundamentally challenge and overturn what we believe. In fact, we find that we have not entirely replaced any major scientific methods or tools we use across fields—from spectrometers to lasers. Likewise, we have not entirely abandoned any major scientific fields—from nuclear physics to biomedicine. They are not subject to radical paradigm shifts; rather, we expand them over time. Scientific discoveries are also highly cumulative: only 1% of over 750 major discoveries have ultimately been abandoned. We also explore the classic scientific method of testing hypotheses by observing and experimenting: is it compatible with the cumulative nature of science—and did science's major discoveries follow it? After examining the inner workings of science, we explore the broader external factors in the next chapter.

Chapter 4: How does the broader environment—demographics, institutions and resources—help support the scientific process, and our powerful new tools? And more intriguingly, who are the scientists behind our major discoveries? By studying science's biggest discoverers and their broader traits, what emerges is a unique portrait of the greatest scientists in history. This also enables spotting what factors only apply to certain breakthroughs—and what connects across them. Strikingly, we find that interdisciplinary scientists, those with broader method training, are behind about half of science's major discoveries. The German Konrad Bloch for example completed degrees in chemical engineering and biochemistry, enabling him to apply new isotopic labelling methods to discover how cholesterol is regulated in our body. We also find that younger, more motivated scientists are the ones behind most of our greatest advances. And surprisingly, most discoverers work at less-known institutions at the time of their discovery. This flips the common assumption about where discovery happens—not just at top-tier institutions. We then broaden the lens from individual discoveries to also study scientific fields in the next chapter.

Chapter 5: How do new scientific fields emerge? Fields embody our greatest scientific advances and accumulated bodies of knowledge. Here we trace the origins of science's major fields including hundreds of fields spanning across science, offering a unique opportunity to uncover the common force driving them. We find that the speed of opening new research domains is mainly determined by how quickly we create new tools: particle detectors launched high-energy physics, microscopy techniques triggered neuroscience and randomised controlled trials kick-started experimental economics. The pace at which science expands is not random. For tools enable a completely new perspective to the world—and without them, the fields would not be possible. Our extraordinary development of new statistical techniques, x-ray devices, microscopes and spectroscopes has fuelled a wave of new disciplines: each enabling over ten new fields to emerge. In fact, about a quarter of fields are themselves the new method or tool—from laser physics and computer science to x-ray crystallography and econometrics. After uncovering how powerful tool innovations spur scientific advances, the key question we need to answer to close the logical circle is how we make tool innovations themselves—the topic we tackle in the next chapter.

Chapter 6: How do we create new methods and tools—that spark new discoveries and fields? To answer this, we begin by tracing the remarkable stories behind science's greatest toolmakers: from Ernest Lawrence's particle accelerator to Theodor Svedberg's centrifuge. In this foundational chapter, we bring together some of the key pieces of evidence. We show that—without powerful tools to detect or measure—more funding and collaborations are not enough, our scientific theories cannot be meaningfully developed or tested, and unexpected or serendipitous observations are not possible. So what if we no longer wait for new discovery tools to emerge by chance? How many big breakthroughs are we missing because we have not yet begun prioritising tool innovation? Here we map out the crucial pathways to create new tools—the discovery engine. We lay out a map of scientific methods—explored and unexplored method combinations that can help uncover and predict untapped opportunities. We introduce the idea of setting up methods labs and hubs—as incubators of innovation—that catalyse tool creation. What emerges is a new general theory of discovery—one that can provide a foundation for a new field targeted to developing methods: the *Methodology of Science*. After tackling the question of the foundations of science in Chapters 1–6, we then take a bigger leap—and explore the deeper origins of how we started science in the next chapter.

Chapter 7: What are the early origins of science and our methods? Is the engine of science today the same as it was in our past? We find that the key turning point in scientific history came with the unexpected invention of tools that enhance human perception: the first microscope and telescope. They opened unimaginable new worlds and sparked most major discoveries of the 17th century. They put science on a no-return path of rapid discovery, triggered wider curiosity and kick-started modern observational and experimental science. *Before* their invention, science was limited by ancient tools and human senses—but their arrival broke those barriers, giving us direct access to microscopic and cosmic realms we never imagined. These extraordinary instruments did not emerge from scientific institutions, a large scientific community or formal scientific method. Instead, they inspired them—*after* observing the extraordinary discoveries these tools made possible. But to understand how we got there, we go further back: how did we develop our ability to do science in the first place? We cannot study what triggered our advances throughout human history as systematically as we can study contemporary discoveries using statistics. But we can trace what has driven dozens of transformational early breakthroughs—from the agricultural revolution to the scientific revolution. So while we, throughout the book, examine scientific methods and tools systematically applied in research—such as computational methods and x-ray devices—in this and the next chapter, we broaden the scope to study more basic, general methods and tools that go far beyond the narrow scientific context. And we turn to how we developed these method-making abilities in the next chapter.

Chapter 8: Can we best explain our success, throughout history, by our unique human capacity for method-making—enabling us to better meet our needs, solve problems and develop vast knowledge? Throughout history, we have continually refined our abilities, to invent navigation techniques using stars, apply plant-based medicines to heal wounds, devise lunar calendars to predict seasons—and eventually develop farming techniques and early mathematical systems. Each method gave us a survival advantage. Eventually, we developed the microscope and telescope that enabled vastly surpassing the limits of our evolved mind—triggering the discovery of entirely new stars, moons and medicines. The evolution of this capacity for method-making and tool-making marks one of the most fundamental changes in human history—and it enables science today. But what made this possible? How did we shift from our unaided mind to developing increasingly cumulative methods? While evolution gradually honed and refined our inherent methodological abilities to observe, think and imagine throughout human history, we eventually began honing, refining and amplifying them into methods and tools. The bigger the innovation in method, the bigger the leap in progress. Our success is a remarkable story of realising the power of leveraging this capacity to conceive better tools that enable us to think, test and predict better. We are complex method-makers who extend our own minds. Bringing together the evidence throughout the book leads us to the question of how we can best develop a general theory of discovery—a question we explore in the next chapter.

Chapter 9: Can we develop a general theory of discovery, backed by different types of evidence and methods across fields? We cannot design an experiment to measure what our tools cannot detect. In science, the tools we use largely determine how and what we can study—and the questions we can ask. Powerful new tools—from super-resolution microscopes to the gene-editing method CRISPR—in turn unlock entirely new types of questions and discoveries that were not imagined before. Here we take a step back and look at the bigger picture, merging and expanding the insights from earlier chapters—with new perspectives. While the search for grand explanations of how science progresses has largely been abandoned since Kuhn's concept of paradigm shifts, we expand the general theory of scientific progress here: *the new methods-driven discovery theory*. This theory is backed by diverse kinds of evidence from five different domains: statistical analysis of science's major discoveries, evolutionary biology, cognitive science, scientific practice, and philosophy of science. Integrating the five different explanations, this unifying theory offers a new understanding and roadmap for discovery. It is commonly about overcoming the limits of our methods, mind and senses with powerful new tools that outthink, outsee and outimagine our own minds. Our tools do not just assist science, they largely define it. This leads us to explore our scientific bounds in the next chapter.

Chapter 10: What are the current limits of science—and what exactly shapes those limits? Can we keep pushing forward into the edges of the universe, atoms, genes and human societies? These are key unanswered questions, but the answers are often less mysterious than they seem. So here we unpack our present boundaries—after we tackle the foundations and origins of science in earlier chapters; and the three puzzles are deeply interconnected. Yet when studying our boundaries, researchers commonly explore *what* we are not yet able to explain—from dark matter in physics, to how life began in biology. But here we explain *how* our current scientific frontiers are shaped by five factors: our *scientific methods* and tools set the present limits of what we can observe, test and discover in most of science today, but our evolved *mind* shapes how we develop those methods—while the *time and place* we live in, our *human* perspective and our *social* context help shape what we study within these limits. Science's boundaries are not fixed, they are the shifting edges of a map we redraw with each major new tool. Still, there are many mysteries we cannot unlock with our current tools today: from the size of the universe and superstrings to how consciousness emerges. But there is a paradox: understanding what drives the current limits of science is the best way we have to expand and redefine these very limits—the topic we turn to in the next chapter.

Chapter 11: How can we push our present limits of science—and make new advances that expand the frontier faster? Inventing tools with cutting-edge innovations—sharper resolving power, greater computational strength, more statistical power—continually break the boundaries at the frontier. From developing the gene-editing method CRISPR to rewrite genes, to creating quantum computers to solve problems we otherwise could never do, the power of our tools is enormous. So how do we actually push our limits and accelerate discoveries? To break new ground in any field, we generally first have to spot where our tools run into bottlenecks—and then tackle them. It is about amplifying our ability to see, measure and theorise about the world with enhanced lenses. Whether better fighting climate change with cutting-edge carbon-capture methods or better mapping the brain's complexity with more advanced neuroimaging and AI analysis, our progress depends on constantly pushing our tools' boundaries. Here we map out the steps we need to take to expand the top ten most influential discovery tools—from x-ray devices to statistical methods. Researchers best equipped at the edge of science are those who can best see the gaps in our tools and what we know—and test new ones to fill these gaps. Here we also explore deeper questions: are there fundamental limits to what we can discover in some domains? And what can the future of science look like—and how can we power it with new tools, including AI tools?

As we go through the chapters of the book, we take a journey across the foundations, boundaries and future of science—by tracing science's major discoveries and turning points. Along the way, we uncover a powerful framework—built on new evidence and a new theory—for understanding discovery and methods. And this lays a foundation for two fields: the *science of science* and the *methodology of science.*

Defining discovery, methods and science—and how we study them

To do science and make breakthroughs, scientists develop and use methods and tools, gather and analyse data, provide explanations or a theory based on the results and, in extraordinary cases, uncover a major discovery—the hallmark of scientific progress. Here we instead go backwards: we begin with science's major discoveries and the discoverers behind them, gather and analyse data on how those discoveries are made and the methods and tools they used, to ultimately develop a theory of discovery based on the common patterns across them.

How do we study and uncover something as complex as scientific progress? Just as there is no one method in biology or physics to tackle major questions, there is also no one method to study science itself. How science and discovery are currently explained can be grouped into five main approaches: explanations by historians exploring rich case studies of individual discoveries,[15,16,59,60] scientometricians tracking high-impact articles and how often they are cited,[3,14,21,23,39,61] psychologists examining discoveries indirectly with experiments on how scientists think,[62–65] computer scientists modelling the discovery process[66–70] and philosophers exploring the concept of what discovery means.[17,25,28,71,72] Scientometricians, computer scientists and psychologists analyse quantitative data, while historians and philosophers offer in-depth, narrative insight.

To unpack how discovery works, we combine the different types of methods and evidence across different fields into a coherent framework. As the backbone of the analysis, we statistically study science's over 750 biggest discoveries—the method we use most in the book; while we also explore the personal motivations behind the discoverers themselves (Chapters 1–6). We draw on qualitative methods and insights from cognitive science and evolutionary biology, and from the history and philosophy of science (Chapters 7–9). We also analyse science's top methods and tools themselves—from lasers to spectrometers (Chapters 10–11). In short, we turn the methods of science onto science itself: applying scientific tools and techniques to study science and its best methods. That means combining large-scale data analysis (to uncover the common patterns across discoveries), rich case studies (to explain the biggest discoveries in detail) and conceptual analysis (to understand scientific concepts deeper). Integrating these mixed methods, we gain a deeper and broader understanding.

So to uncover what drives breakthroughs, we need an integrated approach that overcomes the shortcomings of relying on a single method. Because the current explanations of discovery are built around the particular method that researchers use. The method applied does not just guide research, it shapes—and constrains—the very questions researchers are able to ask and the answers they are likely to find. If we use the common method of scientometricians, or philosophers of science, or historians of science and so on, then we can only look at problems that large-scale data on citations, or conceptual analysis, or individual case studies can address.

Take scientometricians who show for example that the most impactful scientific work appears randomly throughout a scientist's career—a finding based on big data analysis of article citations.(12,35,37) But we are left asking how scientists triggered that high-impact research. Take philosophers like Karl Popper who argue that there is no logic behind how discoveries arise—an idea based on conceptual reasoning.(17,25,28) But we are left without an analysis of discoveries to test that claim. Take historians like Thomas Kuhn who popularised the notion of paradigm shifts shaping science—an insight drawn from few case studies of famous theories.(15,16,73) But we are left with a focus on select theories that does not allow us to spot broader patterns—or generalise about science, but that is precisely what Kuhn still did. This sparked an enormous debate on what actually drives scientific progress that persists today. What is clear is that discovery is far too complex to study or understand using just one method and exploring just one or a few factors—and it is the reason why it is still a puzzle.

Here we do not start with a method and then explore what questions it can tackle. Instead, we begin with the big question: what actually sparks science's major discoveries? And we let that question guide which methods we need. So what is the best way to tackle this question? To answer it, we take the most direct route possible: we open up science's over 750 biggest discoveries across fields and history and—one by one—collect and analyse detailed data for each discovery. This means studying how each breakthrough was made, what methods and tools they used, what the traits of the discoverers are and what kind of broader environment they worked in. This enables us to spot the common thread driving them. The aim has been to provide the most systematic, comprehensive and in-depth analysis of discoveries to date. So instead of just focusing on a discovery in detail, or tracking high-impact papers from high altitude, we merge the different approaches. Throughout the book, we zoom in and zoom out—like turning a microscope on the discovery process itself. What is remarkable is that the broad statistical patterns from two independent sources (all nobel-prize and major non-nobel discoveries)—and the more detailed findings from different methods across fields—are deeply consistent with each other. This kind of cross-disciplinary convergence—that the evidence consistently points in the same direction—is rare in studies on discovery. And it reveals something fundamental and consistent about the hidden patterns driving discovery. It is this kind of broad, evidence-rich approach that enables a bigger picture of scientific progress.

How do we define major discoveries? The structure of DNA, the human genome, the periodic table of elements, gravitational waves, relativity theory and CRISPR gene editing reflect groundbreaking discoveries. These are common examples of major discoveries. A major discovery is defined as a new experimental, methodological or theoretical breakthrough that marks an entirely new way to understand the world, opens new paths of inquiry and later has a proven, lasting impact on science. Here we study science's major discoveries—spanning all nobel-prize and major non-nobel discoveries—that form the foundation of the major fields across the natural and social sciences (as we explore in depth in the next chapter). For both scientists and the general public, these discoveries are often the most exciting aspect of science.

Science in general is often defined as the study of the 'world through observation, experimentation, and the testing of theories' (Oxford English Dictionary).[(74)] But as we go through the evidence in the coming chapters, we will develop a broader understanding that gets at the core of what science is.

What are scientific methods and tools? Scientific *methods* are systematic techniques—such as controlled experimental, statistical and computational methods; scientific *tools* are systematic instruments—such as telescopes, x-ray devices and electron microscopes; both are used to study the world and extend our ability to observe, measure and analyse, and are general-purpose. This means they can be applied to different questions and often domains. In practice, researchers often use the words method and tool interchangeably—and we do too for simplicity. What matters is what they do: both give us better ways to see and think. Together, methods and tools make up what we call our *scientific toolbox*. (To be clear, we are not referring here to cognitive abilities like observation and hypothesising, or theoretical frameworks.)

The origins: the book's methods-driven and interdisciplinary approach

As the book offers a new way to tackle this fundamental puzzle, it is worth briefly explaining how this approach to studying science and discovery came about, before we begin. After finishing my PhD, I worked on applied research projects with governments and the World Bank for five years. It was an exciting mix of research and policy, working on complex, real-world challenges and tackling them by thinking across disciplines. Whether we worked on public health, energy or agriculture, we constantly had to integrate different methods—and pull together perspectives on the environment, human behaviour and social systems to find solutions that work.

That applied, hands-on research led me to two key insights that lay the foundation of this book. First, I realised that the specific method we use shapes what we see. If we apply big datasets and statistical models, our answers are different from if we run controlled experiments, if we employ mathematical methods or if we apply rich, in-depth qualitative methods. The method we use is not a neutral lens—and gives different kinds of insights. Second, I realised how adopting one disciplinary perspective also does not take us very far in understanding and tackling a complex challenge—whether malnutrition, climate adaptation or fostering innovation. In the real world, complex challenges always cut across disciplinary boundaries—and the only way to tackle them is by crossing those boundaries and combining multiple methods. There is no way around it—and this is clear doing applied research. But in academia, research generally stays siloed, shaped by researchers' training in a specific method and specialised in a field. Time, resources and institutional incentives also reinforce that narrow focus.

So when I returned to academic research, I brought this big-picture mindset with me—but this time to study science itself. I realised there is a pattern running through breakthroughs I came across in public health, economics and medicine: each breakthrough seemed to rely on using new method innovations. Developing randomised controlled trials, new statistical methods or novel 'natural experiment' techniques revealed completely new kinds of findings. I began to wonder, could there be a deeper principle at play—that new methods drive new breakthroughs across science? I set out to test it. To my surprise, no scientific study has yet explored the role of new methods and tools in discovery across science, and no study has yet explored the wide range of factors—at the same time—that may play a role in the discovery process. That big-picture approach was lacking.

Over fifteen months, I did extensive full-time research to compile the comprehensive data on science's major discoveries across fields. I read each of the over 750 discovery-making publications to meticulously track down and extract the data. For each of these breakthroughs, I compiled detailed data ranging from the discoverers' age, their education level, their gender and their university at the time of the discovery, to the methods and tools they used, if their discovery was serendipitous, and much more. In total, the study covers 32 different variables for each discovery—and over 20,000 individual data points were collected and verified, one by one. This time-intensive research lays the backbone of this book. It involved building a dataset from the bottom up—detailed to zoom in on any one major discovery, and broad to zoom out across science's major discoveries, enabling rich analysis at different levels. It allows asking and answering new kinds of questions about the role of each of these factors—and crucially, at the same time. At the core: what common driver links across science's major discoveries—and what influences only certain breakthroughs? What patterns show up when we look across centuries, not just decades; or across scientific fields, not just some; or across methods and tools, not just citations or theories, and so on? It offers a new, broader understanding of how discoveries emerge—that earlier research studying one or a few factors has overlooked.

The book is the result of this work that developed over many years—the biggest and most demanding project I have ever taken on. The motivation and input for this book is shaped by these experiences. The different methods I learned working in these different domains enables tackling the complex puzzle from multiple angles. Much of this research has been published in peer-reviewed scientific journals (under a creative commons licence), and this book brings the findings together—and goes far beyond them—by connecting the dots into one big picture.[(75–81,390)] Another book of mine, *Science of Science*, qualitatively studies the foundations of science across 14 different fields—from psychology and philosophy of science, to computer science and biology, sociology and economics of science.[(82)]

Before we get started, let us take a final step back. In a nutshell, we will explore how the story of discovery is not unpredictable and mysterious—and not what most people think: science's major discoveries are not commonly driven by greater funding, larger teams, bigger theories or serendipity.[6,11,15,20] When we examine science's over 750 major discoveries systematically, a different and clear pattern emerges: new tools and methods are the most overlooked factor and commonly the central driver of discovery. What is surprising is that, so far, our methods and tools have been largely created in an unplanned and unstructured way. But if we want to speed up breakthroughs, we have to rethink how science is structured and organised—and shift how universities, journals, scientific awards and funding bodies incentivise science.

PART I
THE DRIVERS OF SCIENCE AND DISCOVERY

In Part I of the book, we take on some of the biggest questions about science: how do we drive new discoveries, scientific fields and science in general? To answer these questions, we have to trace science's major discoveries themselves, map the methods and tools they used, and explore the broad range of traits of the discoverers behind them—spotting the common patterns and insights along the way. Understanding what drives scientific advances will then allow us to tackle other fundamental puzzles about science: what are the deeper origins of science—and what are its current limits and how can we push beyond them? These are the challenges we turn to in Parts II and III.

1
Sparking discovery
How new methods and tools trigger science's major breakthroughs

Summary

Discoveries transform our lives—through breakthrough medicines, technologies and completely new ways to understand the world. Yet one of science's greatest unsolved puzzles remains: how do new discoveries actually emerge? Despite their impact, there is still no general theory explaining how major breakthroughs across science arise. To tackle the puzzle, we study science's over 750 major discoveries—spanning all nobel-prize and major non-nobel breakthroughs—and link them to the methods and tools they used and to the discoverers themselves. We find a surprising and striking pattern: science's major discoveries are sparked by a new method or tool that enables observing, studying and understanding the world in entirely new ways—and commonly reveal what we did not even expect. From new bacteria and viruses to unknown particles and distant galaxies. In fact, we uncover that most discoveries in the last few decades are triggered just a few years after the tool was created—and many at the same time. From new microscopy techniques that exposed the extraordinary structure of our nervous system, to a more powerful particle accelerator that revealed the antiproton. Remarkably, we also find that powerful tools—from new x-ray devices and electron microscopes to spectrometers and chromatography—have each unlocked dozens of major discoveries that were impossible without them. What unites the diverse discoveries that each of these tools made possible is not a shared theory, research teams or more funding, it is using the shared powerful tool—each uncovering groundbreaking findings across different fields. Understanding this tool-driven principle of scientific progress gives us a powerful new way to understand and accelerate discoveries. We can apply this principle widely across science, from physics to medicine, and it can provide a foundation for the science of science—and a new way to predict discovery itself.

Overview

When Theodor Svedberg built the first ultracentrifuge in 1924 at Uppsala University in Sweden, he could not have imagined that his new invention would be used in developing many discoveries across biology and medicine. At its core, the

The Engine of Scientific Discovery. Alexander Krauss, Oxford University Press. © Oxford University Press (2026).
DOI: 10.1093/oso/9780197829790.003.0002

ultracentrifuge is an instrument optimised for spinning samples at ultrahigh speeds at over 20,000 revolutions per minute. It enables us to separate small particles—from cells to viruses—in fluids, gases and liquids. A few years later, when Svedberg's doctoral student Arne Tiselius developed the breakthrough method of electrophoresis in 1930, he could not have predicted that this new technique would be applied in making many discoveries—and even lay the groundwork for the powerful method of DNA sequencing. Electrophoresis allows us to separate different substances by their size and electrical charge—like proteins or DNA fragments. It has helped transform how we understand many diseases and our immune response. Today, gel electrophoresis is used in cutting-edge research from personalised medicine to forensic science. We will return to the human stories behind science's best toolmakers—what drives and motivates them to push the boundaries and make these inventions. But we first zoom out to tackle the big question: how do we actually spark new discoveries?

Ask ten scientists, and we often get ten different answers. Surprisingly, there is no consensus—not across the fields that study science, and not even within them. How can discovery—the most groundbreaking and fascinating part of science—remain so elusive? The two most common answers given across science are: breakthroughs come from more funding (that can enable more researchers and experiments)(6–8) or from research teams and collaboration (that can bring together more expertise).(9–11) Others argue that they happen through interdisciplinary research,(83,84) through greater training and education(44,85) or even through more productive younger researchers.(12–14,86) Still others argue that they come about from paradigm shifts in theories(15) or from problem-solving(15,87,88)—or even from sudden flashes of insight or serendipity.(17–20,29,30) As the book unfolds, we will explore how each of these factors varies by discovery, by field and by researcher—and alone are not enough to trigger breakthroughs.

Of five main disciplinary approaches to explaining discovery, computer scientists are the most ambitious—attempting to develop computer programmes that could generate new breakthroughs.(66–70) Some can only reproduce past discoveries—in abstract or limited ways.(66,70,89) Others leverage large-scale data, new artificial intelligence (AI) methods and advanced statistics to help detect patterns and make predictions—like modelling climate change and identifying new drug candidates.(90–92) Scientometricians track the system and dynamics of science—following the trails of cited papers across metrics like high productivity, collaboration networks and career paths in science.(3,14,39,61) Psychologists explore the minds behind the breakthroughs—tracing what shapes scientific reasoning and motivation.(32,62,64,65) Historians often turn the lens on past scientific theories but also remind us that context matters—with pandemics, changing governments and global challenges at times nudging the history of science in certain directions.(15,16,59,60) And finally, philosophers are often more sceptical—arguing that how discovery happens does not follow a systematic or logical process.(17,25,28,71,72)

Despite the different explanations, we have not yet uncovered what the common driver tying discoveries together is—the unifying thread. Could it be because science's major discoveries have simply not yet been studied in a comprehensive way that lets us spot the common patterns? And because studies have not yet systematically investigated a crucial piece: the role of new methods and tools? In this chapter, we take on this challenge. After all, many see the 'identification of fundamental mechanisms responsible for scientific discovery' as the core question of the science of science—as the holy grail of research on discovery.[3–5,93]

A new way forward for us to solve this fundamental puzzle is taking the most direct path: opening up science's over 750 major breakthroughs for in-depth analysis of who, when, where and how they emerged. By analysing the methods behind them and the broad traits of the discoverers themselves—at the same time—this allows us to compare and assess the different factors together. It enables us, for the first time, to tackle the puzzle at its core: *what consistently sparks science's biggest breakthroughs—the common driver—and what factors only apply to certain breakthroughs*? This powerful new approach moves us beyond the common limits of exploring one factor or small samples of discoveries.[41–44,94] It opens the door to developing a general theory of science's major discoveries across fields. So instead of generally isolating one piece of the puzzle, we can see the full picture. At its core, we explore the wide range of factors from the age and training of the discoverers to their university ranking, whether serendipity played a role, and much more. Surprisingly, we find that discoveries consistently share one common pattern: across fields and time periods, science's major discoverers rely on a new method or tool that triggers their new breakthrough finding. What is remarkable is that this is the only factor that consistently shows up in breakthrough after breakthrough—from chemistry and medicine to astronomy and beyond. We unpack this key finding up front in Chapter 1. In the chapters that follow, we will explore in depth the range of other factors—and why some matter only for specific breakthroughs.

By examining science's biggest discoveries, we will uncover, for example in biology, how powerful methods and tools—from new microscopes and electrophoresis to centrifuges and chromatography—transformed the field and how we understand life. These extraordinary tools triggered discoveries that were previously impossible: from viruses to the structure of the cell. We uncover this striking pattern across science. Before inventing these tools, we faced blind spots and much of nature was simply out of our reach. But after creating them, we could see further, think deeper and probe more complexly into new layers of reality.

How we measure and track science's major discoveries

At this point, readers may just want to get to the discoveries—and the stories behind their discoverers. But before we get there, we first have to lay out the data—this may seem dry, yet it is essential. Much of what follows builds upon it—laying the foundation for studying how science's major discoveries unfold. These cover all 533 nobel-prize-winning discoveries across science—from the first year of the prize in 1901 to 2022. The Nobel prize is widely seen as the most prestigious scientific award for 'the most important discoveries'—with winners selected by leading experts in their field.[95] Yet not every major breakthrough wins a Nobel. Some came before the prize existed, others can fall somewhat outside its scope. To capture as complete a picture as possible, we also incorporated the discoveries featured in all science textbooks that list the top 100 greatest scientists and their breakthroughs—and span across fields and history; in total, seven textbooks were published and added to the dataset.[96–102] After excluding duplicates, 302 discoveries remained—and of those, 74 won a Nobel.

This left 228 discoveries that—when added to the 533 nobel-prize discoveries—brings us to a total of 761 major discoveries that shaped the history of science across fields.

If the Nobel prize had existed earlier, eminent scientists—from Galileo, Newton, Hooke and Boyle to Darwin and Maxwell—would almost certainly have received the prize. And they are all included here among the major non-nobel discoveries. The large overlap in discoveries between each of the science textbooks and the nobel-prize discoveries points to the strong consensus on science's biggest discoveries. It reinforces that we cover basically all of the most influential discoveries across science (as mapped out in online Appendix Figure 1.1). Likely, everyone will find their top 50 or 100 greatest scientific discoveries among these over 750 discoveries.

Still, many leading studies on scientific progress trace patterns in citations simply because how often researchers are cited is easy to access and study. Bibliographic databases already provide the data, eliminating complex manual data collection. But here, with this new approach, we identified the original discovery-making publication for each of the 761 breakthroughs. For each discovery, we extracted the central method or tool they used—outlined in that publication—and linked the data to the broader traits of the discovery and discoverer. We then confirmed the year both the central method and discovery were made. Unlike citation data that can be gathered and analysed in weeks or months, this research demanded much greater time and extensive manual research. This is the first book built on manually reading and extracting data from each of the over 750 discovery papers (as for example most older papers cannot be processed using text recognition). By adopting this deeper approach, we can trace the discovery process as described by the discoverers themselves—and then confirmed with other scientific publications.[95,103–108] This approach enables us to get *inside* each major discovery across science and also explore them at scale—probing them from *outside* or at high altitude as well.

So each discovery—whether experimental, theoretical or methodological—is linked here to the central methods and tools they used (such as specific electron microscopes and statistical methods) and to the characteristics of the discoverers—all at the time of the discovery. The central method or tool is defined as a major method or tool applied in the discovery-making paper. In the paper, it is commonly highlighted by the discoverers (and often also in their Nobel lectures) as key to the breakthrough. We find that the discovery would not have been possible without it. And the method is also later widely adopted, and typically used to generate multiple discoveries, showing its scientific impact. The main source is thus the discovery paper—and for discoveries earning a Nobel that means the prize-winning studies.[95] And just to be clear: throughout the book, whenever we mention a year, we are always referring to when the discovery or method was published—not when the Nobel was later awarded (for interested readers, more details on the data are in the online Methods Appendix).

Five key insights: how powerful new methods and tools trigger discovery

Let us begin with a breakthrough in astronomy. In 1946, the astronomer Martin Ryle revolutionised how we explore the universe with an innovative invention. He did not construct a larger telescope, but linked smaller telescopes kilometres apart that could capture radio waves. Built at Cambridge, this new system opened telescopic observations on a massive scale—as large as the space between the different telescopes. It added power and precision, making it possible to map distant stars and galaxies with unprecedented resolution. And it opened a new window onto the universe that transformed radio astronomy.

Then in 1967, a groundbreaking discovery was made by Antony Hewish and his 24-year-old graduate student Jocelyn Bell Burnell. While scanning the sky, Burnell noticed a peculiar signal—a regular pulse of radio waves, something no star was supposed to emit. It was the discovery of pulsars—neutron stars—sparked by an enormous new radio telescope that Hewish built that same year using that same innovative technique. Before their discovery, neither pulsars nor quasars were known and no theory could have predicted they exist. Jointly, the two published this trailblazing finding in *Nature*, in a paper titled *Observation of a rapidly pulsating radio source*. Yet when the Nobel prize followed, it was awarded only to Ryle and Hewish—overlooking Burnell. But the deeper insight is that without the invention of radio telescopes, pulsars would have remained invisible. We could not otherwise detect the radio waves emitted by galaxies, supernova and pulsars millions of light-years away. In short: they give us access and unlock a part of the universe that was out of reach before them. They are just one of many extraordinary instruments that transform not only what we can detect—but how we can imagine the universe.

By analysing science's over 750 greatest discoveries, we uncover five powerful insights—that we will unpack in the chapter's next five sections:

1) *We spark science's major discoveries by developing a new method or tool—that makes the breakthroughs possible.* New tools like x-ray devices, electron microscopes, spectrometers and chromatography have each enabled uncovering dozens of breakthroughs—from revealing the structure of molecules and vitamins to the mechanisms of cellular life. For each of these tools, the shared thread across the different discoveries they made possible is not a common theory, research teams or additional funding, but it is using the common tool—each uncovering breakthrough findings across different fields.
2) *Our powerful tools trigger new breakthroughs by extending what and how we can observe, detect and measure the world.* We see this clearly in the ten central methods and tools most used to make nobel-prize discoveries: statistical/mathematical methods, microscopes (optical and electron), x-ray methods, centrifuges, spectrometers, chromatography, electrophoresis, lasers, particle accelerators and particle detectors.
3) *There are three ways new methods and tools power new discoveries.* One way is that a researcher develops a method or tool and then uses it in making a discovery (responsible for 25% of science's major discoveries). The second way is that a researcher creates a new method or tool, and another researcher—in the same or a different field—uses it to trigger a discovery (representing 50% of cases). The third way is that a researcher invents a new method or tool that is the major discovery itself (making up 25% of cases)—like the electron microscope and the particle accelerator.

4) *New breakthroughs often follow soon after developing the new enabling tools.* Surprisingly, one in ten major discoveries has been sparked simultaneously—or in the same year—the new key method or tool is created. Two in ten follow within two years. And since 1975, more than half—at 51%—of major discoveries emerge within just four years, highlighting the power of method innovations.
5) *Creating a new method or tool is often more foundational to scientific progress—across different fields and over time—than individual scientific discoveries.* While a single breakthrough can reveal bacteria or detect a planet, the same tool can uncover multiple discoveries: from cells and mitochondria to galaxies and exoplanets that were not even being searched for. Our powerful tools—whether microscopes or spectrographs—often outlive single scientific discoveries in impact. Because many enable multiple discoveries: in *different fields*, at *present* and in the *future*. Many such breakthroughs are exploratory—not theory-driven—and they begin with a new lens to the world.

We spark science's major discoveries by developing a new method or tool—that makes the breakthroughs possible

Digging into the data and papers of science's major discoveries across fields and over time, we find a striking and consistent trend. Each discovery-making paper relied on a new method or tool—and the study, including usually experiments, could only be conducted by using it. The strong link is remarkable: analysing science's major discoveries, we find that new breakthroughs are enabled by applying a new method or instrument for the first time to a problem. That is what commonly makes it a breakthrough—it reveals something we could not see, detect, develop or test before, without that method.

Let us look first at the turn of the 20th century. At the time, physicists focused on the fascinating newly detected subatomic world. The first subatomic particle discovered was in 1897—the electron. Joseph Thomson unexpectedly detected it using an instrument he improved the previous year: a cathode ray tube. At first, many physicists relied heavily on mathematics—using abstract equations to describe the electron. That changed in 1911 with the invention of the first particle detector that could visually show particle trails: the cloud chamber. For the first time, we could actually see and measure the trails left by charged particles. It was built by Charles Wilson working at Cambridge. With this one groundbreaking instrument, we introduced an entirely new kind of evidence to physics: launching particle physics as an observational science (Picture 1.1).[109] Then in 1929, Ernest Lawrence—at just 28—made another extraordinary invention: the first particle accelerator. Working at Berkeley, he designed this new instrument that could propel charged particles to high speeds using electromagnetic fields. Before it, we did not know that many subatomic particles even existed. And without this invention, we would not have discovered the antiproton or uncovered the weak force—that changed how we understand matter and the forces governing physical reality. So even particle physics—often seen as one of science's more theoretical fields—has deep roots in powerful tool innovations that helped make the field possible and transform it into a precise experimental science.

Picture 1.1 *The world's first particle detector that visualised particle tracks—Wilson's cloud chamber, 1911.*

Reproduced from Rolf Kickuth via Wikimedia Commons.

Another great leap came in 1952, when Donald Glaser constructed an entirely new kind of particle detector: the bubble chamber. Built at the University of Michigan, it allowed studying particles at much higher energies than before. Together with particle accelerators, this tool was instrumental in the theoretical discovery of quarks in 1964—fundamental particles that make up protons and neutrons. Our ability to observe and measure the smallest units of matter—atoms—was revolutionised with these powerful discovery tools: particle detectors and accelerators.(110) They opened up new layers of reality that were invisible, unknown and unimaginable before—using our bare mind. With them, we could see the evidence unfold before our eyes. And without them, none of these breakthroughs could have been possible.

In fact, a range of powerful tools and methods that have made discoveries possible were awarded a Nobel prize for their creation: from particle detectors and accelerators, to x-ray methods, spectroscopes and centrifuges; from advanced microscopes, electrophoresis and statistical methods, to chromatography, the laser, the PCR method and many others. Each of these remarkable tools enabled multiple discoveries—experimental, theoretical and methodological—that in turn won a Nobel, as we see in Figure 1.1 (and Appendix Figure 1.3). Each was designed in one field—like physics—but then triggered discoveries in different fields—like biology, chemistry and medicine.

We next zoom further out on this puzzle of how new methods and tools spark breakthroughs. To study it, we trace science's major discoveries—spanning from 1575 to the present—and compare the year each new method or tool was developed with the year the discovery was made using it. It is straightforward to visualise: each new method or tool is shown as a vertical line (|), and each breakthrough made using that method or tool is shown as a dot (●) on the same horizontal line—see Figure 1.2. If a new tool triggered multiple discoveries, we see a cluster of dots aligned with that tool's timeline. Take the first particle detector that visualised particle tracks invented in 1911—marked as a vertical line | . It enabled the discovery of the positron in 1932—appearing as a subsequent dot ● on the same horizontal line. Or take the invention of x-ray crystallography in 1913 that led to the discovery of DNA's double-helix structure in 1953. The history of science reveals a pattern: discoveries cluster after major new tool innovations that are applied for the first time. Over time, the gap between new

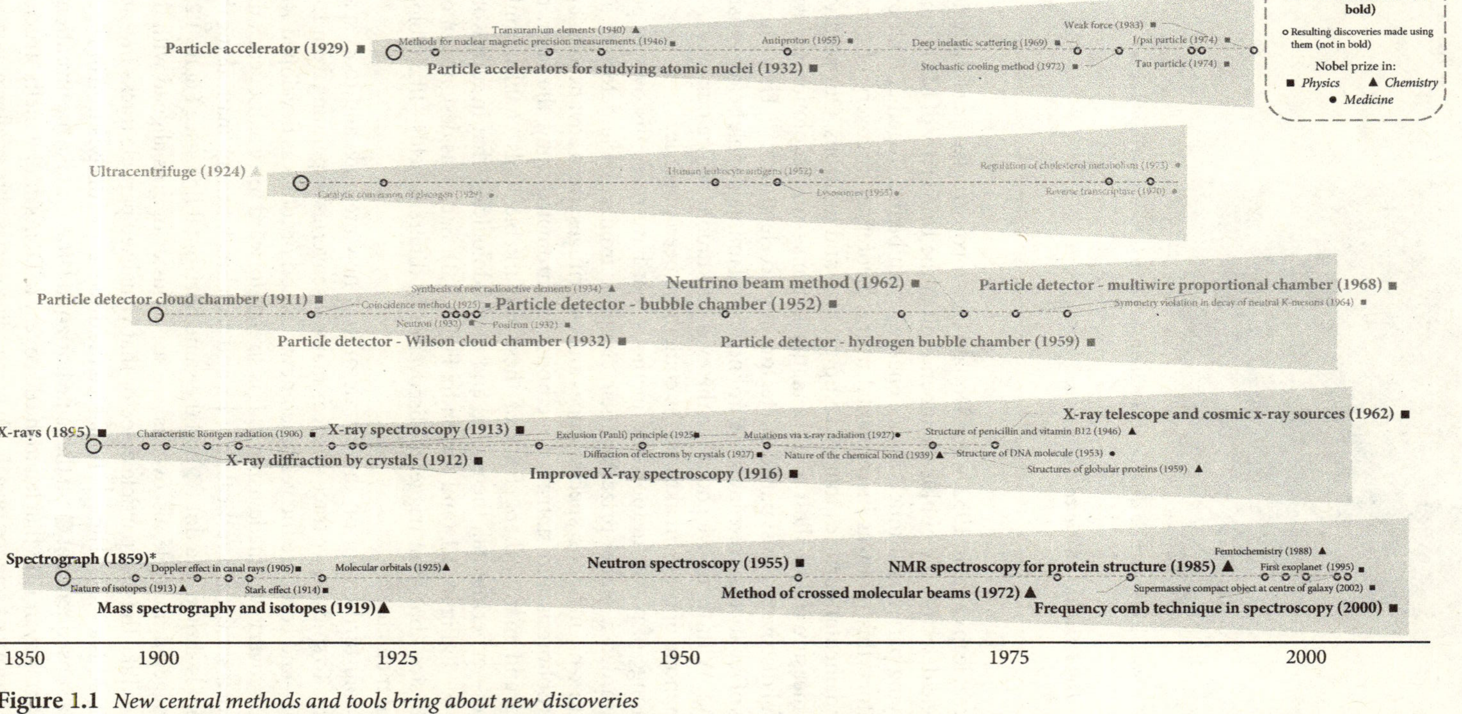

Figure 1.1 *New central methods and tools bring about new discoveries*

*The technical notes below the figures throughout the book are just that—technical notes. They only offer additional details but can be skipped to stay with the main story. The data shown in this figure highlight nobel-prize-winning methods and tools developed before 1930, and each has led to at least four later nobel-prize discoveries. (Those developed after 1930 are shown in Appendix Figure 1.3.) The year listed refers to when the tool was created and when the resulting discoveries emerged. *One exception is the spectrograph, invented in 1859 to study the spectrum of light; although it did not receive a Nobel prize, it was used to make five discoveries and more advanced spectroscopes that were later awarded a Nobel. Due to space limitations, a number of nobel-prize discoveries made using x-ray methods and spectrometers could not be included in this figure; but they are included in all other relevant figures throughout the book.*

tools and new discoveries is reducing. (The horizontal lines are getting shorter over time in Figure 1.2.) We explain why later in the chapter. But this time dimension is important: the earlier we develop transformative tools, the sooner we catalyse new breakthroughs.

Surprisingly, we find this key pattern across science—that new methods drive new discoveries by making novel findings possible. We find that the pattern is consistent across science's major discoveries spanning throughout fields—from physics and chemistry to medicine and economics—and it is consistent across history—from past centuries to contemporary science (Appendix Figure 1.5). It holds across experimental discoveries, such as uncovering the hepatitis A virus with an electron microscope. It applies across theoretical breakthroughs, such as developing Planck's quantum hypothesis based on experiments using spectrometers and bolometers, and on mathematical methods. And it holds across method discoveries, such as creating gas chromatography that vastly extended chromatography. As we move through the chapter, we will unpack this key finding.

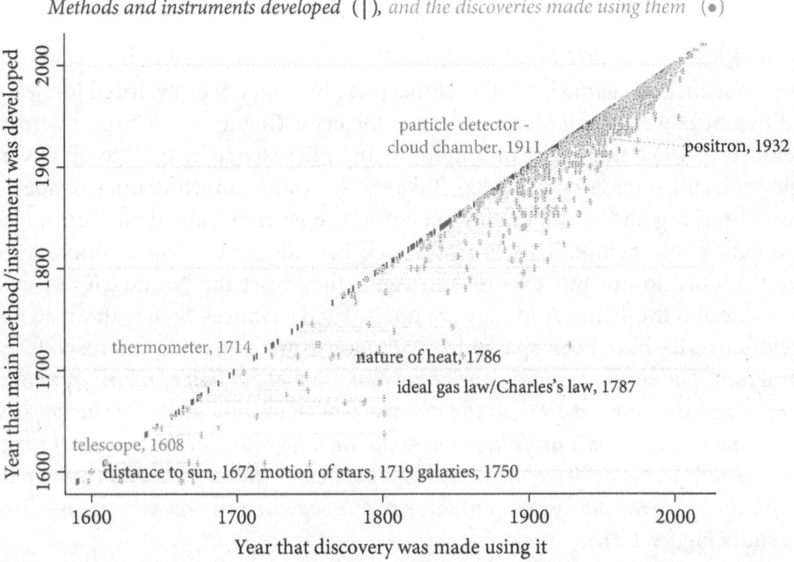

Figure 1.2 *We trigger new discoveries by using a new central method or tool needed to make the breakthrough*

The data reflect science's 734 major discoveries since 1575—including all nobel-prize discoveries across science. Examples are shown for three tools. Each new method or tool is marked as a vertical line (|) that all fall along the diagonal 45-degree line. The year a new method or tool is created is closely linked to the year a discovery followed—with a correlation of 92%. When we split and compare the data—exploring only nobel-prize discoveries and only major non-nobel discoveries over the same time period—the pattern remains strong: a correlation of 88% and 81%, respectively. This highlights that the pattern is not driven by just nobel-prize discoveries but is consistent for both groups of discoveries.

Few tools have transformed science as drastically as the microscope—our prime example of a powerful discovery tool. Microscopes have uncovered about 7% of science's major discoveries—an enormous contribution to scientific progress from a single kind of instrument. This spans from the first microscope in 1590, which inspired later ones such as the achromatic (colour-corrected) microscope in 1841, to scanning tunneling microscopes in the 20th century.(46,47) Each generation of microscopes are not just technical upgrades, they are scientific catalysts—each one revealing deeper layers of reality. It all started with Zacharias Janssen—a Dutch eyeglass maker who created the first microscope using a simple converging lens of glass beads.(46) This groundbreaking invention marked a turning point in tool-driven science: we had—for the first time ever—a tool in our hands that enabled us to look through its lens to uncover an invisible world. Cells, bacteria and mitochondria were revealed without even searching for them. No one hypothesised or imagined they could even exist. But these unexpected discoveries changed how we understand human life itself.

Take the British Robert Hooke who uncovered cells in 1665. By looking through his new 300-power microscope, designed in 1662, he observed a honeycomb-like pattern in a piece of cork: he discovered cells. His book *Micrographia*—the first important work on microscopy—launched and popularised the microscope into wider scientific use.(111) It also helped establish a concept at the core of biology today: the cell as the basic unit of life.

As lens systems improved and magnification increased, optical bottlenecks were broken—again and again.(47,48) The Nobel prize has since been awarded for developing different powerful microscopes—from the crystallographic electron microscope developed in 1962 and the scanning tunneling microscope in 1982 to an advanced single-molecule microscope in 2006. Take the scanning tunneling microscope—that allows observing and manipulating matter at the level of individual atoms, like no instrument could before. This breakthrough helped launch nanotechnology. Such powerful tools do not just extend our sight—they reset the boundaries of what is observable and thinkable. And they are not isolated advances. So far, about 30 nobel-prize discoveries have been sparked using a nobel-prize-winning microscope as the central tool. *The electron microscope alone makes up about half of them: from the discovery of cell structure in 1945 to the discovery of ribosomes in 1955. This pushes the electron microscope—as a unifying tool—into the foreground of these breakthroughs it made possible across fields; while it pushes a particular theory, research team or large-scale funding into the background of these microscope-driven breakthroughs*—Box 1.1 (Appendix Figure 1.3).

Box 1.1 How Ernst Ruska built the electron microscope—catalysing breakthroughs across biology, medicine and physics

Only a handful of breakthrough inventions are as exciting and transformational as the electron microscope. It is one of the ten central methods and tools most used to make nobel-prize discoveries. Before it, many microorganisms, molecules, viruses and nanoparticles were often beyond the reach of any lens—and a mystery. After it, we could observe, study and begin to understand these hidden worlds in the extraordinary detail it offered. The story traces back to the German Ernst Ruska.[47] Ruska said that as a child, he was inspired by his 'father's big Zeiss microscope ... [who] sometimes demonstrated to us interesting objects under the microscope'.[48] So how did he exactly develop the electron microscope? The design is straightforward: he realised that improved magnetic coils can be used as a lens for electron beams—and such an electron lens can be applied to depict images of miniscule objects on a fluorescent screen by pointing the lens towards them. In other words, irradiating extremely small objects with electrons. He then linked two electron lenses to create a primitive microscope—that he kept improving (Picture 1.2).[47,48,112]

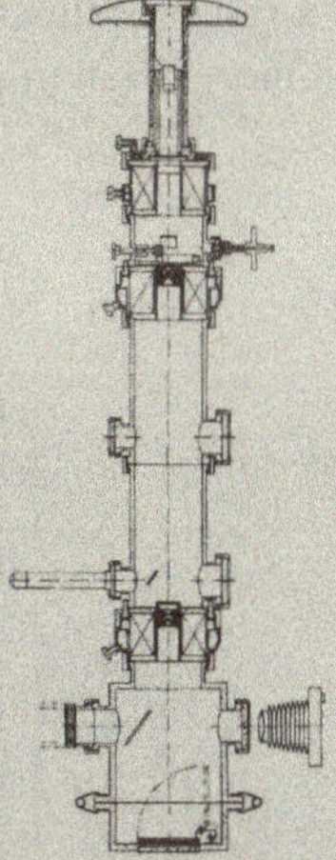

Picture 1.2 *Ruska's instrument, 1933.*

Reproduced from Nobel Prize 1986a.[48]

This new tool could magnify—for the first time—far beyond any conventional light microscope: achieving magnification of up to 12,000 times. When Ruska designed the first prototypes between 1931 and 1933, he was an unpaid PhD student at the Technical University of Berlin. He did get a modest stipend of 100 Reichsmarks a month in the second half of 1933 to improve its design—and he completed it in November. It turned out to be a highly efficient and very low-cost instrument at 500 Reichsmarks in 1933 (about 2800 US$ in 2025 prices).[48] Remarkably, he built the most powerful microscope yet invented at the time without large funding, a large team or a large research lab—a pattern we will see across a number of science's most important methods and tools. After its creation, Ruska went to work at Siemens, where he spent two decades developing commercial electron microscopes. In the meanwhile, his extraordinary tool unlocked over a dozen major discoveries—and it completely transformed biology. Despite the impact,

Ruska received delayed recognition: he was awarded the Nobel in 1986—53 years after his breakthrough. An astonishing fact for 'one of the most important inventions' not just of the century but the history of science.[47] After zooming in on this powerful instrument, we now zoom back out to explore the broad patterns across science.

Next, when we trace science's biggest discoveries over time, a clear trend emerges: in the decades when we develop more major new methods and tools, more major discoveries follow using them. It is not random. It is a consistent pattern that holds across time periods—as visualised in Figure 1.3. Around the turn of the 20th century, the pace of tool innovation accelerated quickly. In fact, the first half of the 20th century stands out as an extremely productive era for major method advances across fields—spanning everything from x-ray crystallography to new statistical methods that transformed fields (explored in more detail in Appendix Figure 1.6).

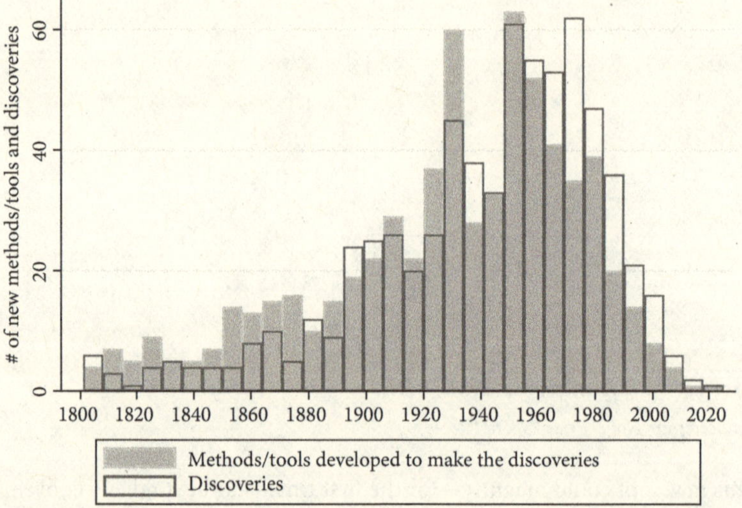

Figure 1.3 *Trends in new central methods and tools closely follow trends in discoveries*

The data show science's 653 major discoveries made since 1800—and include all nobel-prize discoveries—and track how many central methods/tools and discoveries emerged in each time period.

In fact, across all Nobel prizes, there is an average 21-year delay between making the prize-winning discovery and receiving the award. New methods and new discoveries are not decreasing in the past few decades—but trends reflect the delay in the scientific community recognising major discoveries (Figure 1.3). Take biology. The entire human genome—made up of three billion letters of our genetic code—was mapped in 2003. This enormous feat was made possible by powerful new tools such as automated DNA sequencing and sequence-tagged site maps.[113] Then the groundbreaking new method for gene editing, CRISPR, was discovered in 2012 by the French Emmanuelle Charpentier and American Jennifer Doudna. With it, we can change the DNA of plants, animals and even human cells with precision—offering new hope and solutions for treating genetic diseases and cancers.[114]

Or take physics. We detected gravitational waves for the first time in 2015 after upgrading the Laser Interferometer Gravitational-Wave Observatory (LIGO), a massive laser interferometer, in the same year. This proved that space can vibrate and show ripples in spacetime.[115] In astronomy, new high-magnification telescopes have discovered thousands of planets beyond our solar system since 1995—increasing the possibility of finding other life in our universe.[116] Then there is particle physics: the world's biggest particle accelerator, the large hadron collider, was built by CERN on the French-Swiss border, 175 metres underground. In 2012, it revealed a new particle, the Higgs boson, deepening our understanding of the nature of matter.[117] With many new discoveries across fields, we do not see new breakthroughs slowing down—a topic we return to later.

Our powerful tools trigger new breakthroughs by extending what and how we can observe, detect and measure the world

Scientific tools are our lenses, sensors, amplifiers, separators, accelerators, and data-processers—vast extensions of our eyes, hands and minds. Without them, most of the universe would remain invisible and unknown. With them, we can see living cells, weigh single atoms, edit genes, detect black holes, experiment with medical treatments and model climate systems. Our best methods and tools are the foundation of science.

But how do our best tools exactly expand the boundaries of discovery? Let us look at the top ten methods and instruments most used to trigger nobel-prize discoveries. Optical and electron microscopes, x-ray methods and spectrometers vastly extend our visual scope: revealing unimagined worlds of microorganisms, cells, protein complexes and galaxies. Statistical and mathematical methods massively expand our cognitive capacity to reason across vast amounts of data: whether mapping how diseases spread, reconstructing past climate conditions from ice cores or modelling human behaviour. These few general methods and tools are the most universal ones in sparking discovery: each one is used across physics, chemistry, medicine and biology. Then there is chromatography, electrophoresis and centrifuges that give us the power to isolate substances and better understand them—from proteins and DNA to viruses. Particle accelerators and detectors let us collide and track matter at the smallest scales—and peer into the fundamental building blocks of the universe. Lasers give us astonishing precision to measure ultrafast interactions in matter, probe the structure of molecules and calculate vast distances in deep space. Each time we expand our toolbox with a powerful new method, our horizon of possible discoveries moves. As we work through the book, we will dig deeper into the power of these extraordinary tools.

Our powerful tools are a core defining feature of a field. It is difficult to imagine physics without particle accelerators, mathematical methods, lasers and spectrometers. Or biology without microscopes, centrifuges and controlled studies. Or economics without statistics, algebraic equations and equilibrium analysis. Or astronomy without radio and space telescopes (Appendix Figure 1.7). So why do some tools transform science more than others? It is because impactful tools are much better at extending our sensory and cognitive range: giving us greater resolving power, higher sensitivity, more computational strength, greater statistical power and entirely new measurement domains we could not reach before. That is when deeper layers of nature become accessible to us.

Take x-ray instruments—among the most transformative tools of discovery that science has. In fact, the first year the Nobel prize in physics was awarded to the German Wilhelm Röntgen for discovering x-rays. In 1895, Röntgen stumbled on a breakthrough observation: he saw an image of one of his keys on a photographic plate he happened to place next to a discharge tube (a Crookes tube) that was developed in 1875. It was the discharge tube that generated the unexpected image.[118] When he took an exposure of his wife's hand, her bones and wedding ring appeared starkly on the plate; she reportedly said: 'I have seen my death'.[118] He immediately published the finding. This landmark discovery triggered a cascade of breakthrough developments: from x-ray diffraction in 1912 and x-ray crystallography in 1913 to improved x-ray telescopy in 1962 (Figure 1.1). Over 20 nobel-prize discoveries—across medicine, chemistry and physics—have been uncovered with x-ray instruments. What is striking is that using x-ray tools to spark a number of discoveries was not intuitively clear before we had these tools—from uncovering the inner architecture of viruses to revealing the structure of ribosomes and cholesterol. The lesson repeats: before the tool, these worlds were largely hidden; after the tool, they became data.

Box 1.2 How Max von Laue developed x-ray diffraction—triggering advances in chemistry, physics and biology

X-ray diffraction and crystallography opened a new window into matter—and represent another one of the ten methods and tools most used to make nobel-prize discoveries. The remarkable technique would go on to unlock the structure of proteins and viruses. It began with the German Max von Laue. What inspired him to invent the method? Von Laue said that after he joined the University of Munich, his interest was constantly drawn on by 'the influence of Röntgen's work at this University and subsequently by Sommerfeld's active interest in X-rays'.[119] These strange, invisible rays could pass through flesh—but what exactly were they? It was in 1912, at the age of 33, when von Laue conceived the method—and he discussed it with his colleagues while on a skiing trip, but they were sceptical.[119] Yet von Laue went ahead with the design and worked with two technicians at the university, Friedrich and Knipping. X-ray diffraction is surprisingly straightforward: shine a beam of x-rays through a copper sulfate crystal—and then record the pattern they create on a photographic plate. What it revealed was stunning: a pattern of spots in circles (Picture 1.3). These diffraction patterns offered the first proof that x-rays are waves—not particles—and that crystals have regular, repeating atomic structures.[119,120] It marked the creation of a new method that could reveal the structure of matter itself.

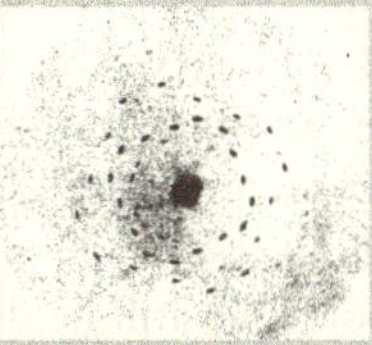

Picture 1.3 *X-ray diffraction pattern from a crystal.*

Reproduced from Nobel Prize 1914.[119]

The x-ray diffraction instrument he developed is simple: an x-ray tube on one side (the left), a small crystal in the centre, and a photographic plate on the other (the right)—as seen in Picture 1.4. These materials were purchased at low costs.[119,121] It was yet another case of science's most important methods and tools designed without large funding or a large research lab. Scientists quickly recognised the importance of this exceptional method—and he received the Nobel prize just two years later, in 1914.

This method immediately opened the way for William and Lawrence Bragg's research on new techniques for analysing crystal structures with x-rays. The techniques spurred new fields like x-ray crystallography, x-ray spectroscopy and molecular biology—and sparked a series of breakthroughs in organic and inorganic chemistry. William and Lawrence Bragg then received the Nobel prize in 1915 for extending these essential x-ray methods. Today, these tools are still at the cutting edge—used to design new materials, study batteries and semiconductors—and even model protein structures for drug discovery.[119,121]

Picture 1.4 *Laue's instrument, 1912.*

Reproduced from Eckert 2012.[121]

Some of the most powerful tools in science are not designed for the fields that end up using them most. Breakthrough methods are often born in one corner of science, are adopted and then spark breakthroughs in entirely different disciplines. What is fascinating is how this happens in ways the original inventor never predicted. Lasers and particle accelerators for example were initially rooted in physics, but they also drive discoveries in chemistry—while they are even applied in biology and medicine. Today, they are used to manipulate biological cells, treat cancer and even guide eye surgeries. Both medicine and chemistry use for instance chromatography, the PCR method and blotting techniques to separate and transfer proteins or DNA—and these methods are even used in forensic science and environmental testing. Both physics and medicine employ for example cathode-ray oscilloscopes and counters to measure radioactivity—whether for cancer treatment or nuclear experiments—and they are even used in environmental science and biochemistry. Such widely used tools spread across a shared interdisciplinary *methods space*—a web of methods and tools that link fields together (Appendix Figure 1.8). Chemistry, medicine and biology share the most tools, followed by physics and chemistry.

The scientific community is largely a tightly knit method community. In fact, the scientific world is far more unified by shared tools than most realise. While theories can differ across fields, methods often move between and across them. This points to a key insight: a method or tool can often be the key unifying thread across discoveries and fields, not just theory. This becomes clear looking at how highly interdisciplinary tools—like electron microscopes, x-ray methods, spectroscopes and Geiger-Müller counters—are each leveraged to trigger discoveries across physics, chemistry, medicine and biology, like no theory can. Yet economics and social sciences largely only share statistical methods with other fields.

Take the radiocarbon dating method—invented by the American physical chemist Willard Libby in 1949 at the University of Chicago. What is radiocarbon dating and how does it work? The method is simple: carbon is a component in biological organisms—including all living animals and plants—and when an organism dies, the carbon-14 content decays at a constant rate; and that is used to measure how much carbon-14 remains in a sample—and to determine its age. To develop the method, he used a Geiger counter, a detector for measuring radioactive decay originally invented in physics. Libby's ingenious method enables us to measure the carbon-14 content of bones, wood, fabrics and fossils—and establish their age with unprecedented accuracy.(122) Libby published the breakthrough method in a paper in Science titled *Age determination by radiocarbon content: world-wide assay of natural radiocarbon.* His technique offered something archaeologists had only dreamed of: a clock for prehistory. It earned him the Nobel prize. Before inventing it, we were heavily constrained in estimating the age of organic material—it was often guesswork. After it, we had reliable data. For the first time, we could date human settlements, extinct animals and climate changes going back tens of thousands of years—with astonishing precision.(122) Nothing has been more important than this method in developing the fields of archaeology and earth science—and reconstructing the history of our species on our planet.(123) Theories and grant funds have not had as much impact as this simple but essential tool across these domains and discoveries.

There are three ways new methods and tools power new discoveries

When we think of discovery, many imagine a researcher using an existing dominant scientific method or tool. Here we dig deeper to ask: what are the actual channels through which our innovations in tools spark breakthroughs? Examining the over 750 discovery-making studies, we find that breakthroughs do not emerge from conventional methods. Instead, they can be categorised into one of three pathways:

- *method-to-discovery by the discoverer*: a researcher develops a new method or tool themselves that they use to make the scientific discovery.
- *method-to-discovery by another researcher*: a researcher develops a new method or tool that another researcher applies—sometimes in an entirely different field—to make the scientific discovery.
- *method-to-method by the discoverer*: a researcher develops a new method or tool that is the major discovery itself, and does so by building on other methods.

Across all nobel-prize discoveries, these three pathways account for 25%, 47% and 28% of discoveries (Figure 1.4). Discoveries are thus not just experimental or theoretical, but 28% were awarded for a method breakthrough (149 times out of all 533 nobel-prize discoveries). So how does this breakdown give us a deeper picture of how science works? It helps understand not just where innovation starts, but how it happens and spreads. The first pathway—where the discoverer makes both the needed method and the discovery—highlights how tool innovation directly spurs breakthroughs. And this strategy is rare in conventional research. Take the experimental physicist Kamerlingh Onnes—who developed an improved liquefier in 1906 along with a sensitive thermometer in 1907 for measuring temperatures as low as absolute zero.[102,124] Using these tools, he discovered liquid helium in 1908 and superconductivity in 1911—the surprising phenomenon where certain materials, when cooled to extremely low temperatures, conduct electric current with zero resistance. Superconductivity did not just advance physics—it changed technology. Today, it powers instruments from magnetic resonance image (MRI) scanners in hospitals to particle accelerators and quantum computers.

Then there is the father and son duo, William and Lawrence Bragg—who designed an x-ray spectrometer in 1912 that enabled them to discover the structure of crystals just a year later. Their groundbreaking paper *The reflection of X-rays by crystals* published in the Proceedings of the Royal Society led to the spread of x-ray crystallography.[125] Crystals, after all, make up an important building block of part of our world—they are in the ice, table salt and sugar we consume. By studying crystals, we can understand the structure of compounds and how the biomolecules of our muscles and bones function.

John Vane, a pharmacologist at the Wellcome Research Lab in England, devised a new bioassay technique in 1964—called cascade superfusion, allowing scientists to study how drugs interact with live tissues in real time. He used his technique to make a life-saving discovery in 1976: prostacyclin, a molecule that prevents blood clots.[126] The breakthrough explained how aspirin works—the most used drug in the world to alleviate pain—and it helped shape modern cardiovascular treatments. Moore and Stein, at the Rockefeller Institute, invented the first automatic amino acid analyzer in 1958—a device that could analyse proteins faster and more precisely than ever before. Later that same year, they used their device to discover the chemical makeup of ribonuclease, a molecule that breaks down RNA.[104] RNA is the molecular messenger that turns DNA into proteins—essential to nearly every function in our body. Each of these scientists earned a Nobel prize for their discovery. In short: many great discoverers do not wait but invent entirely new methods as part of the scientific process—or *methodological process* of discovery. These discoverer-inventors identified the missing methodological link: building the instruments that make it possible to find new answers.

The second pathway we have seen throughout the chapter—a researcher uncovers a discovery using a method another researcher invents. Then there is the third pathway—where the new methods are the discoveries themselves. Again, method breakthroughs (like experimental and theoretical breakthroughs) build on other methods and tools. Method discoveries are remarkably common in some fields: 39% of all nobel-prize discoveries in physics and 36% in chemistry have been awarded for developing a major new method or tool, with medicine and biology at 10% (Figure 1.4b). Take frequency comb spectroscopy, created in 2000. This laser-based technique lets scientists measure light frequencies with extraordinary precision—for example when searching for earth-like planets. The RCT method for poverty alleviation introduced in 2003 enables measuring how effective government policies are. An advanced single-molecule microscope invented in 2006 made it possible to track and image how proteins fold, how viruses invade cells and how drugs interact with receptors—through super resolution. The breakthrough genome-editing technique CRISPR developed in 2012 gives us the extraordinary ability to modify the genes of living organisms and offers us hope for curing inherited diseases.[114] In fact, the share of these method discoveries has been rising over time (see Figure 1.4a). The time around the 17th century, when some of the most foundational scientific tools were born, marks an exception—and we will explore that in detail in Chapter 7.

A striking finding emerges: discoverers are unique in realising—although only for an individual discovery—how to better study a phenomenon with new methods they develop along the way. Some develop a better method or tool they used (pathway one), others make a breakthrough method itself (pathway three). When combined, the two apply to about half of all nobel-prize discoveries (53%). These incredible researchers did not wait for others, they grasped the extraordinary power of improving the way they perceive and measure the world to catalyse discovery. They built lenses, detectors and models needed to find the answers. In chemistry, physics and astronomy, this combined number is higher—more than 60% of discoverers. It shows how powerful it can be when we see not just a problem—but a better way to measure or explore it. It is also a strong sign that science is not just idea-driven or theory-driven—it is tool-driven.

So two key strategies we have to gain knowledge in science are: by tracking changes over time (historical analysis) and comparing different groups in a population to spot differences (comparative analysis). Just studying part of a population at one point in time offers one part of the evidence. When we analyse both, we can uncover the bigger picture: how discovery differs across time and fields (Figure 1.4).

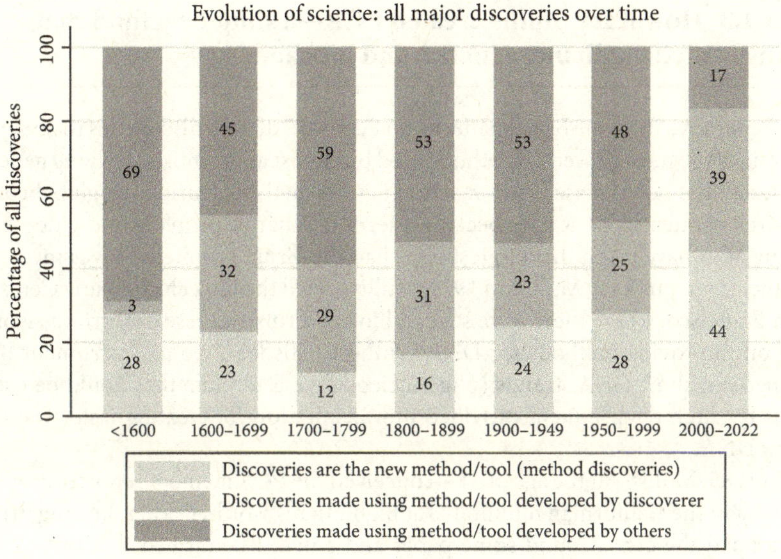

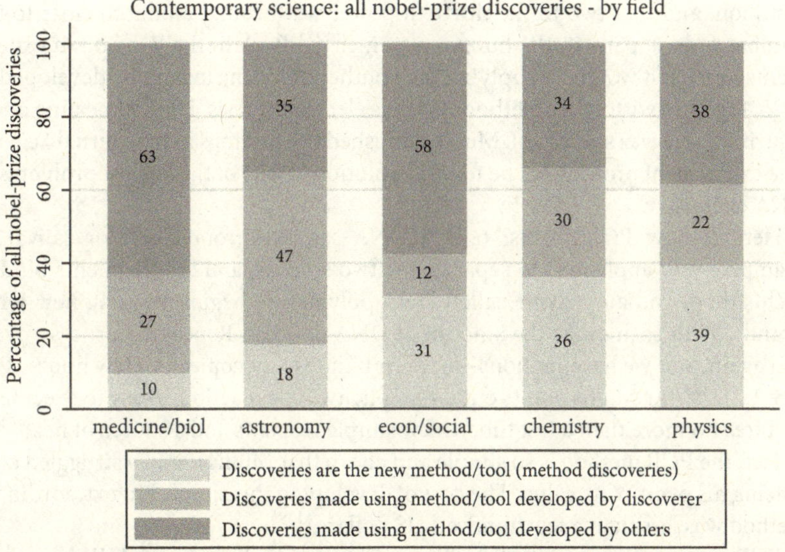

Figure 1.4 *The three pathways of how new methods and tools drive new breakthroughs—across time and fields*

The data reflect science's 761 major discoveries (Figure a)—and all 533 nobel-prize discoveries (which we can think of as a measure of contemporary science) (Figure b). Here we expand the analysis across these five fields in Figure b to include major non-nobel discoveries made over the same time period—bringing the total to 633 discoveries. The results are similar: the share of method discoveries for example is 9%, 12%, 28%, 35% and 36% (left to right). This consistency suggests that the pattern is not just driven by nobel-prize discoveries, but reflects a broader dynamic across contemporary science.

Box 1.3 How Kary Mullis created PCR—a simple method that transformed medicine, genetics and biochemistry

We now move from the broad patterns that connect diverse discoveries to zoom in on one of the most powerful methods used by almost all scientists studying genetic material: the polymerase chain reaction (PCR) method. During the global coronavirus pandemic, PCR tests became a term familiar to people worldwide—and many of us have taken them to test if we had the virus. The method was invented by the American Kary Mullis. In 1979, Mullis joined the biotech company Cetus in San Francisco, where there were several biotech firms and research groups working on improving methods for DNA synthesis. His lab used an instrument that turned out short DNA strands (oligonucleotides) faster than they could be used. This research environment provided him inspiration for creating faster ways to copy DNA.

In 1985, Mullis—at the age of 41—conceived the PCR method, reportedly while driving in the Californian mountains at night. In his words, during the long drive, 'I stopped the car ... found some paper and a pen. I confirmed that two to the tenth power was about a thousand and that two to the twentieth power was about a million, and that two to the thirtieth power was around a billion, close to the number of base pairs in the human genome'.[127] Back in the lab, he ran experiments to see if it worked—applying the Southern blotting technique developed in 1975, together with other methods such as electrophoresis, DNA annealing and a centrifuge. He was successful. Mullis published the findings in the journal *Science*. The experiment proved that he found a solution to one of the biggest problems in DNA chemistry.

Here is how PCR works: take a DNA sample—from blood or saliva for example—and apply heat to separate the two strands, and the fragments bind to each strand, and an enzyme called DNA polymerase begins building new DNA strands. With each cycle, the amount of DNA doubles. Repeat the process, again and again, and we have millions—or even billions—of copies in a few hours (Picture 1.5).[128] As Mullis points out its simplicity, 'The reaction is easy to execute: it requires no more than a test tube, a few simple reagents and a source of heat'.[129] In fact, the PCR method is so simple and cheap that Mullis initially struggled convincing his peers of its value. The cost of developing this simple but extraordinary method was less than a few hundred US dollars.[128]

PCR is now applied in medical diagnosis to detect diseases like HIV infections or identify genetic disorders and even in palaeontology to retrieve DNA from fossils of extinct animals. We also use PCR in police investigations to analyse the DNA of a drop of blood or a hair strand. After creating PCR, Mullis soon left Cetus—receiving a 10,000 US$ bonus for discovering the new method. The company later sold the rights to PCR for 300 million US$. Yet in 1993, Mullis was awarded the Nobel prize—and then turned to his passion, writing. As we uncover with other cutting-edge methods across science, the PCR method is just another example of how we make low-cost but major new methods and discoveries.

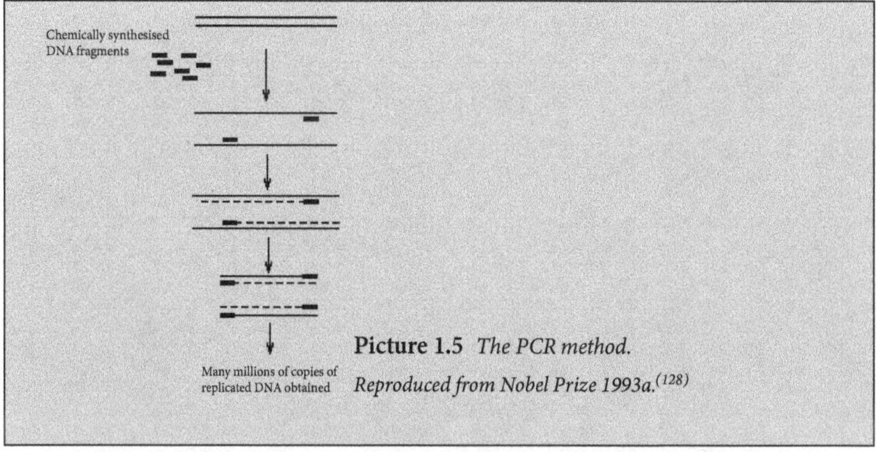

Picture 1.5 *The PCR method.*
Reproduced from Nobel Prize 1993a.[128]

New breakthroughs often follow soon after developing the new enabling tools

How long does it take for a new tool to spark a major discovery? The gap in time is getting shorter over time: in the 1800s, it was an average of 30 years; in the first half of the 1900s, it reduced to 21 and then in the second half to 10 years; and it finally fell to 6 years between 2000 and 2022 (as we see in Figure 1.5a). Since 2000, the average time from tool to discovery has narrowed to 5 years in chemistry—and 1.5 years in medicine and biology. Yet this pattern has taken place in a largely improvised, piecemeal way and without a general awareness or theory—without realising that there is a principle of tool innovation behind discovery. As we would expect, the gap in time is shorter for discoverers who create a new tool (seen in the darker bars in Figure 1.5b). This is important: method-inclined researchers see quicker returns and impact.

The decreasing delay over time illustrates that science—in part—is becoming more efficient. Advances in new technologies like higher-performance computers, quicker data analytics, high-speed imaging and rapid AI and machine learning methods enable running faster experiments, processing data more quickly and uncovering patterns that would have been impossible a few decades ago. But we are only scratching the surface and there is enormous room to make better tools and discoveries and make them faster and smarter—as we explore in depth in Chapter 6.

A key insight emerges here: discoveries generally follow soon after the new key tool is developed. In fact, we find that 10% of discoveries are sparked in the same year—or simultaneously—that the enabling tool is created, 21% emerge within two years, 33% within four years and 53% within ten years. Since 1975, most discoveries—51%—arise within just four years of the new method innovation. This highlights how innovation in tools is inseparably linked to innovation in science (Appendix Figure 1.9).

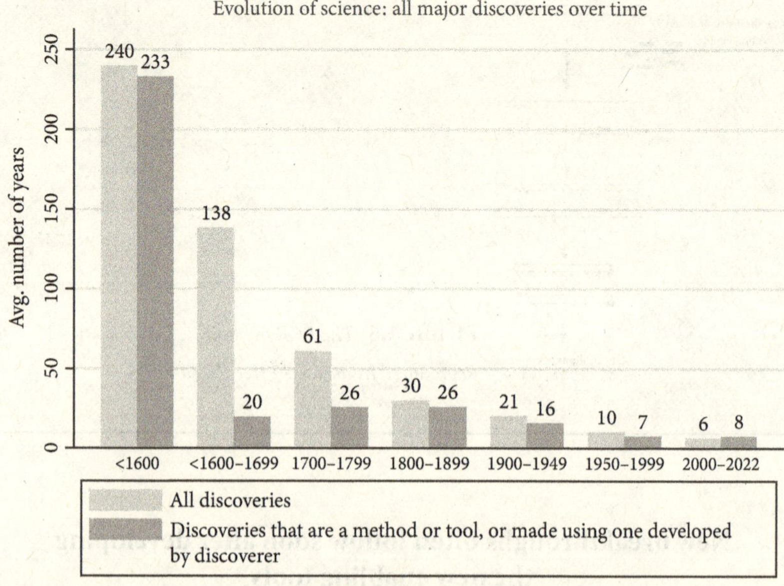

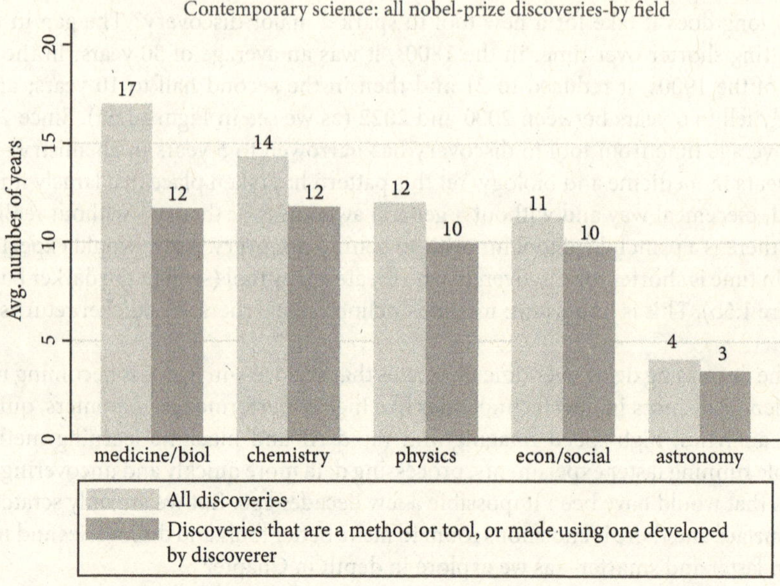

Figure 1.5 *Gap in years between the developed method or tool and the discovery it enabled—across time and fields*

The data show science's 761 major discoveries (Figure a)—and all 533 nobel-prize discoveries (Figure b). Here we expand the analysis and combine all nobel-prize and major non-nobel discoveries during the same time period across these five fields (in Figure b); we find similar results: for example, the share of discoveries that either are a method or tool—or are made using one developed by the discoverer—remains consistent at 14%, 12%, 12%, 10% and 3% (left to right).

What patterns emerge across fields? Astronomers are by far the quickest in exploiting new tools: more than half (54%) of nobel-prize discoveries are triggered within two years—and a remarkable 88% within ten years. Take the discovery of a supermassive compact object at the centre of our galaxy, the Milky Way. It was uncovered in 2002 by the German Reinhard Genzel and American Andrea Ghez. The groundbreaking finding was sparked by a new adaptive optics spectrometer constructed the same year, together with the world's largest telescopes.(130) Or consider the discovery of the first exoplanet. It was revealed in 1995 using a new fibre-fed echelle spectrograph built in 1993 at the Haute-Provence Observatory in Southern France.(116) This astonishing breakthrough redefined how we view our own planet—and our very place in the vast universe. Some fields are slower at translating new tools into new breakthroughs. In medicine and biology, 51% of nobel-prize discoveries emerged within 10 years; in economics, it is 54% (Appendix Figure 1.9). But how we can narrow the tool-to-discovery gap across fields is a key question we tackle later.

What is one of the most direct signs of the power of our tools in driving science? When the moment a tool is invented, it sparks a breakthrough. Take the spectrograph, created in 1859 at the University of Heidelberg. As soon as it was built, it enabled discovering atomic light signatures: that each chemical element gives off a unique light signature. This groundbreaking instrument allows us to produce properties of light in the electromagnetic spectrum, measure the mass of atoms and molecules and identify what planets and stars are chemically made of. The spectrograph became a cornerstone of modern astronomy, chemistry and physics. Take the microscope designed with a silver staining technique in 1873. It immediately led to discovering the structure of our nervous system—and the first view ever inside our previously invisible nerves and how they could function (Picture 1.6).

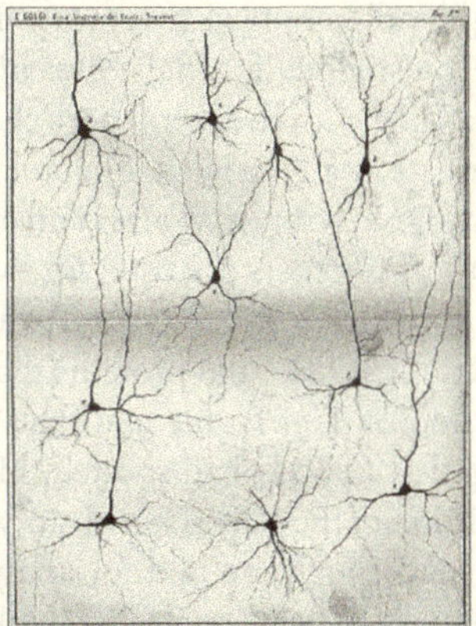

Picture 1.6 *The first image of a nerve cell, depicted with a new silver staining technique. Reproduced from Wikimedia Commons.*

The mass spectrograph invented in 1919 immediately sparked the mapping and identification of different isotopes—atoms of the same element that have different masses. Some are stable, others radioactive. Isotopes help power medical imaging, cancer treatment, carbon dating and even natural resource exploration. The frequency variation method designed in 1924 immediately made the experimental detection of the ionosphere possible—a layer of our upper atmosphere at the edge of space. It is home to many of our satellites and influences our radio and GPS signals. An atom trap—called the time orbiting potential trap—developed in 1995 along with laser cooling immediately enabled the discovery of an ultra-chilled state of matter: Bose–Einstein condensate. This extreme state of matter allows producing nanotechnology and manipulating matter on an atomic scale. A powerful expression cloning strategy created in 1997 immediately led to discovering our elusive receptors that detect temperature. It revealed how our body translates heat and cold into electrical signals in the brain. This fascinating breakthrough by the American David Julius at the University of California and the Lebanese Ardem Patapoutian at Scripps Research in California uncovered the molecular basis of touch, pain and temperature.

These are just a few examples of the fastest route to discovery: creating a tool that instantly reshapes what we can detect, measure or understand. And it is not rare. *The immediate method-to-discovery link makes up about one in every eight discoveries since 1975—and about three in every ten discoveries are uncovered within two years. That is not just remarkable, it is key to understanding the nature of discovery* (Appendix Figure 1.9). In some fields, it can take a year or longer to run experiments with a new tool, write up the study, go through the peer review process—in one or more journals—and publish the breakthrough findings.

Discovery time lags exist across science: defined as lags in making breakthroughs when a new method already exists for years but is not immediately applied to tackle a specific problem. They are important missed opportunities to spark advances using new methods not yet exploited—within and across fields—over longer periods. Take the very innovative technique called optical tweezers—initially created in physics in 1970 and published in the prestigious journal *Physical Review Letters*. This laser-based tool can grab and hold (like with tweezers) particles using beams of light. But it was not until 1987 that scientists finally applied it in biology and medicine. Suddenly, researchers could manipulate DNA strands, viruses and even living cells with extraordinary precision. Today, optical tweezers are used for everything from studying the structure of proteins to running diagnostic tests for diseases like malaria.[131]

Take the discovery of cyclin in 1982—the proteins that control the cell cycle. The discoverer, the British biochemist Tim Hunt, said that 'I did a very simple experiment ... It is very surprising that nobody had spotted this before; it could have been done any time after the invention of the slab SDS polyacrylamide gel [electrophoresis] in about 1970'.[132] That breakthrough method for analysing proteins was invented by the Swiss molecular biologist Ulrich Laemmli—published in a landmark article in Nature, *Cleavage of structural proteins*. It has played an enormous role in helping understand cancer and designing drugs. The tool was ready; but no one had yet used it that way.

The same pattern appears again and again across science. In development economics, one of the greatest transformations in understanding the causes of poverty and low education came from leveraging a unique experimental method: randomised controlled trials (RCTs). Employing the method, Michael Kremer, Abhijit Banerjee and Esther Duflo received the Nobel prize for having 'introduced a new approach to obtaining reliable answers about the best ways to fight global poverty'. Yet the RCT method had long been developed in medicine[(133)]—with over half a million RCTs already conducted up to 2003,[(134)] when economists used it for the first time.[(135)] Economists did not reinvent the RCT method but applied it to economic questions. In fact, the earliest experimental trial is often traced back to 1747, when Scottish physician James Lind proved that oranges and lemons are effective in treating scurvy (a severe lack of vitamin C) among sailors at sea.[(136)]

So why do we have large discovery time lags across science? When there is a delay of more than several decades, the new tools are often developed in other fields and not immediately known in one's own field. Think of optical tweezers and the RCT method already long employed in other fields. Because researchers are rarely trained to look beyond their own field for methods, many important discoveries are unnecessarily delayed. So some discovery delays stem from artificial borders across fields and their methods. Yet methods are still the binding constraint—but disciplinary isolation reinforces that constraint and creates a longer gap. Other discovery delays stem from scientists trained to focus on findings and theories, not always on testing and experimenting with a wider range of tools that enable them. The key is that whether there is a long time lag or not, the main spark of discoveries is generally *the new tools never applied before to the specific problem*—even if created in one field and adopted later in another. So discovery delays often represent some of the most overlooked opportunities and low-hanging fruit in science. Because new tools can help cure diseases and improve society, every year or decade they lie unused is a year or decade of missed impact. Closing these gaps accelerates breakthroughs—from life-saving treatments to advances in clean energy. Ultimately, none of these transformative discoveries would have been possible without these remarkable tools—hidden in plain sight in other fields (Table 1.1).

But with time lags for years—or even decades—for some discoveries, what is then a *new* tool? A new tool does not just need to be invented, it also has to be used. Innovation comes from application, not just invention. We find that 70% of discoveries in the last half-century are made within ten years of creating the enabling tool—and 90% within twenty years (Appendix Figure 1.9b). Even if a tool was invented decades ago, the first time it is ever applied to a new problem is what commonly unlocks breakthroughs. Yet the longest delays are mostly in the early history of science, when the scientific community was smaller, communication was limited and researchers were scattered across vast distances—significantly slowing progress. Take the telescope. It was invented in 1608 and then improved, yet major breakthroughs like discovering the distance to the sun in 1672, the motion of stars in 1719 and galaxies in 1750 came much later. Even longer time lags occurred before the 1600s. In these exceptional cases, the tool had never been used before to tackle the problem, but with the new lens, the discovery was then possible. Scientific progress relies not just on invention but access: on getting the right tool in our hands.

Table 1.1 Discovery time lags: delays between the new tool and a discovery (that slow real-world impact)—examples across different time periods

Making the discovery is within *x* number of years of developing the method used	***Method or tool developed* → the *discovery made* using it**			
	1600–1900	*1901–1990*	*1991–2020*	
0 years (same year)	pneumatic trough, improved (1774) → oxygen (1774)	electron tube, improved (1913) → laws governing impact of electron on an atom (1913)	time orbiting potential trap (1995) → Bose–Einstein condensate (1995)	*Impact of new methods and thus discoveries directly realised*
	spectrograph, improved (1859) → atomic light signatures (1859)	frequency variation method (1924) → ionosphere (1924)	spectroscopy, improved adaptive optics (2002) → supermassive compact object at galaxy's centre (2002)	
	microscope, with silver staining technique (1873) → structure of nervous system (1873)	electron image processing, improved (1990) → electron cryo-microscopy for bacteriorhodopsin (1990)	LIGO detector, improved (2015) → observing gravitational waves (2015)	
1–5 years	telescope (1608) → Jupiter's moons (1610)	starch-column chromatography (1948) → synthesis of polypeptide hormones (1953)	fibre-fed echelle spectrograph (1993) → first exoplanet (1995)	
	microscope, improved (1662) → cells (1665)	particle accelerator, Bevatron (1954) → antiproton (1955)	neutrino detector, improved (1996) → neutrino oscillations (1998)	
	quadrant electrometer, improved (1897) → chemistry of radioactive substances (1899)	spark chambers, improved (1959) → symmetry violation in decay of neutral K-mesons (1964)	in vivo coimmunoprecipitation assay (2000) → oxygen-sensing mechanism in cells (2001)	
Time lag	thermometer, mercury (1714) → Fahrenheit scale (1724)	metal-oxide-semiconductor (1960) → CCD sensor (1970)	electron density map, improved (1991) → spatial structure of potassium ion channel (1998)	*Impact of new methods and thus discoveries postponed*

6–10 years	leyden jar (1745) → nature of electricity (1752)	molecular spectroscopy (1939) → magnetic moment of electron (1947)	telescope, Hubble space (1990) → accelerating expansion of universe (1998)
	galvanometer (1820) → Ohm's Law (1827)	ultraviolet spectrophotometer, improved (1954) → configuration of ribonuclease (1961)	microinjection, improved (1991) → RNA interference (1998)
11–20 years	battery (1800) → electromagnetism (1820)	chromatography (partition) (1941) → carbon dioxide assimilation in plants (1952)	laser, titanium sapphire (1982) → stimulated emission depletion microscopy (1994)
	isomer counting (1875) → linkage of atoms in molecules (1893)	transistor (MOSFET) (1960) → quantized hall effect (1980)	Northern blotting technique (1977) → toll-gene in immune system (1996)
	discharge tube (1875) → x-rays (1895)	electrophoresis (SDS-PAGE), improved (1970) → cyclin (1982)	atomic force microscopy (1986) → production of graphene (2004)
21–50 + years	eudiometer, improved (1777) → atoms (1803)	superconducting quantum interference device (1964) → superconductivity in ceramic materials (1986)	...
	galvanometer (1820) → Joule's law (1841)	x-ray crystallography (1913) → structure of DNA (1953)	...
	microscope, improved (1826) → antiseptic medicine (1867)	ultraviolet spectrophotometer (1941) → crown ethers (1967)	...

Methods or tools—and the discoveries they enabled—are based on science's 761 major discoveries—with examples shown by time period and time lag.

We face time lags in not only generating discoveries but also realising the importance of some. When the Moravian-Austrian monk Gregor Mendel statistically studied peas, his research went largely unnoticed for decades, but eventually contributed to how we understand inheritance and genetics. When the British Charles Babbage and Ada Lovelace designed the first mechanical computer and computer programme, and when the British Ronald Fisher vastly expanded statistical methods,[137] little did we know that most of science—including governments and our everyday lives—would rely so heavily on computers and statistics. We do not always immediately grasp, and cannot yet imagine, the important impact that discoveries eventually trigger. The same goes for the Wright brothers. They tested flying gliders in a small town in North Carolina that eventually paved the way for airplanes and how remarkably connected we are today around the world.[101] Yet such seminal breakthroughs revolutionise our lives and how we understand life itself.

Let us look beyond the timing of tools to an important question: have scientists been able to predict some discoveries before they happened? Some discoveries have in fact been anticipated because the right tools and knowledge were already in place. Yet predicted breakthroughs make up only a small fraction of major scientific advances. Take the discovery of DNA's structure. DNA was first isolated in 1869, using chemical extraction methods and new advanced microscopes capable of observing cell nuclei. Once DNA was detected, we knew—and could predict—it must have a structure. And once x-ray crystallography was invented in 1913, we also had the necessary method at our disposal that could reveal molecular shapes. It was only a matter of time. Eventually, researchers used Rosalind Franklin's crucial x-ray images to uncover DNA's double-helix structure—in 1953.[1]

A similar story played out with the human genome. After inventing DNA sequencing in 1977, it was no longer a question of *if* we could decode the entire human genome, but *when*. A powerful existing method—gel electrophoresis—played a crucial role in separating DNA fragments in the sequencing process. What followed was a competitive race across research teams for finding the best way to sequence the whole genome. The publicly-funded Human Genome Project was launched under geneticists James Watson and Francis Collins. A large privately-funded initiative was led by biotechnologist Craig Venter at his genomics firm. In 2000, President Clinton and Prime Minister Blair intervened, requesting open access to the data to coordinate efforts. An integrated method for genome mapping developed in 2001 sped up the discovery process—and the finish line came into view. It ultimately enabled achieving the incredible feat in 2003: mapping the entire human genome. This milestone transformed modern biomedicine—allowing us to better diagnose and treat diseases—and also led to major advances in forensics and even our understanding of human evolution.[113]

A particularly exceptional case is gravitational waves, predicted a year after Einstein published his general theory of relativity in 1915. Gravitational waves were predicted to move through space at the speed of light—and they were, at first, a hypothetical phenomenon, speculative for Einstein himself. Einstein's theory was made possible through advanced mathematical methods and experimental findings, especially by Michelson and Morley on the speed of light in 1887 that applied the recently invented interferometer. Even those discoveries that at times seem—at first glance—to be largely theoretical breakthroughs rely heavily on recently developed powerful tools (Chapter 6). And later, we also first had to create an instrument sensitive enough to physically detect the waves. A century later, the LIGO detector was constructed—and it was updated in 2015, and within just days, it captured gravitational waves. The detected waves were about 1.3 billion light-years away, meaning the signal had travelled for over a billion years before reaching the earth.(115,138,139)

Another example is the periodic table of elements—created by the Russian chemist Dmitri Mendeleev in 1869 while working at Saint Petersburg State University. He achieved the breakthrough using an early classification of chemical elements in 1829 and the mathematical law of octaves developed in 1865. He also indirectly relied, beforehand, on several new elements recently identified using the new spectrograph in 1859. And afterwards, the spectrograph was also used to identify some of the predicted elements where empty places remained in the periodic table. It was just a matter of detecting the missing elements we knew existed. The elements, as a whole, make up the world we live in and the foundation of chemistry. So uncovering these exceptional discoveries is often a question of developing or applying the missing method—filling in the missing pieces (as seen in Figure 1.6).

Creating a new method or tool is often more foundational to scientific progress—across different fields and over time—than individual scientific discoveries

Many discoveries are almost inevitable once we create a new tool that enables seeing and measuring further. These breakthroughs range from how we discovered capillaries in 1661, cells in 1665 and bacteria in 1674 after inventing the microscope in 1590. To how we found mitochondria in 1898 once we designed an improved achromatic microscope in 1841. These breakthroughs span from how we uncovered Jupiter's moons in 1610, Saturn's rings in 1655, galaxies in 1750 and Uranus in 1781 after constructing the telescope in 1608. To how we revealed that the universe is expanding in 1929 after making the needed telescope in 1917 and how we detected the first planet outside our solar system in 1995 after building a high-resolution spectrograph in 1993. These extraordinary new tools sparked these discoveries that no one had ever imagined—by extending what we can observe. And they sparked the later theoretical explanations to describe them. These discoveries were triggered by what the tool in our hands made possible. And they were highly likely once we invented the needed tool. Before these tools, we did not know these even existed—the same with microscopes later triggering breakthroughs from viruses and chromosomes to nerve cells and tissues.

It is difficult to argue that—for example in astronomy—one discovery (such as Jupiter's moons) may be more influential for scientific progress in the field than other discoveries (such as galaxies or the expanding universe). Yet none of these discoveries

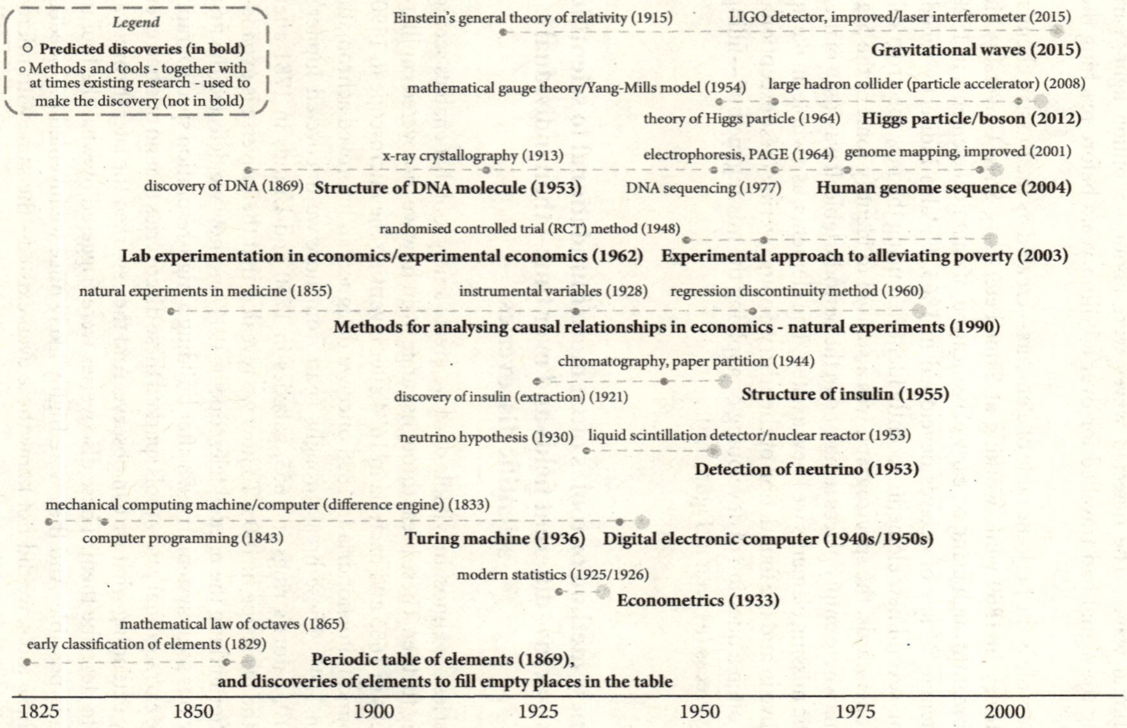

Legend

○ **Predicted discoveries (in bold)**
∘ Methods and tools - together with at times existing research - used to make the discovery (not in bold)

Einstein's general theory of relativity (1915) — LIGO detector, improved/laser interferometer (2015) — **Gravitational waves (2015)**

mathematical gauge theory/Yang-Mills model (1954) — theory of Higgs particle (1964) — large hadron collider (particle accelerator) (2008) — **Higgs particle/boson (2012)**

discovery of DNA (1869) — x-ray crystallography (1913) — **Structure of DNA molecule (1953)** — electrophoresis, PAGE (1964) — DNA sequencing (1977) — genome mapping, improved (2001) — **Human genome sequence (2004)**

randomised controlled trial (RCT) method (1948) — **Lab experimentation in economics/experimental economics (1962)** — **Experimental approach to alleviating poverty (2003)**

natural experiments in medicine (1855) — instrumental variables (1928) — regression discontinuity method (1960) — **Methods for analysing causal relationships in economics - natural experiments (1990)**

discovery of insulin (extraction) (1921) — chromatography, paper partition (1944) — **Structure of insulin (1955)**

neutrino hypothesis (1930) — liquid scintillation detector/nuclear reactor (1953) — **Detection of neutrino (1953)**

mechanical computing machine/computer (difference engine) (1833) — computer programming (1843) — **Turing machine (1936)** — **Digital electronic computer (1940s/1950s)**

modern statistics (1925/1926) — **Econometrics (1933)**

early classification of elements (1829) — mathematical law of octaves (1865) — **Periodic table of elements (1869), and discoveries of elements to fill empty places in the table**

1825 1850 1900 1925 1950 1975 2000

Figure 1.6 *Predicting some discoveries: examples of breakthroughs anticipated before they happened*

The timeline shows when the central methods and discoveries were made—and the moment when the breakthrough could be predicted. These predicted discoveries (shown in bold) received a Nobel prize, except for the digital electronic computer, the Turing machine and the periodic table of elements.

would have been even conceivable without first inventing the telescope—created by a Dutch lensmaker, Hans Lippershey, in his eyeglass shop in 1608. He designed the first telescope to help magnify distant objects by mounting lenses within a tube. A tool he thought could 'likely to be of utility to the state' especially for military or naval use. And he had no connection to science.[(140)] From a new highly sensitive spectrograph that enabled discovering many exoplanets since the 1990s, to a new radio telescope that sparked the discovery of pulsars and quasars in the 1960s, this pattern repeats itself. It highlights the fundamental role of a single instrument: one that sparks multiple unintended discoveries and makes it often the more foundational discovery itself. What makes them so powerful is that a single tool triggers a cascade of breakthroughs. Many researchers who would have gotten their hands on the new microscope, optical telescope, spectrograph or radio telescope could have made these discoveries that were not searched for—guided more by empirical observations than theory. They were also made without a large research team.

Discovery often comes down to which researchers get access to the new tool first. This often pushes our tool innovations into the foreground, and discoverers and other factors into the background of discovery. We trigger many major scientific advances not through testing a clear hypothesis, but through exploratory research using a new tool innovation—a topic we turn to in the next chapter on serendipity.

Exploring science's major discoveries, we uncover the striking rise and importance of the most prominent tools in science: hot streaks of discoveries cluster after we develop powerful new tools (as mapped out in Figure 1.7). The discoveries we unlock bundle together not by chance but by extending our toolbox—from x-ray methods and spectrometers to statistical methods. For each of these tools, the shared link across the diverse discoveries it enabled is not the same theory, teams or funding—it is the tool itself that made them possible. Each one unlocked findings across different disciplines. What commonly makes new tools the crucial driver of progress is not just that each of these tools has made multiple discoveries possible, but that many do so across different fields and in ways their inventors never imagined—and can also do so in the future (Figure 1.1 and Appendix Figure 1.3).

The causal effect: how new tools trigger new discoveries

We have seen how discoveries follow the invention of powerful new tools. Now let us dig deeper: how does this cause-and-effect relationship work? There are five pieces of evidence that show how the two link together.

First, science's top-ten most influential methods and tools were not designed with different discoveries in mind—but they still made them possible. Charles Townes did not invent the maser (the precursor of the laser) to unlock the structure of molecules. Ernst Ruska did not design the electron microscope to trigger major advances on viruses and nanotechnology. Max von Laue did not create x-ray diffraction to enable breakthroughs on the atomic structure of complex proteins or in materials science. And yet, each of these tools triggered more than ten such major discoveries. In fact, these ten general-purpose tools are responsible for 21% of all nobel-prize discoveries—over 100 of the 533 breakthroughs they could not have predicted. And

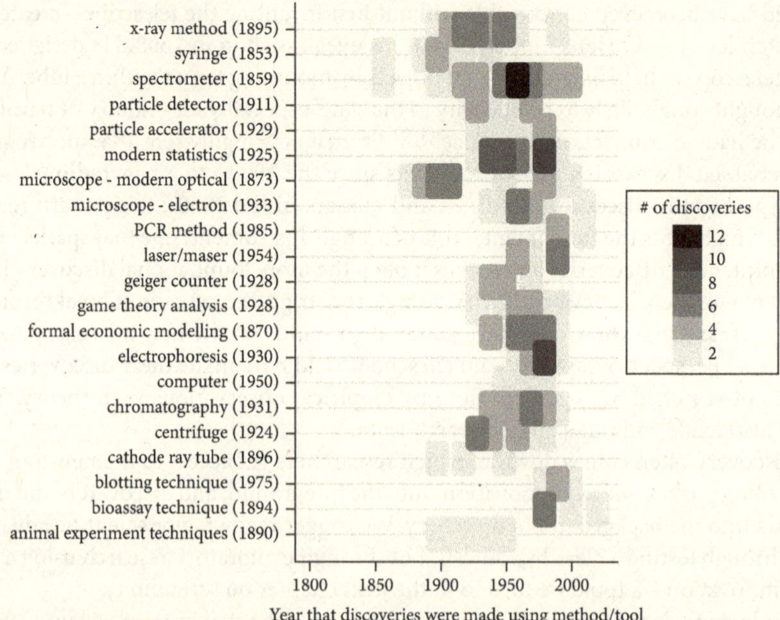

Figure 1.7 *Timeline of developing and using central methods and tools—each enabling clusters of major discoveries, since 1800*

The data reflect science's 308 major discoveries made since 1800—including all nobel-prize discoveries—using central methods and tools that each triggered four or more discoveries. The year shown marks when each method or tool was first developed—but all have since been vastly expanded. For modern statistics, 1925 is used as when the method was created—as it was the year Ronald Fisher, the father of the field, published 'Statistical Methods for Research Workers'. It marked the first full-length book on statistical methods and played a critical role in establishing and spreading modern statistics.[137] *For animal experiment techniques, these became standardised and widely adopted around 1890, including producing animals for experimental use.*

remarkably, about 55 methods and tools have led to two or more nobel-prize discoveries and make up about 80% of all those breakthroughs (426 of 533). This leads to a key insight: each powerful new tool unlocks *multiple* questions and findings we cannot predict. *And because each of the ten most powerful discovery tools uncovered multiple breakthroughs across different fields, their impact often outweighs the varying role of serendipity, funding or specific theories.*

Second, all top-ten tools, and most of the 55 most used tools in science, were invented in one field but went on to trigger discoveries in other fields. Lasers and electron microscopes came out of physics, but ended up transforming biology, chemistry and medicine. The tools' inventors did not anticipate the different breakthroughs across disciplines (Figure 1.1 and Appendix Figure 1.3). This is one of the strongest signs of a causal relationship: when scientists working in completely different fields pick up a tool and spark major discoveries using it—with supporting factors like collaboration and funding varying by context. It is not necessarily about who holds the tool, but what the tool can do across fields. Third, not just new tools, but extensions to science's best x-ray devices, lasers and centrifuges, improvements to our best statistical, microscopic and electrophoretic methods, have consistently brought about new breakthrough advances. Fourth, we have not found a major breakthrough that did not rely on a *novel* method or tool (Figure 1.2). If we want to uncover something new, we need a new lens, detector or technique to reveal it. There is no way around it. Fifth, many discoveries have not been guided by a hypothesis or theory, but by designing a new tool and doing exploratory research with it. In fact, serendipitous discoveries commonly depend on using a new tool that enables the surprising breakthrough observation. We explore this in depth in the next chapter.

We can think of a causal relationship between new methods and new discoveries using a kind of experimental logic: a quasi-experiment. How does this causal mechanism work? Before we develop the new enabling tool (*the treatment*), we cannot trigger the breakthroughs (*the outcome*); but after the tool is created, the breakthroughs follow—often soon. The odds of making many discoveries are zero beforehand, but jump when we develop the tools. The timing matters. The invention year is the *baseline*, and the discovery year is the *endline*. The breakthroughs—like uncovering DNA's structure with x-ray crystallography—were not possible without the enabling tools (*the counterfactual*). The breakthroughs are not bound to occur specifically when they do—but they cannot happen *without* the right method in hand.

Inventing a powerful new method is like getting the clock ticking towards uncovering a discovery that was previously out of reach. The biochemists Archer Martin and Richard Synge invented chromatographic methods that enabled breakthroughs they could not have imagined: from the structure of insulin and sugar nucleotides to carbon dioxide assimilation in plants. The physicist Max von Laue developed x-ray diffraction that led to x-ray crystallography and discoveries he could never have predicted: from the structure of vitamin B12 and penicillin to the DNA's double-helix, and so on.

Some major tools began with no specific purpose, like the first maser. Or with no scientific use in mind at all, like the first microscope and telescope. Or are serendipitously found, like x-rays. The physicist Charles Townes famously described his new maser—invented while at Columbia University—as 'a solution looking for a problem' since it had no clear relevance at first.[141] The first microscope and telescope were both constructed not by scientists but eyeglass makers as an extension of eyeglasses and simple magnifying glasses. At the time, no one imagined they would become foundational for modern biology and astronomy—with the telescope for example first thought to be useful for navigating (Chapter 7).[46,140] Röntgen discovered x-rays by unexpectedly seeing a glowing photographic plate near a discharge tube.[118]

This gives us a powerful way to understand cause and effect in science: if a tool is not designed to make specific discoveries, and yet still made them possible—often in completely unrelated fields—then the tool itself is the key cause. This is clear looking at science's most used tools. Using quasi-experimental reasoning, this lack of an initial relationship to the outcome (not originally intended as tools to make different discoveries) is important to identify the causal effect of the new tools in catalysing the discoveries they commonly were not even designed for. The microscope was not invented to discover bacteria, and a sensitive spectrograph was not created to reveal planets around distant stars—yet both tools enabled such transformative discoveries. This lack of an initial intended relationship helps isolate the new tool as the causal spark (the independent variable) in driving the new breakthrough (the outcome) that is otherwise not possible—see Figure 1.8. With the tools already developed, it also reduces other explanations—like just funding or research teams as the key triggers. Our tools directly unlock a major advance by providing a completely new perspective. Without them, supporting factors like money[6–8] and collaborations[9–11] are not enough. And without something to detect or measure, we cannot generally create and test theories (as we lay out in depth in Chapter 6).

It is not only discoveries that are surprising but also some tools are unforeseen. While eyeglass makers unexpectedly developed the first microscope (Zacharias Janssen) and the first telescope (Hans Lippershey) as instruments we could later use in science, they caused unexpected discoveries: from cells and bacteria, to the motion of stars, and galaxies (Figure 1.8). Why are these unintended discoveries? Because they were not just unintentional for the inventors—who had no idea the tools would be used in science[46,140]—they were often also unintentional for the scientists who made the discoveries using them. Their unexpected observations were not guided by theory. The theories only came after, built to explain what the tools had revealed. Again, *before these tools existed, the chance of making such discoveries was zero: there was no way to see exoplanets, or pulsars, or bacteria*. And then, suddenly, there was. That is as causal as science can get. New tools do not just *help* us discover—they *cause* discovery. This causal link holds across time, across fields—from medicine to chemistry—and across expected and unexpected discoveries. And we turn our lens to these unexpected breakthroughs in the next chapter.

With this consistent pattern across discoveries, can the powerful role of methods and instruments in science seem self-evident? Yet, what is clear is that there is no consensus among researchers on what powers scientific progress. That is striking, since all scientists share a common goal: discovery. But the current explanations of discovery do not place powerful new methods at the centre of major breakthroughs—as the key spark. Rather, they focus on broad supporting factors like funding, teamwork, chance and serendipity. In practice, science is generally done with *existing* methods and tools. That is the default—conventional research using current microscopes, statistical models and imaging techniques. But the engine of discovery is observing or measuring something we could not before. And that requires a new tool. That is the key novel insight here: discovering that science's major breakthroughs come from applying a *new* method or tool (not existing ones) to a problem for the first time. These tools expand the edges of what we can see and understand. They fill in blind spots to studying the world in ways not possible before. This is the missing piece of the puzzle in understanding how we cross a boundary we could not cross before—and it has been overlooked until now.

With new methods needed (and not just conventional ones), how do we actually develop them? Scanning science's over 750 major discoveries, we uncover four common ways. First, *we create entirely new methods* not yet conceived before—like electrophoresis that separates molecules. That technique made discovering DNA sequencing possible. [142] Second, *we extend methods in novel ways* that also represents a new method not yet used—like the ultracentrifuge that vastly expanded

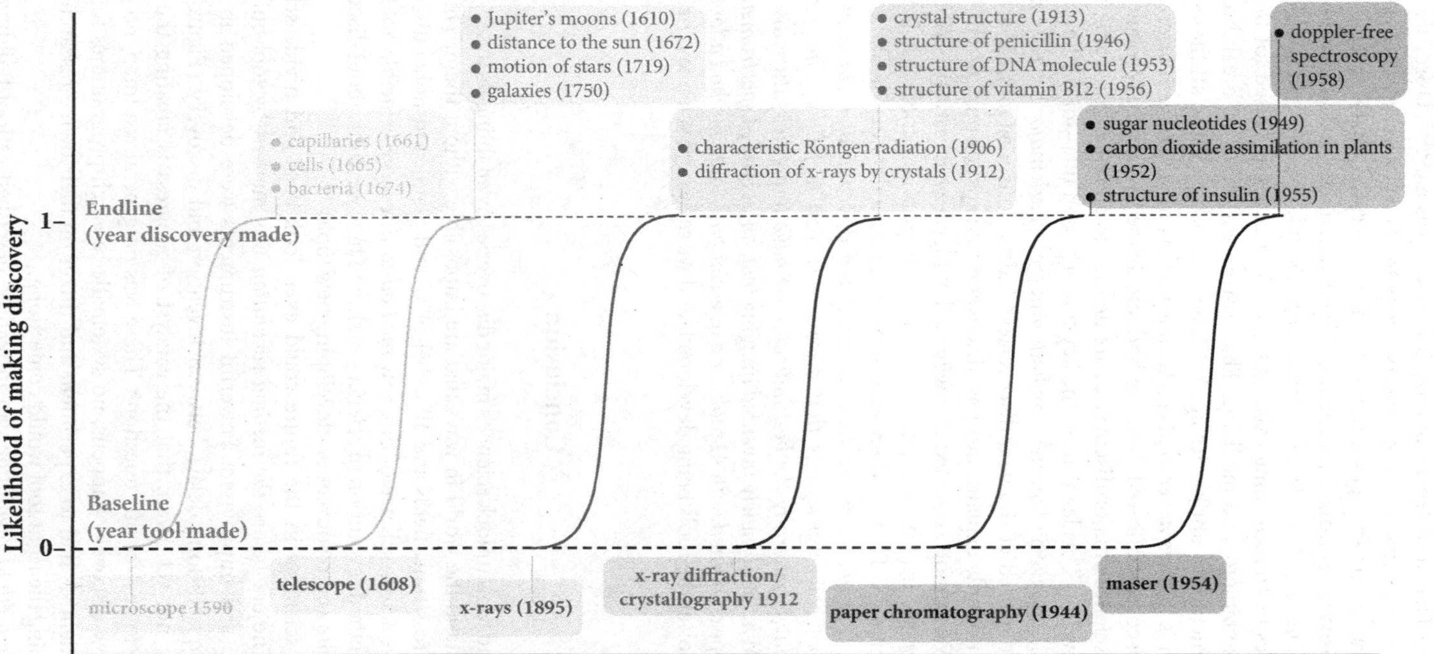

Figure 1.8 *The causal power of new methods and tools—unlocking discoveries they are not even designed for*

Each of the tools—and the discoveries they enabled—earned a Nobel prize, except for those made with the first microscope and telescope. Six examples of tools are provided.

early centrifuges. That tool led to uncovering blood plasma.[143] Third, *we combine methods in novel ways* that also reflects a new method not yet leveraged—like x-ray crystallography that merges x-ray methods with crystal analysis. That technique sparked discoveries from the structure of penicillin to DNA's double-helix.[120] Fourth, *we adopt new methods developed in other fields*—like the electron microscope created in physics by transforming existing light microscopes and used for the first time in fields like biology and medicine. That instrument unlocked breakthroughs from the cell structure to antibiotics against tuberculosis.[112] What all these cases have in common is that *we never applied the new method to the problem before.* And when applied, discovery followed. New methods are defined as created through these pathways—and science's major discoveries are defined as all nobel-prize and major non-nobel discoveries. This leads us to the key finding—and title—of the book: *The engine of scientific discovery: how new methods and tools spark major breakthroughs.*

So can we better predict the next discoveries? Indeed, this *new tools-driven discovery principle* not only explains our past discoveries, it expands our current limits of prediction. We cannot know exactly *who* will achieve *what* path-breaking discovery. But when we invent, upgrade, combine or adopt a new tool, that is when breakthroughs happen. That is when we can predict *where* the next breakthroughs can come from, and *when. By tracking the pace and direction of method innovations, we can spot powerful signals that discovery is soon to follow. The speed of tool-building today is generally the best indicator we have to predict future discoveries we can spur. Whether a newly invented imaging tool, an upgraded spectrometer, an advanced telescope combined with AI tools, or a new computational method adopted in another field.* Look at the tools being developed, and we see where science is heading next.

Conclusion

New methods and tools unlock science's major discoveries by enabling us to see, measure and understand the world in ways that are impossible without them. This key pattern holds across time, fields and all nobel-prize and major non-nobel discoveries. An awareness across science of this new *method-to-discovery principle* holds the potential for a method revolution in science—a shift in how we think about discovery and focus our time and resources on developing new tools.

Imagine that scientists in the future could even look back and divide science into two eras. The era *before the method revolution* was marked by tool-building that was ad-hoc and improvised. Powerful instruments were developed in scattered ways, often with long delays between invention and discovery (Figure 1.5). We were largely not yet aware that the catalyst of progress is creating the very methods we use to answer our questions. There was no general roadmap, no guiding theory of how discovery happens, no systematic research programmes for tool development. Scientists had to experiment and tinker on their own—researcher by researcher—hoping the right tool would come along.

The era of *the method revolution* can instead be described as tool-building that would be structured, planned and targeted. We would all have become aware of this powerful principle of new methods propelling science. Future scientists may look back at the advent of this revolution as the point when the scientific community as a whole began to strategically focus on refining, combining, restructuring and inventing tools that accelerate the pace of new breakthroughs. No longer would tool innovation be the side project of a few tool-inclined researchers who deviated from their initial academic training in established fields. Method invention would be at the centre of how science is done. It is a shift from asking *what* to study to asking *how* we can study better. It is about pushing science's current boundaries—from gene editing and brain mapping to exoplanet detection and climate modelling—with our hard-won tools. And we lay out how in Chapter 6. As the quantum revolution transformed physics in the 1920s and the cognitive revolution reshaped cognitive science in the 1960s, the method revolution could accelerate science at an unprecedented rate.

This principle of method innovation reflects a general rule we can apply across science. The powerful principle has proven very effective in explaining and uncovering new advances—and can better predict them. The pace of tool innovation is our best predictive signal of discoveries. And this principle can also help us lay the foundation for the science of science.

We will continue to advance our understanding of the world mainly at the pace of diversifying our toolbox. What is key here is that we need to shift part of our attention and research to refining the very methods and tools we use as a central part of the discovery process. This would open yet unexplored terrain, opportunities and research areas. Scientific progress would then no longer largely remain an uncoordinated and unorganised result of few method-curious scientists. It would be like moving from breakthroughs made by trial and error up to now, to sparking discoveries using systemic controls based on this methods-powered principle.

Ultimately, tomorrow's biggest discoveries will be made by those who create and apply the best new methods and tools today. The future of science—our next leaps forward—largely lies in tackling our tools' bottlenecks: the limits of how we measure, observe and test the world.

2
Engineering serendipity

How new tools make unexpected discoveries possible—and highly likely

Summary

Uncovering how new tools trigger new breakthroughs leads us to explore the most puzzling and mysterious feature of discovery: serendipity. Many think breakthroughs are serendipitous—an unexpected and fortunate observation. From the discovery of x-rays to superconductivity. In fact, many believe discovery itself is unexplainable and unpredictable precisely because of serendipity. But what role do chance and serendipity actually play in discovery? Are they what we commonly think they are? And has discovery been more random than it needs to be, just because we have not been looking in the right place? By studying science's over 750 major discoveries, we uncover a surprising and powerful pattern that has been overlooked until now. Discoveries often labelled as serendipitous are actually sparked by using—for the first time—a powerful new tool that makes the unexpected observation possible that we are not even looking for: an improved microscope revealed cells, a discharge tube uncovered x-rays, gel diffusion revealed the Hepatitis B virus and a cutting-edge spectrograph detected the first planet outside our solar system. What seems like chance is often just using an entirely new lens—just as in each of these discoveries. In other words, breakthroughs that seem serendipitous follow shortly after we create the new tool—with other researchers able to arrive at the same findings applying these same discovery tools. While a serendipitous moment was reported in about one in six major discoveries, there is a pattern behind the surprise: using new tools to make the breakthrough finding commonly through exploratory research. Why is this important? It changes how we think about discovery: we can generally better understand individual *serendipitous discoveries* as *tool-triggered discoveries* that multiple researchers can make. And most crucially, developing powerful new tools helps reduce chance and predict new discoveries—whether we label them serendipitous or not. So here we reframe discovery: because new tools are what commonly unlock surprise, we can actually design and accelerate discovery—shifting it from a more passive outcome of serendipity.

The Engine of Scientific Discovery. Alexander Krauss, Oxford University Press. © Oxford University Press (2026).
DOI: 10.1093/oso/9780197829790.003.0003

Overview

Are many discoveries serendipitous and unpredictable—or can we actually learn to better anticipate them? Are they more accidental—or more designed and engineered? Many think discoveries emerge as a sudden serendipitous moment, or a flash of insight, that we cannot study and measure systematically.(17–20,29,30,34) The term serendipity itself originally comes from a Persian fairy tale, *The Three Princes of Serendip*, who were 'always making discoveries, by accidents and sagacity, of things they were not in quest of'.(144) Seeing discoveries as arising serendipitously, defined as a moment when we make an observation or finding we are not searching for, is attractive. It is a concept that captures the imagination of both scientists and the general public.(19) The narrative of serendipity is compelling and appears in science textbooks, in studies of scientific discoveries(30,35,145) and in science news outlets.(33) An editorial in *Nature* argues that 'data are not available to track it in any meaningful way ... [so largely] academic research has focused on serendipity in science as a philosophical concept ... [and so] giving curious minds free rein to explore nature may well be the best way to generate discoveries'.(19) It seems natural to think we cannot study serendipity: after all, how could we analyse something that happens by chance? So is serendipity—by its very nature—unpredictable?

Some scientists, psychologists and philosophers have tried to study the elusive and difficult-to-measure role of serendipity. Scientists examining the careers of scientists suggest that serendipity plays a key role in breakthrough moments.(30,35,36) Studies of hundreds of thousands of scientific publications suggest that a scientist's most impactful research may be randomly distributed across their career—and appear at unexpected moments.(12,35,37) Scientists stress 'the profound unpredictability that pervades many aspects of scientific discovery'.(4) Psychologists explore how scientists reason, arguing that unexpected discoveries are surprisingly common in science(146)—and can be linked to creativity.(32) Even philosophers like Karl Popper argue that 'there is no such thing as a logical method of having new ideas ... every discovery contains "an irrational element"'.(17,28) The influential physicist Henri Poincaré also famously said that 'It is by logic that we prove, but by intuition that we discover'. And Nobel laureate Carlo Rubbia highlighted that 'Scientific discovery is an irrational act'. Other researchers also believe that thinking that 'the process of discovery has a distinct logic may have been vastly overstated'.(20) Yet arguing that discovery is irrational or unpredictable does not actually explain it at all—and leaves no direct way to intentionally spark it. It is also misleading—and can waste researchers' time and resources.

A keyword search of 'serendipity' up to 2025 in the journals *Nature* and *Science* reveals about 2000 hits. Yet despite the interest on the topic, no large-scale study of science's major discoveries yet exists on serendipity.(19) Researchers exploring the topic commonly focus on the unexpected observations made by discoverers—not what sparks those surprising observations across discoveries.(17–20,29,30,34) So have we been looking at the wrong side of the coin?

Here we test that hypothesis: is there a logic to how discoveries emerge, whether they seem serendipitous or not?(17,28) We will see that we can in fact establish a common cause that precedes new discoveries and makes them possible. But to do this, we need to systematically analyse science's major discoveries and serendipity at scale. Only then can we uncover a common hidden pattern that links seemingly random discoveries together.

So which discoveries do scientists describe as classic examples of serendipity? Galileo famously discovered the moons of Jupiter *serendipitously* in 1610, just like Hooke did when uncovering cells in 1665 and Röntgen when revealing x-rays in 1895.(147) But these scientists were only able to unlock these unexpected discoveries by *deliberately* using a powerful new tool: Galileo's recently invented telescope in 1608, Hooke's improved microscope built in 1662 and Röntgen's discharge tube created in 1875. These extraordinary tools enabled these three surprising observations they were not even looking for—rather than just a flash of insight. Take Arno Penzias and Robert Wilson who *unexpectedly* discovered cosmic background radiation in 1964—providing strong evidence for the Big Bang theory and transforming cosmology. But they *actively* applied a new highly sensitive radio receiver—a 20-foot horn antenna built in 1961—that made the surprising finding possible in the first place.(147,148) Even Robert Hooke himself pointed to serendipity when saying that: 'the greatest part of invention being but a luckey bitt of chance'.(149) But he was using his new advanced microscope to unexpectedly discover cells—the basic building blocks of life. Many also think of Kamerlingh Onnes' groundbreaking discovery of superconductivity in 1911 as unintentional.(150) But Onnes used a liquefier he created in 1906 and a sensitive thermometer he built in 1907 to run systematic experiments. These tools allowed him to measure temperatures as low as absolute zero and ultimately made discovering liquid helium and superconductivity possible.(102,124)

What do these classic serendipitous discoveries have in common? Could it be that they all happened when scientists used—for the first time—a new tool? They were not expecting what they saw, but could only make the surprising observations because they were using a tool that had never been used in that way before. So what if that is how serendipitous discovery generally works? Could the key to understanding serendipity be the tools we use to explore the unknown—rather than chance? What if the reason why discovery is commonly thought of as mysterious and unpredictable is because researchers have not yet studied it in a comprehensive way spanning across science's discoveries and fields to reveal what they share in common?

Framing discoveries as serendipitous makes for exciting stories of individual discoverers. We are fascinated by discoverers, often seen as geniuses. It seems natural to focus on the personal side of science, where discovery can seem random. And we are more intrigued by surprise than design. From the perspective of an individual discoverer, a moment of surprise can seem very unsystematic. But what if framing discoveries as triggered by serendipitous moments makes them seem much more blind than they actually are?

By studying serendipity among science's major discoveries, we uncover a key pattern here: breakthroughs that seem serendipitous are often made soon after we develop the new enabling tool. We will see how these tools make the surprising observations possible—and make them increasingly likely (before the discovery) and replicable by other researchers (after the discovery). The irony is that serendipity is seen as illustrating no structure in the discovery process. But by analysing these moments of serendipity across science and identifying the pattern driving them, we find that they actually reveal strong evidence of the mechanism behind discovery. The key to these unexpected moments is generally the tool—one that allows us to look at the world in a completely new way. Yet serendipity suggests a passive process—something that happens to us rather than something we can actively shape. But tool development highlights a systematic strategy that is far from passive: scientific progress depends on designing better tools that maximise the odds of uncovering new observations. It shifts the focus to creating the conditions where surprising moments become possible in the first place—and increasingly likely.

Thinking of discoveries as serendipitous is not only unhelpful in creating new breakthroughs, it also offers little guidance for policy on how to best fund science and incentivise researchers to spur innovation (Box 2.1).[19] So what if the unpredictability associated with discovery does not stem from researchers, scientific institutions or even timing? What if it comes from not yet realising that the very tools we use—and their blind spots—are largely shaping what we can see, and what we miss? What if the best and fastest strategy we have to generate new discoveries—whether they appear serendipitous or not—is to innovate and improve our tools? Could that be the missing key to reducing chance in science and sparking more new unexpected discoveries—from CRISPR to exoplanets?

Box 2.1 The mystery of discovery: Karl Popper's view on the illogical nature of breakthroughs

The Austrian philosopher of science Karl Popper published his seminal book in 1959: *The logic of scientific discovery*. Ironically, there he argues that identifying a logic of how scientific discovery emerges is not possible.[17] Popper's core argument is that discovery arises through a scientist's unexpected intuitions and chance observations—moments that cannot be predicted because each discovery involves an irrational element.[17,28] In his view, chance and serendipity play a strong role in discovery and make it unexplainable. As he points out, Einstein also argues that for universal laws of science: 'There is no logical path ... leading to these ... laws. They can only be reached by intuition.'[17] The consequence for Popper is to shift the focus from how discoveries happen to testing whether theories and discoveries hold up to scrutiny. In fact, for Popper, 'The question how it happens that a new idea ... [or] scientific theory [arises] is irrelevant to the logical analysis of scientific knowledge.'[17] For him, scientific knowledge grows when scientists propose theories and then test and attempt to falsify (disprove) them through experiments.

Other philosophers of science—from Reichenbach to Carnap—also think discovery is a deeply subjective process and highlight its unpredictable nature.[17,25,28,151] For them, discoveries are a mystery of human psychology.[17,25]

These philosophers do not focus on the process of discovery, but on assessing an output—scientific theories. After all, testing theories is easier to measure and evaluate than how we create them. In short, they neglect the fascinating process of how discoveries actually emerge. As we reveal here, we can ultimately uncover measurable features of discovery—and establish a logic of how we trigger breakthroughs.

Measuring serendipitous discoveries

In statistical studies, we generally cannot make a census or gather data on all individuals in a population because the numbers are too large. Instead, we take a sample—a smaller subset of the population—and use it to draw conclusions. Yet, collecting data as comprehensively as possible is crucial to draw general insights about discovery and the role of serendipity. Here we explore science's over 750 biggest discoveries—from those made by Marie Curie, Nikola Tesla and Charles Darwin, to Louis Pasteur, Rosalind Franklin, James Maxwell and Richard Feynman. This big-picture approach enables us to move from unpredictable features of a few select discoverers to a more predictable pattern across the total population of major discoverers.

To explain the extent and nature of serendipity, a discovery is categorised here as involving a serendipitous moment—including an unexpected observation or finding—if the discoverers described it as such in their original discovery publication. Discoveries are also classified as including a serendipitous moment if the discovery is described as such later in prize ceremonies and interviews of the discoverers—within nobel-prize documents,[95] one of six encyclopaedias of science[103-108] or the seven mentioned science textbooks. In fact, the description for most discoverers comes from two sources, Nobel prize speeches[95] and the discovery-making paper. Through this search strategy, we confirmed that we captured all well-known and classic cases of serendipity found in scientific publications. Discoverers' reported serendipitous moments are the only way to study whether a discoverer experienced such a moment as only they had access to the entire discovery process.

It is important to note that serendipity is not luck. Far from it, serendipity depends on our ability to use the needed tool to make the discovery and recognise it is a discovery. Luck means blind chance or passive stumbling—that we do not find. A serendipitous moment reflects one element in the broader discovery process that involves leveraging scientific methods and tools and often systematic research and experiments. We can also assess how effective tools are by applying both general criteria like empirical power and how generalisable they are—and tool-specific criteria like resolution, sensitivity, speed and scope. Tool development is precisely where science is most intentional and design-driven in response to bottlenecks in our tools—while serendipity, in contrast, is not about planning and strategy.

Uncovering the elusive role of serendipitous moments in scientific discoveries

How common is a serendipitous moment in science's major discoveries? About one in six. In total, 84% of discoveries have been sparked using just tools and methods and 16% using tools and methods and involve a serendipitous moment. So for discoveries with a serendipitous element, there is one thing they have in common: using a new method or tool—whether an electron microscope, an advanced x-ray method or sophisticated computer simulations—that generally made the unexpected breakthrough possible. No major discovery—nobel-prize or major non-nobel breakthrough—has been made by accident, without tools and methods, or by non-researchers.

Surprisingly, when we explore trends over time, we find that discoveries with a serendipitous moment are becoming less common (as we see in Figure 2.1a). We also uncover that 15% of nobel-prize discoveries involve a serendipitous moment while 18% of major non-nobel discoveries. So the results for both groups of discoveries are comparable and robust—and not just driven by one set of discoveries. But do these overall trends mask important differences across fields? Indeed, discoveries with a serendipitous moment are concentrated most heavily in astronomy—both for nobel-prize and major non-nobel discoveries. In fact, they represent about a quarter of nobel-prize discoveries in astronomy (Figure 2.1b). And yet, what is key is that all of these astronomical breakthroughs were made possible by newly developed instruments. These tools enabled us to spark the surprising observations about the universe in a way we never could without them. Take pulsars for example—unexpectedly discovered in 1967 using an immense new radio telescope built the same year at the Mullard Radio Astronomy Observatory. The serendipity was not in stumbling upon something by chance—it was in deliberately using this new tool that allowed revealing something completely new and surprising in the universe.

Vision-enhancing tools trigger unexpected discoveries most often

What kinds of instruments uncover unintended discoveries? We find a striking pattern: most often serendipitous breakthroughs come from using tools that enhance our vision—and enable seeing what was previously invisible. Following these come tools designed to separate substances—like chromatography, electrophoresis and centrifuges; and then come other tools—such as thermometers and Geiger-Müller counters. In fact, about one third of discoveries with a serendipitous moment were made alone using our vision-enhancing tools: microscopes, telescopes, x-ray methods, spectroscopes, cathode ray tubes and other particle detectors (see Figure 2.2). Each vastly expands our visual range to the world—and sparked the unexpected. The types of questions we can answer with these tools are questions about 'what is out there'. These are often very different from the questions we can solve with mathematical and statistical methods, commonly used after we know 'what is out there'. Discoveries sparked using statistics and mathematics are less likely to involve

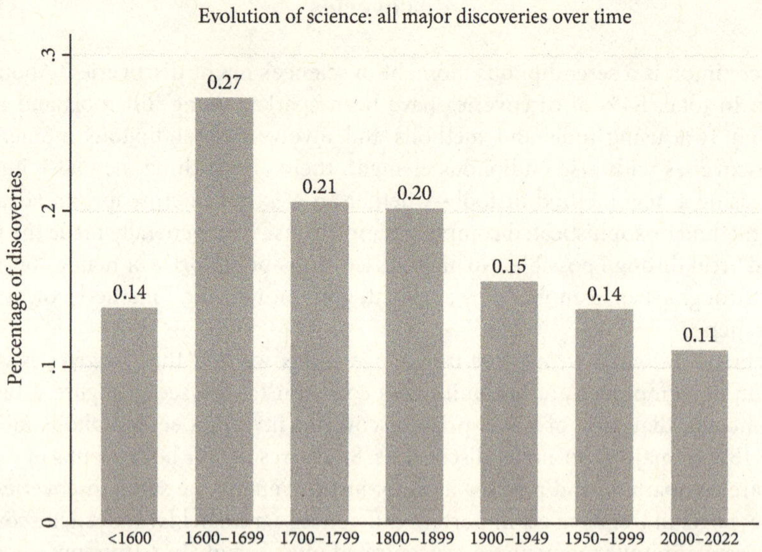

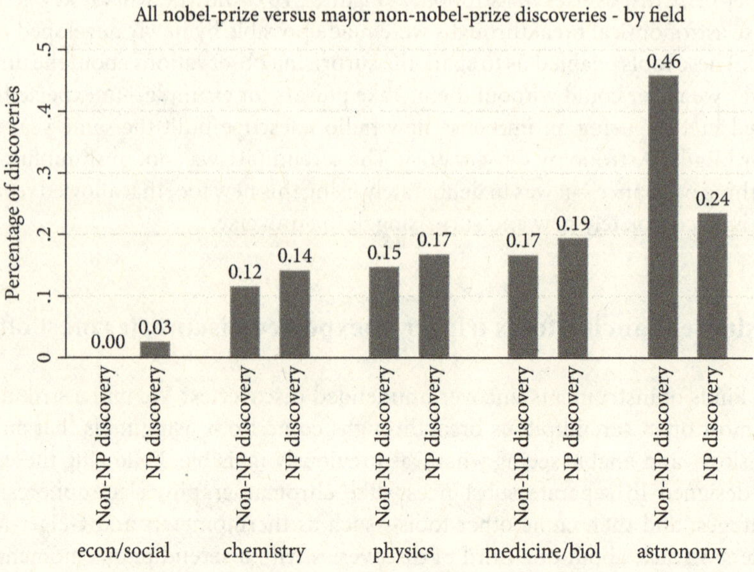

Figure 2.1 *Discoveries sparked with a serendipitous moment are decreasing over time—and are most strongly concentrated in astronomy*

The data reflect science's 761 major discoveries (including all nobel-prize discoveries) (Figure a), and all 533 nobel-prize discoveries compared to 228 major non-nobel discoveries—as an independent control and replication analysis (Figure b). NP stands for Nobel Prize. When we combine all nobel-prize and other major discoveries over the same time period across these five fields, we find that the aggregate shares are 2%, 14%, 16%, 19% and 37%.

serendipity, as they tend to be more deliberate and require much cognitive effort—and not just making a single new unexpected observation. We find this in fewer serendipitous breakthroughs in economics—that all used statistical methods or algebra. This is an important insight: for discoveries with a serendipitous element, we can distinguish between those we uncover slower and with more analytical effort and those we unlock faster and with less effort—especially in more visually driven fields.

Astronomy is a fascinating example where vision-enhancing tools have been the catalysts for big breakthroughs. In fact, we have sparked two thirds of major discoveries with a serendipitous moment in astronomy by using optical tools—a new telescope or spectroscope. In such cases, exploring the unknown leads to discoveries that seem profoundly serendipitous, simply because we had no way to predict what we are going to find in unknown terrain—like in a dark room before we suddenly install a new light. But the only way we achieve them is by designing and leveraging new tools of discovery.

This brings us to a crucial insight: when a serendipitous moment arises in the discovery process, it is often the new systematic discovery tools that enable the surprising

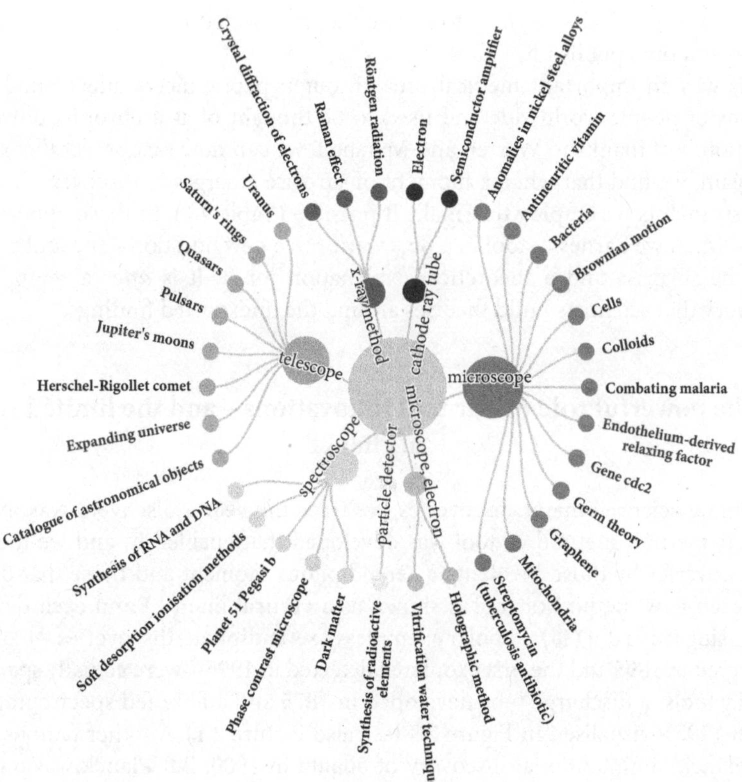

Figure 2.2 *Discoveries with a serendipitous moment made using vision-enhancing tools*

finding through exploratory research. What is more is that these tools allow other researchers to arrive at the same results, sometimes without yet a theoretical understanding about what was discovered. So serendipity is generally about using the right tool at the right moment to unlock surprise.

Serendipitous discoveries triggered by new tools and methods—concrete examples

We next explore discoveries with a serendipitous moment that happen right at the time of uncovering the discovery, and those that occur during the scientific process before discovery. The most popular examples focus on serendipity that arises at the moment of discovery itself—simply because they make for more exciting stories. A well-known example is the Australian Robin Warren's discovery of bacteria as the cause of gastritis and peptic ulcers. Warren was not looking for bacteria. In fact, he was studying gastric samples when he unexpectedly spotted them.[29] Yet working as a medical doctor at Royal Perth Hospital, he and his colleague Barry Marshall had to use a high-power electron microscope and fibre endoscope to be able to see the bacteria. They had the right instruments at the right time. But they also needed method training and medical knowhow to realise the unexpected finding is important and novel—and then publish it.

This was an important medical breakthrough: peptic ulcers affect hundreds of millions of people worldwide and used to be thought of as a chronic, untreatable condition. But thanks to Warren and Marshall, we can now easily treat them. Again and again, we find that when a moment of surprise emerges in discovery, it is often the instruments we employ that make it possible (Table 2.1). In these opportunistic cases—when we harness a tool in a new way or to a new question—the tool precedes both the surprise and a theoretical explanation for it. It is *after* a serendipitous discovery that scientists build theories around the unexpected findings.

The powerful role of our tool innovations—and the limited role of chance

Examining science's major discoveries, we trace the year a discovery was made by the year the new method or tool was developed that enabled it, and we then split the discoveries by those involving a serendipitous moment and those that did not. Here each new method or tool is shown as a vertical line (|) and each discovery made using it as a dot (●). Popular examples of serendipitous discoveries—like x-rays uncovered in 1895 and the first exoplanet detected in 1995—were actually sparked by two key tools: a discharge tube developed in 1875 and a fibre-fed spectrograph created in 1993—visualised in Figure 2.3 (see also Picture 2.1). Another famous case is Max Planck's unintentional discovery of quanta in 1900. But Planck was only able to make the surprising discovery by intentionally building on the famous experiments on blackbody radiation made possible by precise tools like spectrometers and bolometers—and a model of linear oscillators developed in 1886.[53,160]

Table 2.1 *Serendipitous moments in nobel-prize discoveries—sparked by tools (12 examples)*

	Discoveries (and the *tool or method* used to uncover them)	**Moment of serendipity arising:**
Physics (incl. astronomy)	*Diffraction of electrons by crystals* Clinton Davisson had an accident in his lab in 1925 when a liquid-air bottle exploded that broke an experimental tube. But continuing the experiment, he surprisingly found that the distribution of the scattered electrons had changed—by applying *x-ray diffraction* (developed in 1912) along with other tools such as an electron gun and galvanometer.[152] Davisson stumbled upon the wave-like nature of electrons, providing strong evidence for quantum mechanics. *Bridgman pressure apparatus* Percy Bridgman studied phenomena under high pressure in 1905 when the tool he used unexpectedly broke. And he immediately began designing a new *sealing device* that was much more efficient—and gave rise to high-pressure physics.[153]	*during the scientific process before discovery*
	First exoplanet Michel Mayor and Didier Queloz were not actually searching for planets, but for brown dwarfs. But in 1995, they detected a star wobbling (that was too small to be a brown dwarf) by using a *fibre-fed echelle spectrograph* (constructed in 1993)—it was the first planet discovered outside our solar system.[102,116] *Optical tweezers (using lasers to move objects)* Arthur Ashkin accidentally left samples of viruses open overnight in 1986. He studied these larger particles using an improved method for *optical trapping*—that he created in 1978 and involved a microscope with a laser beam. This enabled him to discover that the particles did not actually move around freely when close to the laser beam but were trapped in the light beam.[131] The breakthrough allows scientists to manipulate single molecules and cells.	*at discovery*
Chemistry	*Method for genome editing (CRISPR)* Emmanuelle Charpentier was studying the bacteria streptococcus pyogenes and made the unexpected, groundbreaking discovery of an unknown molecule in 2012: tracrRNA. She was using genome-editing methods and recreating the bacteria's genetic scissors—and she had to apply improved *differential RNA sequencing* (developed in 2010) together with other methods such as the PCR method and electrophoresis.[114] *Methods for mass spectrometry (soft desorption ionisation methods for studying,* e.g., *proteins)* Koichi Tanaka used trial and error and made in 1988 'a monumental blunder, which was followed by a series of fortuitous, arbitrary decisions [when he] used the glycerin instead of the acetone' to observe ions in gas form. It was uncovered by applying an improved *mass spectrometer* (built in 1986) along with other tools such as laser ionisation and a microcomputer.[154] This breakthrough method has become essential to analyse complex proteins.	*during the scientific process before discovery*

continued

Table 2.1 continued

	Discoveries (and the ***tool or method*** **used to uncover them)**	**Moment of serendipity arising:**
Chemistry	*Conductive polymers (plastic that conducts electricity)* Hideki Shirakawa mentioned to Alan MacDiarmid how he added 'by mistake a thousand-fold too much catalyst' to discover an organic polymer that gleamed like silver in 1977. And he also used an upgraded *bolometer* (created in 1961) together with other tools including x-ray crystallography and an electron microscope.[155] This discovery has revolutionised electronics and solar cells. *Water channels (transport water in our cells)* Peter Agre, 'while working on a completely different problem, stumbled across a protein in red blood cells' in 1992. It turned out to be the water channel: the discovery of how water moves across cells—critical to everything from kidney function to brain health. And he made the breakthrough using *immunoblotting* (invented in 1979) along with electrophoresis, a controlled study and video microscopy.[156]	*at discovery*
Medicine	*Cause of gastritis and peptic ulcers* Robin Warren was not searching for bacteria and his colleague Barry Marshall—when cultivating the bacteria in 1982—accidentally left the agar plates too long in incubation, but by using *fibre endoscopy* (developed in 1979) and an electron microscope, they could identify the bacteria they unexpectedly observed.[29] *CDC genes and cell division* Leland Hartwell, 'through a very amusing bit of serendipity' with his student, happened to come across photo-microscopy methods. These were the missing link in studying the cell cycle and enabled discovering the genetic control of cell division in 1971—by applying *time-lapse photo-microscopy and a Coulter particle counter* (created in 1960) along with a centrifuge and statistics.[157] The discovery advanced our understanding of cancer.	*during the scientific process before discovery*
	Immune response Baruj Benacerraf surprisingly found that different guinea pigs had different responses to the same antigen. The discovery revealed the cause of different immune responses in genes in 1966—by using *starch block electrophoresis* (designed in 1955) together with a scintillation counter and controlled study design.[158] *Hepatitis B (infectious virus affecting the liver)* Baruch Blumberg 'unexpectedly discovered an infectious agent for hepatitis B while researching blood proteins' in 1965. The observation was uncovered applying *agar gel diffusion* (developed in 1949) and immunoelectrophoresis, a controlled study and statistics.[159] Discovering the hepatitis B virus and the development of the vaccine has saved countless lives.	*at discovery*

The data are based on nobel-prize discoveries—with four examples highlighted in each of the three fields.

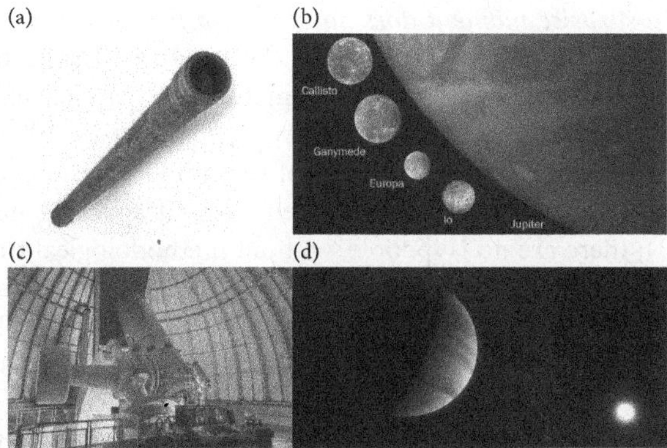

Picture 2.1 *The first telescope constructed in 1608 that Galileo used to serendipitously discover the moons of Jupiter in 1610 (a, b). A new echelle spectrograph developed in 1993 that Mayor and Queloz applied to serendipitously discover the first exoplanet in 1995 (c, d).*

(a, b) Reproduced from NASA; NASA/Kevin Gill. (c, d) Reproduced from José Rodrigues; ESO/M. Kornmesser/Nick Risinger.

When we study science's major discoveries since 1575, we see that the year we developed methods and tools is closely correlated with the year we made the discoveries—for both serendipitous discoveries and for other discoveries. We uncover an extraordinary trend: the time window between the new method and new discovery is nearly identical for both groups of discoveries. The discoveries with a serendipitous moment emerge basically within the same average time as other discoveries (Figure 2.3 and 2.5). This is an important finding because serendipitous discoveries follow the almost identical tool-driven timeline as other discoveries—regardless if we label them serendipitous or not. *The strikingly similar distribution between the two groups of discoveries provides strong evidence: new tools—not just serendipity—are the driving force behind science's discoveries that make the unexpected possible.* Because we must invent and apply new methods to bring about major discoveries—whether unexpected or not—means that breakthroughs have a clear method-driven structure. This includes the small share of theoretical discoveries often driven by new mathematical methods.

A prominent study published in Science, *Quantifying the evolution of individual scientific impact*, found that 'the highest-impact work in a scientist's career is randomly distributed within her body of work'—what the authors call a *random-impact rule.*[12,35,37] The influential finding is based on analysing citations over scientists' careers including nobel-prize-winning careers. Here we study directly how science's biggest discoverers actually make groundbreaking research. This leads us to a very different finding: major high-impact research does not emerge randomly at any point in a scientist's career, it specifically arises when they apply a new method or tool that enables a new way to see the world—what we can call a new *methods-impact rule* (Figure 2.3).

Experiencing a surprise moment does not diminish the importance of new tools—it increases it, because tools generally enable the surprise. Despite the unexpected moment, we still need to learn and implement methods to spark those breakthroughs. So do we find the strong role of serendipity as many claim?[30,35,36] When we look at the data, we instead find a soft role of serendipity that is powered by hard tools: there are no major discoveries achieved without applying new methods or tools (Figure 2.3); there are no laypeople, without methodological training, making nobel-prize discoveries (Figure 4.1b); and only 1% of nobel-prize discoverers are not working at a university at the time of their discovery (Figure 4.3). Winning a Nobel prize is not like winning the lottery—it has predictable features. It is something we can understand by systematically studying how we generate cutting-edge research. Discoveries soon follow after new tools and if they did not, they would be distributed more randomly—as a hypothetical distribution illustrates (Figure 2.4).

The odds of uncovering serendipitous breakthroughs are zero before we develop the tools of discovery

There is another side to the coin: researchers have focused on exciting serendipitous moments in discoveries, but unexpected scientific tools used to spur them have not been studied. Here we explore all 149 nobel-prize-winning method discoveries—all major methods and instruments awarded the prize. Interestingly, these major new tools are less likely to involve a serendipitous moment—about one in eight—compared to scientific discoveries they enable—about one in six. By examining these major tools, we find they almost always emerge from specific method constraints. Our tools are developed deliberately. This is an important distinction: one between intentional tool development and the serendipitous observations they enable.

Only few nobel-prize-winning tools emerged somewhat unexpectedly—such as the x-ray method[118]—or were invented without a specific application in mind like the maser (built just to amplify beams of microwaves).[141] The first microscope and telescope were also invented with no scientific purpose.[46,140] Yet others then harnessed these surprisingly effective tools. Such a tool can power a kind of meta-serendipity: it opens up the possibility of multiple discoveries that no one could have predicted before the tool became available. Serendipity generally comes after the tool is developed and applied.

What is key here is that some powerful tools are not originally designed for the specific breakthroughs they later enabled—they were never meant for that purpose. And yet, the tools still caused the new discoveries they were not originally intended for—and would not have been possible without them. Again, this disconnect between a tool's original purpose and its later impact provides a kind of quasi-experimental insight: because the tools' creators had no idea what breakthroughs they would eventually unlock, we can directly attribute the breakthroughs to the tools as the causal spark. It also helps reduce alternative explanations as the key trigger, like serendipity (Chapter 1). Yet some researchers think serendipity 'is one of the important factors in scientific discovery'.[30,35] But what is clear is there is no chance of serendipitous

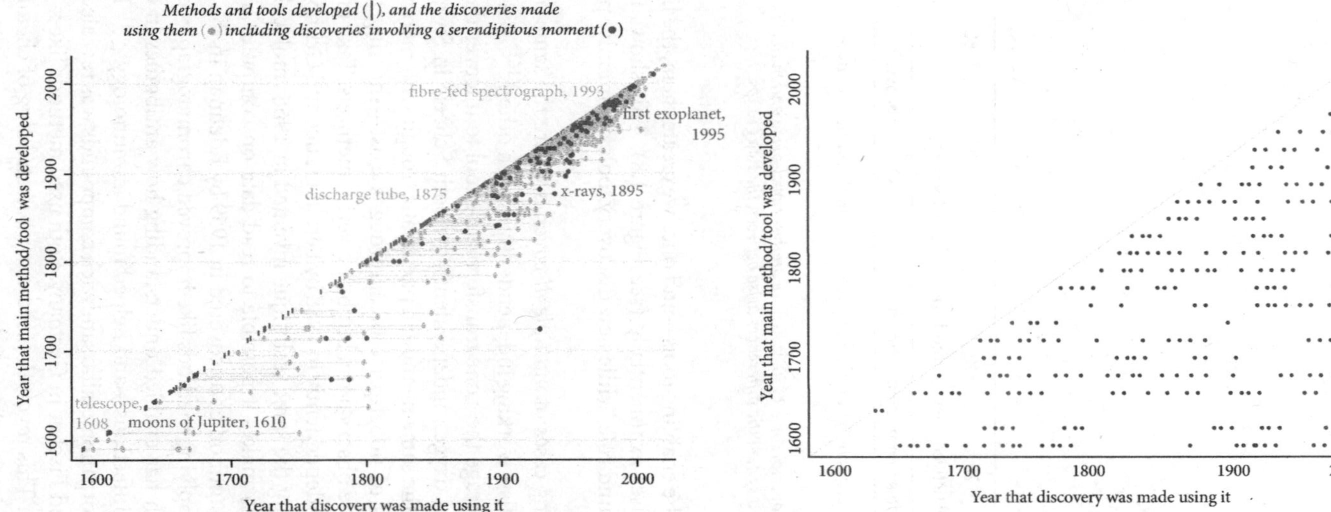

Figure 2.3 *New methods and tools drive science's major discoveries (actual distribution)*

The data reflect science's 734 major discoveries since 1575, including all nobel-prize discoveries (Figure 2.3). Figure 2.3 extends the analysis in Figure 1.2 by capturing discoveries with a serendipitous moment. The year of the method or tool's invention is closely correlated with the year of the serendipitous discoveries—explaining 90% of the variation, and for other discoveries, 92% of the variation.

Figure 2.4 *A hypothetical distribution of discoveries if they occurred through a stronger role of serendipity*

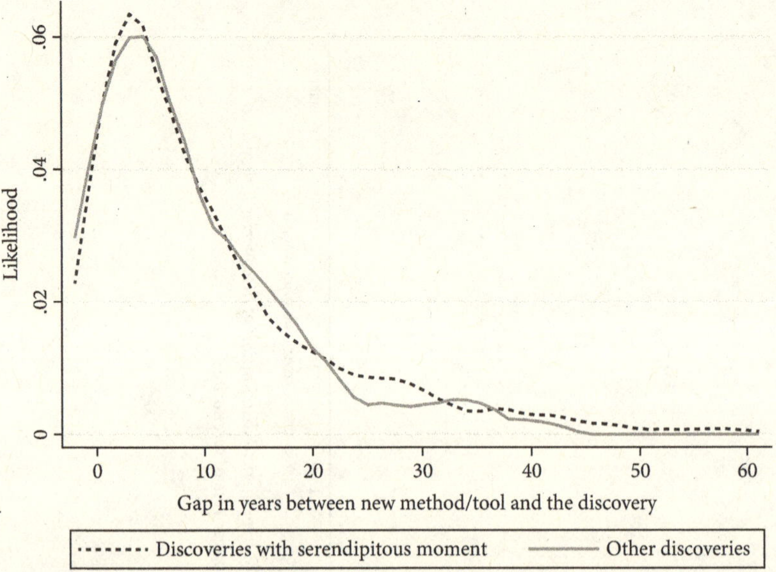

Figure 2.5 *The average gap in years between the new method or tool and the discovery (actual distribution)*

The data reflect all 346 major discoveries since 1950. When we analyse nobel-prize and major non-nobel discoveries separately, no systematic differences arise between the two groups.

discoveries before we generate the enabling tools—and once we create them, the likelihood of surprising findings rises exponentially (see Figure 1.8). This shows that chance plays more of a background role, while new discovery tools commonly spark chance moments.

Science also reveals fascinating cases of *multiple discoveries* where the same phenomenon is uncovered by scientists working independently of each other.[(161)] But we find that they often actually leverage the same transformative tool to uncover them. A classic example is discovering oxygen independently by both Scheele in Sweden and Priestley in England, using the same method: a pneumatic trough with mercuric oxide.[(162)] Numerous examples of nobel-prize breakthroughs exist—showing that they do not just arise by a chance observation, but by powerful methods. Take giant magnetoresistance discovered independently in 1988 by Fert in France and Grünberg in Germany—both implementing the same technique invented in 1968: molecular-beam epitaxy. The breakthrough makes it possible to read data on computer hard disks.[(163)] Take the structure of antibodies uncovered in 1959 by Edelman in the US and Porter in England—both applying methods like improved chromatography created in 1956. This breakthrough has enabled understanding how antibodies in our blood defend us against deadly infections—and helped found immunology.[(164)] The chemistry of compounds called organometallic sandwich compounds was revealed in 1952 by Wilkinson in the US and Fischer in Germany. Both used methods like x-ray crystallography created in 1913. This revolutionised our understanding of chemical bonding and spurred new fields in chemistry.[(165)]

Even the theory of the Higgs particle was developed independently in 1964 by Higgs in Scotland and Englert and Brout in Belgium. Both used mathematical gauge theory formulated in 1954, while also building on recent empirical breakthroughs in particle physics—and motivated by key insights into how particles interact. This field discovered many fundamental particles over the previous decades using particle detectors and accelerators (Chapter 1). The existence of the particle was then confirmed using the world's most powerful particle accelerator, the large hadron collider.[117]

The *scientific toolbox* view of discovery

When most think of major scientific advances, the moment of discovery comes to mind—the final output. Other aspects of the discovery process simply do not catch the eye like a flash of insight and final discovery. Behind breakthroughs, there is a process not only of developing methods but also of designing experiments, collecting and analysing data, and replicating results—all powered by the tools we create. These features are just not as appealing, but we can trace them in the discovery process. They make evidence—and discoveries—measurable, reliable and replicable. Our *scientific toolbox* is the bridge that connects us to the world and how we measure, experiment and generally experience those moments of surprise. By leveraging the tools of a scientific community, we—and science as a whole—transcend individual intuitions and chance moments. *What is fascinating is that, independent of serendipity, other researchers often apply the same tools to arrive at the same findings or discovery. This deflates what seems like an individual's (or lone genius's) unique serendipitous moment to a common tool of discovery used by multiple researchers* (see the conceptual framework in Figure 2.6).

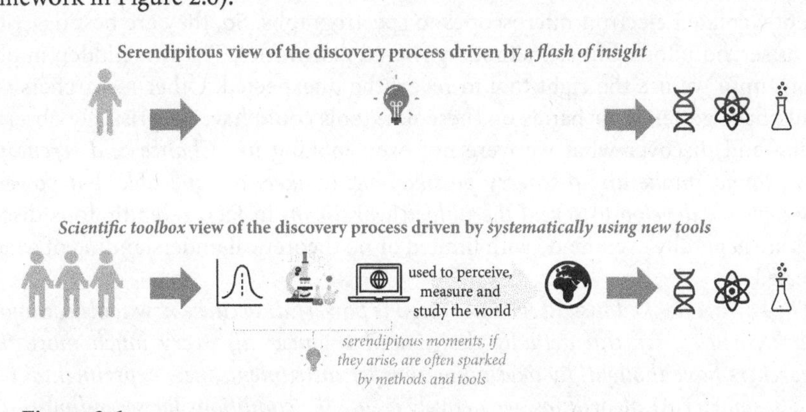

Figure 2.6

The French microbiologist Louis Pasteur said, 'chance favours only prepared minds'.[29,166] And Nobel laureate Albert Szent-Gyorgyi expanded on the idea, 'A discovery is said to be an accident meeting a prepared mind'. The idea is that without training, a serendipitous discovery can go unrecognised.[20] But when we examine science's major discoveries, what becomes clear is that: *chance favours those applying new methods and tools.* The scientific toolbox view explains how we discover by using new tools that open a new lens to the world and it integrates the possibility of chance surprising us by using these tools. Think of breakthroughs described as serendipitous—like discovering cells and bacteria after inventing the microscope, or uncovering Saturn's rings and Uranus after creating the telescope. With these remarkable new tools, other researchers could have made the surprising new observations (whether we call them unexpected discoveries or not). *If we do not know what factor to look at, then the surprise seems random—but once we begin turning our attention to the tools that commonly spark surprise, much of the randomness of discovery fades.* In many ways, those were basically *inevitable discoveries* once we came up with the needed tool and put them in the hands of researchers. This insight is important: it moves the spotlight from individual moments of serendipity and places it on the innovative tools that drive surprising findings.

Conclusion

Most people think serendipity is the most puzzling and the least predictable aspect of science and discovery. After all, it is about coming across something we did not expect. Discovery generally means uncovering a surprising finding, but surprise generally means we cannot anticipate it. So if serendipitous discoveries arise by chance, how could they possibly be predictable or replicable? What we have uncovered here is that surprising new findings commonly follow a hidden pattern. In fact, serendipitous discoveries across science and history—from revealing viruses to the first exoplanet—are sparked not by chance, but generally by applying a new tool to a problem: from recently created electron microscopes to spectrographs. So, they are best described not as serendipitous, but as methods-powered discoveries. They are hidden in plain sight—until we use the right tool to reveal the unexpected. Other researchers who would have gotten their hands on these new tools could have surprisingly observed them—and discover what we were not even looking for. *Chance and serendipity alone cannot make any discovery possible, highly likely or replicable, but powerful new tools we develop to reveal the hidden world can.* In fact, serendipitous discoveries are generally even made with limited or no theoretical understanding of what is revealed.

The key overlooked insight we uncovered is powerful: because new tools commonly unlock surprise, we can actually design and engineer discovery much more than researchers have thought. By developing sharper instruments, new experimental techniques, smarter AI algorithms, we actively create the conditions for serendipity—and multiply the odds of surprise with ever more powerful tools. New tools are the levers that uncover the unknown—whether we are surprised by what we find or not.

Take the breakthrough gene-editing method, CRISPR. The method was not the original focus but emerged serendipitously, using improved methods such as differential RNA sequencing. The method itself then opened up entirely new research areas that were not expected by the method's creators: from finding cures for genetic diseases to developing new crops. This extraordinary method did not just enable a single breakthrough, it creates the foundation for a series of breakthroughs.

When we shift our attention from serendipitous moments in the discovery process to the new tools that enable them, we realise that it is new tools that ensure discoveries are replicable. Different researchers using the same tool can arrive at the same findings. At first glance, serendipity seems to offer the weakest evidence of a logic of discovery, but it actually provides the strongest evidence we have. The fact that discoveries can be replicated using the same tools shows that science is far less random than we think. It often does not make much difference if one labels discoveries serendipitous or not: we do not role dice to generate discoveries but we pick up and apply new tools that make our major breakthroughs possible. What has been more random is who gets their hands on the tool first. So with powerful tools, we move from a random, unpredictable approach to discovery to one where breakthroughs are structured, engineered and repeatable.

Let us picture the landscape of scientific discovery, where about 750 major discoveries make up the foundation of science, about nine million scientists are actively doing research worldwide(167) and roughly '80 to 90 per cent of all scientists that have ever been, are alive now'.(168) The odds of uncovering a major discovery in one's lifetime are thus very low. This reflects a lower-bound estimate of about a 1-in-11000 chance. But there is an unnecessary role of chance that still plays out in many discoveries—and in the speed at which they emerge. It is the chance of someone coming across the right tool, or improving one, without yet a theoretical understanding of how new tools trigger new breakthroughs.

While chance may arise in discovery, giving the impression it is unexpected and unplanned, it takes a new tool to realise that chance moment—and spark any major discovery. In fact, we can accelerate—and anticipate—unexpected moments by making new tool innovations and applying them in new terrain. In short, fewer things are left to serendipity and chance if we create new tools in a targeted way: more new tools systematically foster more new tool-powered discoveries—whether we label them surprising or not. *Here we reframe discovery as an active, design-driven process, offering a productive alternative to common misunderstandings of passive serendipity. Serendipity is especially surprising if not aware of this tool-driven principle of discovery.* We can more accurately label most cases of serendipity as methods-driven—or even methodipity, a fortunate and unexpected new discovery sparked by a new method.

Just as science uncovers patterns and stability in the world, we uncover patterns and stability in the discovery process itself. Unlike common belief, there is a nature and logic of discovery and it is driven by our new tools. With this new *methods-driven discovery rule*, we can demystify much of the discovery process. In 1873, the Polish Boleslaw Prus predicted that 'Someday a science of making discoveries and inventions will exist'.(169) But systematic data were lacking at the time. Over 150 years later, we now have systematic data on science's major discoveries and how they are triggered with a new method or instrument (Figure 2.3). So we can lay out here a science of discovery: one that explains how developing new methods and tools is a necessary condition for major discoveries, and commonly the key trigger. The more we innovate, the faster discoveries will come—and the more we can predict and guide future breakthroughs.

It is time we replace the classic image of the *romanticised discoverer*—someone who has a sudden eureka-like insight of luck—with a new image: the *tool-driven discoverer*, who builds on powerful methods and existing research to unlock the unexpected. We need a fundamental rethinking about scientific breakthroughs. While discoveries receive the most attention in science—by journals, university departments, hiring committees, awards and funding bodies—the tools that spark those discoveries are not celebrated in the same way. In fact, research into developing new tools remains heavily underdeveloped, and we will dig deeper into how this limits scientific progress in Chapter 6.

The fact that we break new ground with recently invented tools significantly decreases the role of chance in discovery. The more we rely on new tools, the less room there is for chance. *By strategically developing new tools—and testing them in unexplored areas—we reduce randomness and serendipity in both expected and unexpected discoveries.* The future of science is in our hands—quite literally, through the tools we create. Expanding our toolbox allows us to better predict and speed up the pace of new breakthroughs following a new tool advance. In fact, shifting greater focus of research, labs and institutions towards devising new tools is commonly the most efficient path to discovery we have. Researchers with an open eye (for unexpected techniques) and with an exploratory mind (for pursuing and applying them) are commonly the ones who spark our future discoveries. If we want to make the next breakthrough, the best strategy we generally have is straightforward: develop and apply a new tool—or adopt one from another field—that gives us an entirely new perspective to a problem.

To wrap up, we return to two prominent researchers and the stories they tell about scientific progress. We test here two of the most influential explanations of science to date: Karl Popper's explanation that discovery has no logical structure but is shaped by chance observations and intuitions (Chapter 2) and Thomas Kuhn's explanation that science does not progress cumulatively (Chapter 3). These different views sparked intense debate and intellectual rivalry. But by systematically examining how discoveries emerge, we uncover a very different picture of science: discovery does in fact follow a predictable logic, and scientific progress is cumulative—with both driven by our powerful and expanding toolbox. In the next chapter, we turn to how our discoveries—expected and unexpected—fit into the overall picture of scientific progress.

3

Revolutionary paradigm shifts or cumulative progress

Rethinking scientific progress and the scientific method

Summary

Understanding how powerful tool innovations unlock breakthroughs leads us to fundamental questions about how science evolves: do we abandon some major discoveries—including breakthrough tools? Is science best seen as a series of sudden paradigm shifts—or rather as deeply cumulative? In the landmark book *The structure of scientific revolutions*, Thomas Kuhn developed the most influential and cited explanation of science yet proposed. For Kuhn, science does not progress cumulatively but is driven by abrupt paradigm shifts: radical upheavals—like the Copernican revolution—that overturn what we believe and can redefine entire fields. Since then, this core question about scientific progress is widely debated, but no consensus and comprehensive analysis yet exist. By studying this question systematically, we uncover a remarkable pattern: three key measures of scientific progress—science's major discoveries, methods and fields—each illustrate the deeply cumulative evolution of science rather than sudden ruptures. First, no major *scientific methods or tools* we use across fields have been entirely abandoned or subject to paradigm shifts—from statistics and microscopes to centrifuges. Next, no major *scientific fields* have been entirely replaced—from genetics and computer science to chemistry. Instead, we refine and extend them over time, often over centuries, and they represent vast bodies of cumulative knowledge. Finally, *scientific discoveries* are also highly cumulative: only 1% of over 750 major discoveries have ultimately been abandoned. Our methods and tools provide the strongest and most striking evidence of science's deep continuity. So we offer a new answer to this foundational question in science and the philosophy and history of science by drawing on methods from statistics and empirical sciences. This brings us to questions about the classic scientific method of testing hypotheses by observing and experimenting—a principle that has remained unchanged for centuries. Is it compatible with the cumulative nature of science? Do science's major discoveries follow it? Or do we need a new understanding of scientific methodology?

The Engine of Scientific Discovery. Alexander Krauss, Oxford University Press. © Oxford University Press (2026).
DOI: 10.1093/oso/9780197829790.003.0004

Overview

Does science mainly go through revolutionary paradigm shifts that fundamentally challenge and overturn what we believe? Or is science mainly a deeply cumulative system, where our current methods and discoveries connect back to our past methods and discoveries and extend forward to enable future methods and discoveries? The question is important because it shapes how we approach scientific research and science education and policy. Should we focus on conducting bold, disruptive science or on refining existing science? Two dominant ways have shaped how researchers study the evolution and progress of science. One way is taken by the highly influential historian of science Thomas Kuhn who offered the most well-known explanation of science to date. Using select case studies, he explored how individual theories may go through extraordinary paradigm shifts: fundamental changes in the theories of a field.[15,16,73] Many still find this classic view of how science unfolds compelling—and many still research his view of non-cumulative science.[16,170–177] Kuhn's seminal book still dominates debates on scientific progress, accruing thousands of citations each year (Google Scholar).

The other way to study scientific progress is taken by scientists using big data to analyse patterns in publications. These scientists argue that scientific articles—on the whole—may be becoming less revolutionary over time.[39] They highlight that bold, high-risk innovation is becoming rare as researchers focus more on established knowledge.[45] These scientists suggest that smaller teams are more likely to generate disruptive ideas than larger teams.[10] They also stress that the sheer volume of publications in today's scientific system can lead us to overlook new transformative ideas.[23] Researchers adopting big data trace science through article citations—the most widely used measure of the impact of discoveries.[3,4,14] But this approach does not uncover other important factors to understanding scientific progress.[38,40] Citations are driven in part by path dependency—researchers with limited time often just cite more cited research—that can impede the spread of new groundbreaking ideas. Focusing on already well-known established research (impact) can hinder cutting-edge breakthroughs (novelty).[178] Citations do not capture the large impact of most major discoveries throughout history, simply because we only began to track citations widely in the second half of the 20th century.[179] Citations also do not capture scientific fields or our powerful tools well.

Most studies—using citations or not—explore a sample of major breakthroughs or publications. Some do suggest that science has cumulative aspects by relying on case studies.[17,180–183] But no comprehensive analysis yet exists that answers the foundational question: is science—namely science's major discoveries, methods and fields—fundamentally cumulative or revolutionary? Here we aim to help tackle this long-standing debate.

Here we move beyond commonly studying just a sample of discoveries or theories often within a field (as Kuhn did mainly in physics)[15,41,43,44] or using a sample of article citations.[10,12,22,23,39] Instead, we systematically analyse the nature of scientific progress through science's major discoveries across fields. In probing the evolution of science, we uniquely shift the focus here to also study an overlooked aspect: our scientific methods and tools. This enables us to draw completely new insights about the cumulative nature of science's major discoveries and methods across fields.[15,16] Introducing this novel method perspective, we find an extraordinary continuity in the methods that we refine over time and enable generating new theories, breakthroughs and fields. A shift from analysing at the individual to the aggregate level helped transform our understanding in numerous fields. It is how Boltzmann and Maxwell developed statistical mechanics, and how Austin Bradford Hill's randomised controlled trials paved the way for cumulative meta-analyses that reshaped biomedical, agricultural and behavioural sciences.

We will explore specific fields. In genetics for example, a set of major advances was only possible using our cumulative methods including microscopes, x-ray methods and electrophoresis we continually upgrade. These sparked a series of transformative discoveries, including heredity, DNA sequencing and mapping the human genome, that keep expanding genetics, improving our health and reducing diseases we face.[113,184] In computer science, a set of major advances was only achieved using our cumulative methods including mathematical and statistical methods and transistors we constantly improve. These led to a series of discoveries, including the Turing machine, information theory (often called the Magna Carta of the digital age) and microchips, that constantly extend computer science and enable the computers, smartphones, internet and artificial intelligence that we use and define our modern society.[102] In the field of electricity, a set of major advances was only feasible using our cumulative methods including galvanometers, batteries and electric generators we continually enhance. These triggered a series of discoveries, including electromagnetism, the theory of electromagnetic radiation and alternating current, that collectively enable the world of electric motors and power plants we rely on daily. The story of genes, computers and electricity is a remarkable story of cumulative evolution—a story of how we stand on the shoulders and methods of giants. And they are representative of major fields across science. So even if new breakthroughs can at times challenge and disrupt research, science is still overall cumulative.

Box 3.1 Thomas Kuhn's explanation of science—driven by revolutionary paradigm shifts

Kuhn published the most widely read and best-selling book on scientific progress to date, *The structure of scientific revolutions*, in 1962.[15] Written while at the University of California, Berkeley, this pioneering book shaped popular understanding of the history of science. The history of science can be seen as a cycle in which established ideas and facts are doubted, new problems and evidence then lead to new revolutionary ideas and facts (and replace the established ones), which eventually over time are also doubted once problems and anomalies with them become apparent, and the cycle begins again. For Kuhn, the process of science reflects revolutionary paradigm shifts, in which we reject current assumptions and theories and adopt entirely new ones. He argues that a paradigm shift 'is far from a cumulative process ... Rather it is a reconstruction of the field from new fundamentals' and 'cumulative acquisition of novelty is not only rare in fact but improbable in principle'.[15] 'Scientific revolutions are thus disruptive episodes of fundamental reconfigurations, through which scientific knowledge develops in a noncumulative way'.[16] Revolutionary science is not cumulative for Kuhn because scientific revolutions (major breakthroughs) replace our current views and theories entirely—and here we focus on this central hypothesis (not on everyday normal science).[15]

A cumulative (or disruptive) advance in science can lead to a cumulative (or disruptive) advance in technology; and the same applies from technology to science.[185] Kuhn's central and bold thesis is: not just some but '*All significant breakthroughs are break-"withs" old ways of thinking*'. But there is a paradox here—changes in theories are the result of a new breakthrough and not what brings about the new breakthrough. Changes in theories cannot explain how science's breakthroughs arise in the first place. This is the key gap and what we want to understand.

With a degree in physics and a PhD in history of science, this laid the background for Kuhn's hypothesis of paradigm shifts. He studied select theories especially in physics up to the early 20th century. Yet he still made sweeping claims about all of science.[15] By focusing on the iconic examples of radical changes in theories of physical reality from Aristotle to Newton to Einstein that span over two millennia, these cases may seem to partly support his hypothesis. For Kuhn, the stark shift from the Ptolemaic earth-centred theory of the universe to the Copernican sun-centred theory characterises the classic paradigm change—and he focused much research on it.[15,73] Yet by examining the paradigm shift hypothesis with over 750 major discoveries, we uncover that the dramatic shift from Ptolemy's model (developed in the year 150) to Copernicus' model (developed in 1543) presents one of the few exceptional cases when we abandoned a central model—though it was in early science.[186] And it was cutting-edge new methods and tools that supported and confirmed Copernicus' model.

Measuring revolutionary and cumulative scientific progress

Examining science's major discoveries and methods across fields offers us a unique opportunity to probe the fundamental nature of scientific progress. This approach captures science's major *theoretical, experimental and methodological breakthroughs*—rather than just focusing on select *theoretical breakthroughs*.[15] While there is no clear cut-off for what counts as a small discovery, there is strong consensus within the scientific community on science's major discoveries (Chapter 1). To assess scientific progress, we classify discoveries, methods and tools across fields into three categories: those that have been *updated* (extended) with new evidence, those that have *not been updated*, and those that have been entirely *replaced* (abandoned or subject to a paradigm shift for being incorrect). Discoveries, methods and tools across fields are defined as contributing to cumulative scientific progress if they have not been abandoned. The classification is based on descriptions within the six mentioned encyclopaedias of science, Nobel prize documents or the seven mentioned science textbooks. The description for about four-fifths of major breakthroughs is drawn from two sources—Nobel prize documents (such as prize summaries and press releases)[(95)] and Encyclopaedia Britannica[(103)] where entries are written by scientific experts as established knowledge (and so are not self-reported by discoverers).

One measure of scientific progress: scientific methods and their cumulative nature

We begin by first testing the fundamental hypothesis of paradigm shifts in science by examining scientific methods and tools—a yet unexplored perspective. If science were truly defined by abrupt revolutions, we would expect to see major tools discarded over time. But what does the evidence show? Take the ten central methods and tools most used to make all 533 nobel-prize discoveries: statistical/mathematical methods, spectrometers, (electron) microscopes, x-ray methods, chromatography, centrifuges, electrophoresis, lasers, particle accelerators and particle detectors. These powerful instruments are the backbone of science that have shaped entire fields. And remarkably, each has been continuously refined, upgraded and expanded to increase power, accuracy, precision and efficiency, at times over centuries. None have been abandoned. Instead, we apply them widely across disciplines, enabling us to spark dozens of new discoveries (Figure 1.7). Among all 149 nobel-prize-winning method discoveries—the major methods and tools awarded the prize—we find a striking trend: 99% have been updated, 1% have not been updated and none have been abandoned. These extraordinary tools like the PCR method, electron microscope, radiocarbon dating method and electrophoresis have been fine-tuned over time.[(95)] In fact, we find no major methods or tools we employ across fields—from calculus and controlled experimental methods to telescopes and thermometers—that have been entirely discarded. Our most prominent scientific methods and tools do not undergo abrupt breaks but are deeply cumulative (Figure 3.1c).

Microscopes—our classic example of a powerful discovery tool—have evolved cumulatively over centuries. From the earliest microscopes relying on light and lenses to electron and scanning probe microscopes, we keep extending the field of microscopy and so our ability to observe the microscopic world (Picture 3.1). Our best light microscopes today do not compete as an entirely different paradigm to the first light microscope, with the same function of visualising miniscule objects. Our methods of arithmetic today do not represent a break from those the Sumerians developed, but build on it. Major methods and instruments used across fields do not go through competing method paradigms. We instead expand our best tools over time into even more powerful tools of discovery, forming the foundation of science and our ability to do science. In fact, our best tools are the very building blocks making up a continually stronger and more refined foundation of science. (So a possible paradigm shift that would abandon them would so fundamentally change how we conceive science that we could likely no longer call what we do science.)

Think of statistical methods for example. Over hundreds of years, they have been developed and expanded by pioneers like the German Carl Gauss, French Pierre-Simon Laplace, British Karl Pearson and Ronald Fisher and many others. Statistical

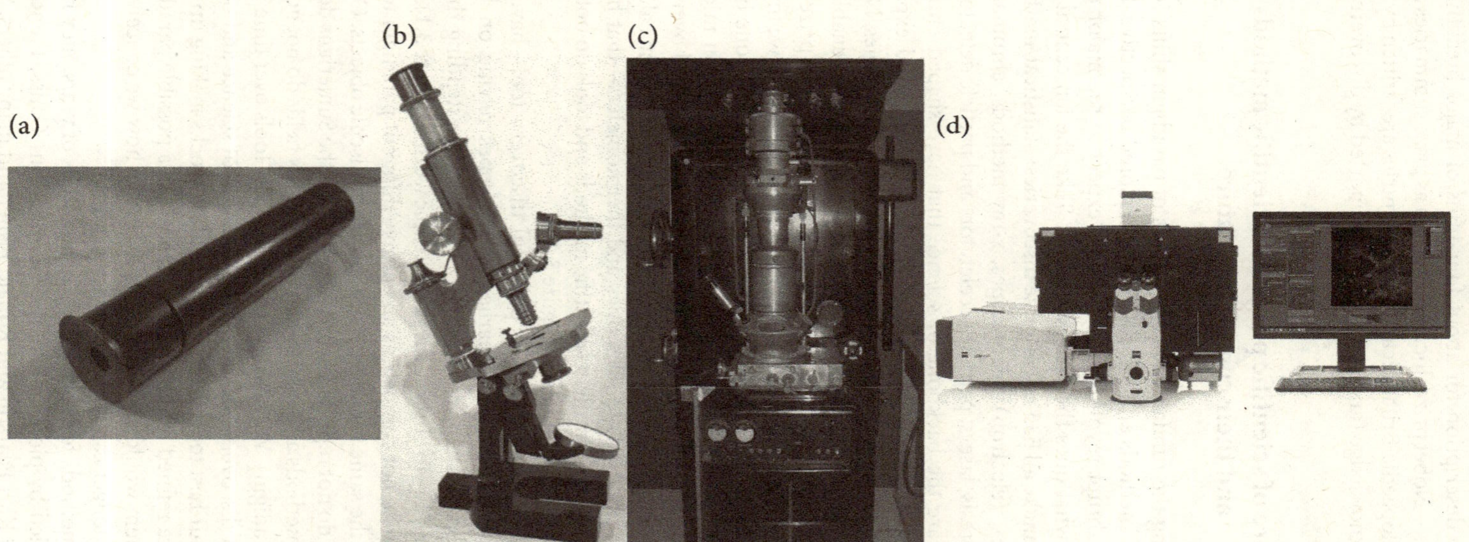

Picture 3.1 *First microscope developed in 1590 (by Zacharias Janssen), compound microscope with apochromatic lenses in 1873 (Ernst Abbe), electron microscope in 1933 (Ernst Ruska) and super-resolution microscope in 1994 (Stefan Hell). Reproduced (left to right) from US Department of Health and Human Services via Wikimedia Commons; Timo Mappes via Wikimedia Commons; J Brew via Wikimedia Commons; Zeiss Microscopy via Wikimedia Commons.*

methods share a common purpose: to extract meaningful patterns from data and make inferences. Today, our cumulative statistical methods test hypotheses, analyse vast data and make systematic predictions in fields from physics and biology to economics. Statistics has arguably received most attention in ongoing debates on improving how results are reported and published.[187–189] The replication crisis, the push for open science and developments in Bayesian statistics all contribute to upgrading statistical methods. Far from replacing the foundations of statistics, these reforms only strengthen them.

Some tools, like mercury thermometers and barometers, are expanded into new tools, like electronic thermometers and barometers, offering greater accuracy and safety. Although cathode ray tubes have largely been physically phased out of physics and electronics labs, their core function—visualising electrical signals—remains unchanged today. Cathode ray tubes were gradually outdated with more advanced digital oscilloscopes, fulfilling the same purpose but with greater precision and efficiency.

Looking at the bigger picture, a key trend emerges here: many methods and tools are each applied in sparking five or more discoveries, often across different fields, underscoring the cumulative tool-driven nature of science (Figure 1.7). In short, with our best tools deeply intertwined across fields, they are the scaffolding of scientific progress (Appendix Figure 1.8).

A second measure of scientific progress: scientific discoveries and their cumulative nature

We next test the fundamental hypothesis of paradigm shifts in science by probing scientific discoveries. Analysing science's over 750 major discoveries, we find a pattern: about 83% have been extended using new methods and evidence, about 16% have not been extended, while only in a few exceptional cases—1%—has a discovery been abandoned. What were once the leading discoveries of the time have generally been built on and updated with new methods and more accurate evidence, making up over four in five discoveries. The discovery of DNA sequencing in 1977 by the American Walter Gilbert and British Frederick Sanger was for example vastly extended with improved electrophoresis and sequencing machines.[184] The discovery of the nature of isotopes in 1913 by the British Frederick Soddy was updated once the neutron was discovered, made possible by new particle detectors.[102] The discovery of hormonal treatment for prostatic cancer in 1940 by the Canadian Charles Huggins was expanded with new methods that paved the way for more sophisticated hormone therapies and improved patient outcomes.[95] Far from abrupt paradigm shifts that render past knowledge obsolete, the vast majority of nobel-prize-winning discoveries has been built upon with new and more cutting-edge methods and later discoveries.

So about one in six discoveries has stood the test of time and has not (yet) been extended but remains largely unchanged. The nobel-prize-winning discoveries of the first exoplanet, the detection of the neutrino and the isolation of fluorine are examples. Often, these are one-off breakthroughs that establish the existence of a new phenomenon. The electron and the double-helix structure of DNA are other such foundational discoveries in the sense that other breakthroughs build on them but do not directly revise them.

Ultimately, only 1% of discoveries have been entirely replaced by new methods and evidence, reflecting only eight superseded discoveries among science's over 750 major discoveries. Surprisingly, three of them were awarded a Nobel prize, but their findings were later abandoned. One case is the initial findings of the Danish Johannes Fibiger in 1913 that ingesting a roundworm caused cancer in rats. While his work was initially celebrated, later methods and research revealed that the real cause was actually a lack of vitamin A. The worm larvae only led to tissue damage where cancer could then develop.(190) Another case is the development of leucotomy by the Portuguese Egas Moniz as a treatment for mental illnesses in 1935. Yet this surgical technique, involving cutting into the brain's prefrontal lobe, also could cause major personality changes as an unpredictable 'side effect'. So it was abandoned as new methods and medications for treating mental illness emerged in the 1950s. Today, it stands as a cautionary tale of early surgical interventions in mental health.(191) The third case comes from economics: the economic theory of portfolio management by the American Harry Markowitz in 1952. Yet he later 'warns the reader that the 1952 piece should be considered only a historical document—not a reflection of my current views about portfolio theory', since he later discarded his initial theory because of methodological errors and changes in his views about mean and variance.(192)

Even in these eight exceptional cases, the discoveries, theories and methods applied served as reference points—stepping stones enabling us to build on and supersede them (Table 3.1). So in these few cases, scientific progress can still be conceived as cumulative, since mismeasurements and mistakes also contribute to the overall picture by triggering correction. Science is self-correcting. Revolutionary leaps that represent a complete rupture from what we currently know are rare in science—unless we search for exceptional differences over millennia as in the most famous example of a paradigm shift from Ptolemy to Copernicus.(15,186) The last step or discovery often seems to be the most impressive or revolutionary. But it is only possible by building on methods and resulting discoveries that come before.

Comparison across fields reveals insight into the evolutionary nature of science. Astronomy stands out again with the highest share of nobel-prize discoveries that have not been updated, making up about four out of ten discoveries in the field (as we see in Figure 3.1b). Pulsars (neutron stars) and the accelerating expansion of the universe for example, once discovered, have not been updated given the very nature of these discoveries. Yet we have expanded nearly all discoveries in economics and social sciences. Why? For these fields deal with highly complex and fast evolving systems. The key here is that we update or replace discoveries commonly by using a new method or tool that offers a new perspective and evidence. New tools provide the lens through which we can revisit and revise earlier findings. Overall, less than 1% of nobel-prize discoveries have been abandoned and slightly more at 2% for major non-nobel discoveries across history, highlighting comparable results across the two groups. This provides us with evidence against the popular hypothesis that revolutionary paradigm shifts dominate science.(15)

Let us take a closer look at the classic and most widely discussed paradigm shift: the transition from the model of our universe from Ptolemy to Copernicus.(15) The

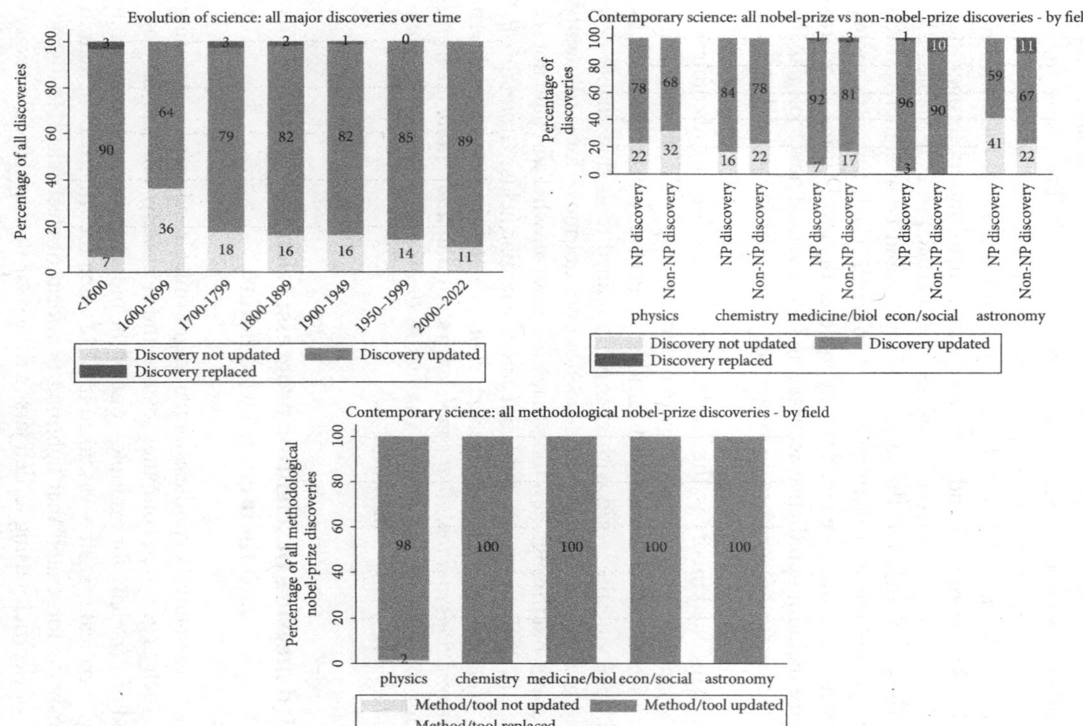

Figure 3.1 *Discoveries are most likely to be cumulatively updated—across time and fields*

The data reflect science's 761 major discoveries (including all nobel-prize discoveries) (Figure a); all 533 nobel-prize discoveries compared to 100 major non-nobel discoveries made over the same time period (Figure b); and all 149 nobel-prize-winning method discoveries only (Figure c). NP stands for Nobel Prize. When we combine all nobel-prize and other major discoveries over the same time period across these five fields in Figure b, we find that the aggregate shares of replaced discoveries are 0%, 0%, 2%, 2% and 4%.

ancient Greek Claudius Ptolemy's geocentric model proposed that the Earth was the centre of our universe, with the sun revolving around it. The Prussian-Polish Nicolaus Copernicus' heliocentric model proposed the opposite: that the sun is the centre of our universe, around which the Earth orbits.(186) For over a millennium, Ptolemy's model reigned. Then Copernicus came and used observational data—collected with tools like the quadrant and astrolabe—and mathematical calculations. These were laid out in his book *De revolutionibus orbium coelestium* that later shook the foundations of astronomy and was published just a few months before his death.(186) Yet the shift from Ptolemy's model to Copernicus' model occurred gradually, once evidence backed by new tools supported the heliocentric model. At the University of Padua in Italy, Galileo refined and confirmed Copernicus' model through new discoveries. Galileo's newly invented telescope played a crucial role: it enabled discovering the phases of Venus and the moons of Jupiter and compellingly proved that celestial bodies could orbit something other than Earth. The heliocentric theory gained greater acceptance once we could test hypotheses rigorously using standards of modern science including predictive accuracy and we invented new tools like the telescope.(193) Once invented, we were no longer blind to incorrect models but could disprove them.

The older an abandoned theory is, the less likely it was developed and confirmed using rigorous scientific methods and tools that cumulatively build on each other over time. What seems like a sudden, radical idea we generally make possible by measuring and observing the world with ever more refined tools that enable new perspectives and greater precision. What if science does not move forward mainly through theory shifts, but through new tools that allow us to see what we could not see before—and not just in theory-heavy fields but across scientific fields? *Groundbreaking discoveries—whether we call them paradigm shifts or not—are commonly driven by new and better methods and tools that enable us to study and understand the world in new ways. Our tools do not just endure changes in theory; they commonly cause them—by providing a new lens to the world.*

A third measure of scientific progress: scientific fields and their cumulative nature

We now test the fundamental hypothesis of paradigm shifts in science by exploring scientific fields, such as genetics, computer science and electricity. Take the development of the field of electricity for example—one of the most transformative forces in modern society. A critical first step was taken in 1752 when the American Benjamin Franklin famously demonstrated that lightning is a form of electricity. He discovered the nature of electricity using leyden jars (a device for storing static electricity) invented a few years earlier in 1745. The breakthrough, explaining how electricity flows from lightning through a metal kite, sparked new questions. It led to the Italian Luigi Galvani developing an animal electricity theory in 1791 by also using leyden jars—a theory of how electricity causes frogs' muscles to contract. Building on this theory and its limitations, the Italian Alessandro Volta created the first electric battery in 1800—a steady source of electric current that marked a transformative advance.

Table 3.1 Discoveries updated, not updated or replaced (examples of discoveries)

	Cumulative science		Non-cumulative science
Field of discovery	**Discovery updated**	**Discovery not updated**	**Discovery replaced**
Physics	Structure of atoms based on quantum theory, 1913 Electron microscope, 1933 Electroweak theory, 1964	Electron, 1897 Detection of neutrino, 1953 First exoplanet, 1995	Ptolemaic geocentric theory, 150* Animal electricity theory, 1791* Steady-state theory of universe, 1948*
Chemistry	Method of electrophoresis and adsorption analysis, 1930 Sandwich compounds, 1952 Computational methods in quantum chemistry, 1969	Isolation of fluorine, 1886 Conductive polymers, 1977 Fullerenes, 1985	... ...
Medicine	Acquired immunological tolerance, 1949 Radioimmunoassays (RIA) technique (for peptide hormones), 1959 RNA interference, 1998	Double-helix structure of DNA molecule, 1953 Reversible protein phosphorylation, 1956 Regulation of neurotransmitter release, 1990	Biogenetic law, 1866* Spiroptera carcinoma causes cancer in rats, 1913 Leucotomy as treatment for mental illnesses, 1935
Economics/ social sciences	Econometrics, 1933 Economic models of causes of poverty, 1954 Integrated assessment model of climate change, 1994	Importance of exchange rate regime, 1963 Expectations-augmented Phillips curve, 1967 ...	Hereditary genius ability, 1869* Theory of portfolio management, 1952 ...
Share of science's major discoveries:	*83%*	*16%*	*1%*
Share of all nobel-prize discoveries:	*85%*	*14%*	*1%*

We provide three examples for each category and field based on nobel-prize discoveries—except for five of the eight 'replaced' discoveries in the last column (*) as only three nobel-prize discoveries have been abandoned. The last column reflects the traditional view of non-cumulative science, yet even these cases contributed to correction and show some path dependency. Major discoveries in chemistry have not been replaced. In economics, only two nobel-prize discoveries have not been updated.

For it enabled us to better study electricity and produce electricity-powered technology. Volta's breakthrough powered new discoveries, with the Danish Hans Ørsted applying the newly created battery to discover that an electric current generates a magnetic field in 1820. Building on the work of Volta, Ørsted and Faraday, the French André-Marie Ampère was then able to develop the law of electromagnetism in 1827 by using a magnetic conductor he invented in 1822. This mathematical theory explained the relationship between electricity and magnetism. In the same year, the German Georg Ohm introduced the mathematical concept of electrical resistance, establishing Ohm's Law, by using a galvanometer.

The next giant leap came in 1831 when the British Michael Faraday created an electric generator using electromagnetic induction. This was key in laying the foundation for electrical power generation. Through Faraday's work and using mathematical methods, the Scottish James Maxwell could then take the pivotal next step to unify what was known about electricity and magnetism into a theory of electromagnetic radiation in 1865. Maxwell's highly influential theory revealed that electric and magnetic waves are on the same electromagnetic spectrum—governing everything from radio waves and light to x-rays. These collective methods and knowledge fuelled the Serbian Nikola Tesla's development of alternating current (an induction motor) in 1883. This fascinating breakthrough brought electric power plants and electricity to our cities on a mass scale.[(99)] Building on the accumulated tools and evidence, the British Joseph Thomson ultimately unexpectedly discovered the electron in 1897 using a cathode ray tube he designed the previous year.[(194)]

These pioneering scientists had an expanding and powerful toolbox at their disposal. A key insight emerges: each of these scientists since 1820 applied the newly developed galvanometer, battery and/or electric generator to spark their discoveries. The fact that major discoveries could not be made without first generating these innovative tools and resulting discoveries offers compelling evidence of the deeply cumulative nature of scientific progress. Even the most revolutionary ideas arise from a chain of incremental steps (Table 3.2).

Taking a broader step back, the German physicist Albert Einstein then built upon these methods to be able to achieve a critical breakthrough in 1905: the special theory of relativity and the famous equation $E = mc^2$, reshaping the field of physics.[(99)] While Einstein's theory is often seen as a classic example of just a theoretical discovery, it was deeply rooted in the existing methods and experimental findings of his time. We can trace the methods he heavily relied on directly or indirectly: advanced mathematical methods—including Maxwell's equations—and Michelson's measurements of the speed of light using the newly invented interferometer in 1881. Without these tools, he could not have formulated his theory (a topic we return to in Chapter 6). Einstein's innovation lies in interpreting and logically restructuring the existing experimental results and methods. New tools, such as particle accelerators and atomic clocks, enabled later experimental tests that confirmed and expanded his predictions with extraordinary precision.[(99)]

Strikingly, genetics and computer science, like electricity, reveal how we develop complex knowledge through deeply interlinked methods, tools and resulting discoveries that build on each other and often span more than a century (Table 3.2).

Table 3.2 Cumulative nature of science across fields: Discoveries and the methods we use to develop them build on each other over time

Discoveries in the field of genetics	Heredity (Mendel 1865)	Composition of nucleic acids (Kossel 1897)	Role of genetics in evolution (Dobzhansky 1937)	Genes regulate definite chemical events (Beadle and Tatum 1941)	Mobile genetic elements (McClintock 1951)	DNA as bearer of genetic information (Hershey 1952)	Structure of DNA molecule (Franklin, Crick and Watson 1953)	DNA sequencing (Gilbert and Sanger 1977)	Genes causing genetic diseases (Collins 1989)	RNA interference (Fire and Mello 1998)	Human genome (Watson and Venter 2003)	Method for genome editing—CRISPR (Charpentier and Doudna 2012)	...
Central method or tool used to make the discovery	statistics, probability (1814)	hydrolysis (1888)	statistics, improved (1919)	x-ray analysis, improved (1913)	microscope, improved (1940)	isotopic labelling (1923)	x-ray crystallography (1913)	electrophoresis, PAGE (1964)	chromosome jumping technique (1987)	micro-injection, improved (1991)	genome mapping, improved (2001)	RNA sequencing, improved (2010)	...
Discoveries in the field of computer science	Computing machine/ mechanical computer (Babbage 1833)		Computer programming (Lovelace 1843)	Incompleteness theorem (Gödel 1931)	Turing machine (Turing 1936)	Transistor (Shockley, Bardeen and Brattain 1947)	Information theory (Shannon 1948)		Compiler for developing computer codes (Hopper 1951)	Integrated circuit/ microchip (Kilby 1959)		World wide web (Berners-Lee 1989)	...
Central method or tool used to make the discovery	mathematics (automated logarithm tables) 1822		computing machine (Babbage 1833)	mathematics, improved logic (Hilbert 1928)	mathematics, improved logic (Hilbert 1928)	cathode-ray oscilloscope (1897)	statistics/probability theory, improved (1923)		computer, Mark I (1944)	transistor (1947)		computer network (internet) (1983)	...
Discoveries in the field of electricity	Nature of electricity (Franklin 1752)	Animal electricity theory (Galvani 1791)	Electric battery (Volta 1800)	Electro-magnetism (Oersted 1820)	Law of electro-magnetism (Ampere 1827)	Ohm's Law (Ohm 1827)	Electro-magnetic induction (generator) (Faraday 1831)	Self inductance—principle of electricity (Henry 1832)	Theory of electromag-netic radiation (Maxwell 1865)	Alternating current (induction motor) (Tesla 1883)	Evidence of electromag-netic radiation (Hertz 1887)	Electron (Thomson 1897)	...
Central method or tool used to make the discovery	Leyden jar (1745)	Leyden jar (1745)	alternating disks (zinc and silver) (1800)	electric battery (1800)	magnetic conductor (Ampère 1822)	galvano-meter (Oersted 1820)	galvanometer (Oersted 1820)	electromagnet, improved (1831)	mathematics, electromag-netic theory of light (1846)	electric generator/ dynamo (Faraday 1831)	induction coil (1836)	cathode ray tube, improved (1896)	...

The data are based on 761 major discoveries. A number of other discoveries also played an important role in these fields.

The history of electricity, genes and computers exemplifies how we apply interconnected tools to trigger interconnected discoveries in fields across the sciences. In fact, no major fields—from biology and nuclear physics to medicine and mechanical engineering—have been entirely discarded. Our highly connected system of science is the outcome of these deeply collective feedbacks.

In chemistry, assembling the periodic table of elements has been an enormous collaborative and cumulative effort over time, forming the foundation of the field. In biology, most of what we know builds on the theory of evolution and the mechanisms of evolution, laying the backbone of the field. Yet it would be odd if we did not see a big change in theory in the classic examples between Ptolemy and Copernicus, or between Aristotle's, Newton's and Einstein's view of physical reality spanning two millennia. These changes did not emerge from nowhere. Scholars in the 16th and 17th century, including Copernicus and Newton, turned to Ptolemy's geocentric astronomy and Aristotle's laws of motion as the very theories they tested, disproved and built on.[(195)] Several centuries later, Einstein then relied on Newton's laws of motion and theory of gravity as a reference point to build on. Yet Newton's classical mechanics still remains useful today to describe everyday, macroscopic objects. Even the rare historical cases of abandoned discoveries reveal elements of cumulative knowledge we used to go beyond them. By studying discoveries systematically, we can overcome what seems to be discontinuity in some select theories in the past, especially about physical reality, and view them as stepping stones.

A conceptual framework of the cumulative nature of science

While an individual hypothesis or theory of a scientist can be tested, challenged and even abandoned, the broader methods and scientific fields represent our extensive bodies of knowledge consolidated over time. We discard many of our ideas, hypotheses and even some theories, but we do not abandon our major methods applied across fields or our major fields—as depicted in Figure 3.2. Testing the paradigm shift hypothesis with over 750 major discoveries including science's major theories, we uncover that the few abandoned discoveries were mostly theories—not experimental or methodological breakthroughs. These discarded theories often lacked robust evidence, reinforcing the fact that scientific progress fundamentally relies on the right methods to collect robust evidence (Table 3.1).

So the history of science tells a cumulative and unified story—one driven by our collective method and tool advances. Shifting our attention from isolated hypotheses and select theoretical discoveries to science's major discoveries, methods and fields is a more systematic way to assess the cumulative nature of scientific progress. For they make up the foundation of science and how we do science. So cumulative knowledge exists on a spectrum: from unestablished ideas and hypotheses, then experimental findings and theories (experimental and theoretical discoveries), and ultimately to well-established methods and fields.

Within the vast landscape of science's major theoretical, experimental and methodological discoveries, Kuhn narrowed his lens on theories. Within that scope, he zoomed in mainly on physics. And he honed in even further on its early development

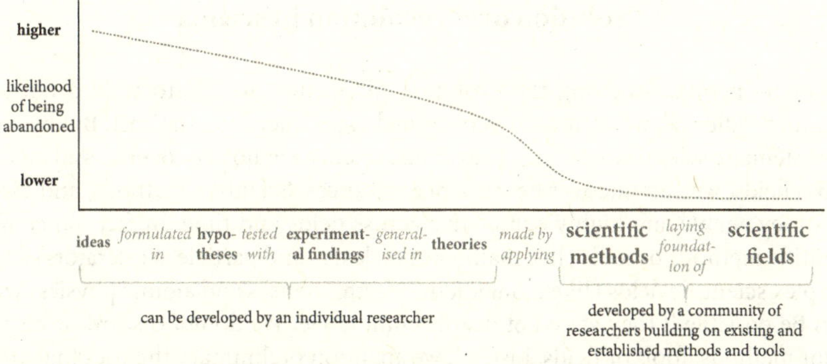

Figure 3.2 *Cumulative nature of science: the spectrum of cumulative knowledge*

up to the early 20th century. He finally zeroed in on a handful of theories often spanning centuries where evolution would be more likely. Yet if we want to draw general insights about scientific progress across the history of science, we cannot be selective but need a comprehensive approach spanning the history of science itself—the aim of this chapter. Kuhn's landmark book *The structure of scientific revolutions* popularised the idea that science undergoes dramatic paradigm-shifting upheavals—like political revolutions. Yet Kuhn did not establish a universal structure across scientific shifts, and they are generally slow and incremental rather than abrupt and revolutionary. A more accurate title, given Kuhn's select case studies, could have been *An account of theory change mainly in early physics.* Yet a more fitting description of science would actually be *Cumulative scientific progress embodied in scientific methods and fields.*

We keep reworking the details of our knowledge as we generate new methods and collect new evidence. Evidence and explanations are works in progress—valid until we update them using better methods that provide better evidence and explanations. Science is about creating new tools that enable us to revise our best theories and explanations of the world in light of new evidence. This is the nature of scientific progress. Science and scientific methods thrive on iteration: they are a bootstrapped (correcting) process of constant improvement, refinement and synthesis. Quantum theory and the mechanisms of evolution, statistics and microscopes, chemistry and computer science are all continually refined over time as we come across new methods and challenges.

Ultimately, we tend to measure a discovery's significance by its impact—how useful it is in helping us solve problems and better understanding the world. Some discoveries and methods seem timeless, like discovering the electron, the theory of evolution and the sun-centred theory of our solar system, and statistics, microscopes and x-ray methods that will be around in the future. For they form the backbone of entire fields. Others seem less timeless, like discovering the Coronavirus vaccine, partition chromatography and cardiac catheterisation (a technique for inserting a catheter into the heart). Yet each piece contributes to our ever more cumulative body of methods and knowledge that we build on—and account for science.

Evolution over revolution in science

For many people, studying the evolution of science over history does not seem scientific—after all, we cannot run controlled experiments on the past. But by applying systematic scientific methods to examine science's major discoveries and methods across fields, we can uncover how science advances. Scientific methods, and discoveries they enable, are highly cumulative across fields and time. In fact, no complex scientific methods or tools (like mathematics, lasers and particle accelerators) and no complex scientific fields (like biomedicine, earth sciences and atomic physics) would even be possible if they were not deeply cumulative. We do not discard or disprove major methods, tools or fields. Instead, we abandon preliminary theories that are not grounded in rigorous methods and are often more provisional and speculative than our established methods—that largely lay the foundation to establish our theories and fields.

If we instead scan the history of science and focus on a few select theories (like some historians and philosophers of science like Kuhn) or explore a random sample using article citations (like some scientists using big data), we can find what looks like discontinuity between those individual cases. But testing this fundamental question with rich evidence provides a very different, cumulative answer. The idea of grand paradigm shifts, which govern science and overturn central theories with entirely new theories, applies to only about 1% of major theoretical, experimental and methodological breakthroughs. If revolutionary science were pervasive, one may even expect the over 750 biggest discoveries in history—spanning all nobel-prize and major non-nobel breakthroughs—to serve precisely as the strongest evidence of disruptive breaks. Yet we find the opposite: the most groundbreaking of discoveries in history have been extended over time, reinforcing the deeply cumulative progress of science. *New and continually improved methods and tools better explain scientific progress than new revolutionary, paradigm-changing ideas and theories that generally result from these improved methods and tools of discovery.*

Scientific progress is, at its core, fundamentally brought about and grounded on the tools we develop to explore, test and refine our understanding of the world—not just by changes in theories. The common emphasis on theoretical shifts reflects a final output but overlooks the crucial methodological process we take to create, replicate and refine the output. Traditionally, many think discoveries and theories are the heart of science, while methods are a temporary bridge that, once we develop the discoveries and theories, no longer receives our attention. *We need to also place methods at the heart of science, with the discoveries and theories we develop using methods seen as the temporary output of science, until we update them with new evidence using our expanding and cumulative methods.* Most theories thus remain provisional, evolving as we improve our tools. This explanation of *methods-led science* offers a more realistic picture of how knowledge advances—a more coherent alternative to the idea of winner-takes-all paradigm shifts. It aligns with actual scientific practice and enables us to better understand and accelerate scientific progress.

Science advances cumulatively regardless of how we theorise about scientific change, so the debate among historians and philosophers of science on whether some exceptional theories may evolve through paradigm shifts has little impact on our scientific advances. Our vast and interconnected web of cumulative methods, knowledge and technology keeps growing regardless of the theoretical debate. For some, the debate on paradigm shifts is, at best, a negligible or non-existent problem. Yet Kuhn would insist that we are not progressing towards the truth, in the philosophical sense—for example from Ptolemy to Copernicus. But we uncover a key insight from the debate: we can in fact speed up the pace of cumulative scientific progress by shifting how we conceive scientific change—to expanding our cumulative *scientific toolbox*. In the final section of this chapter, we turn now to the evolution of the classic scientific method.

Redefining the scientific method: as leveraging sophisticated scientific methods that extend our mind

The classic scientific method—observing, experimenting and testing hypotheses—has remained unchanged for centuries; so can it be compatible with the deeply cumulative nature of science and its ever-evolving methods? Does it accurately capture the way we gather knowledge—and then revise our knowledge as we get our hands on new evidence? Or do we need to rethink and expand this classic method into a broader, overarching method of science that better reflects how science actually works?

The classic scientific method is defined in science textbooks and dictionaries as 'the collection of data through observation and experiment, and the formulation and testing of hypotheses.'[196–198] Many science textbooks present the scientific method to students as a simple sequence of steps: observe, experiment and test hypotheses.[199–202] A study of major science institutions like the National Science Foundation (NSF) and National Institutes of Health (NIH) also revealed that they primarily endorse this hypothesis-driven method rather than exploratory methods that lack predefined hypotheses.[203] So this approach, as a unifying method of science, is embedded in science dictionaries, textbooks and institutions—often stating that we follow, and should follow, the method. It is commonly traced back at least to Francis Bacon who popularised the concept in his book *Novum Organum* in 1620. Bacon emphasised that what we know comes from evidence and experimenting (in his words, 'twisting the lion's tail' and observing what happens); and is the only way to actually do science.[204] His book not only laid the foundation for philosophy of science but also fundamentally shaped how generations of scientists conceive the practice of science.[199,200,202,203]

But science has advanced far beyond its early history, so does this traditional view still hold? Before hypothesising about what science's general method is and what it should be, we need to first take a step back and examine the evidence on how science is actually conducted. Surprisingly, the classic scientific method has not been systematically analysed using scientific methods themselves, as many assume that it cannot be subjected to scientific study. But by studying science's major discoveries, we can tackle the basic question: to what degree is the classic scientific method actually applied in making science's groundbreaking research?

Think of Einstein's theory of special relativity that reshaped physics in the 20th century and how we understand space and time. Darwin and Wallace's theory of evolution by natural selection transformed biology and how we comprehend the historical origins of our species. Franklin, Crick and Watson's discovery of the double-helix structure of DNA redefined genetics and how we conceive the way genetic information of living organisms is stored, copied and passed along. These scientists fundamentally changed the way we view the world, but they themselves did not always directly carry out experiments to uncover these path-breaking discoveries.

Examining science's major discoveries, we uncover an unexpected finding: 25% of breakthroughs since 1900 did not apply the traditional scientific method (all three features)—with 6% of discoveries made without observation, 23% without experimentation and 17% without testing a hypothesis. Expanding our lens and analysing science's over 750 major discoveries over history, we find that the traditional scientific method (all three features) is applied in making 71% of discoveries (with individual shares seen in Figure 3.3). Some hypotheses are tested through systematic experiments, while others rely solely on observation—such as an astronomical observation or in observational studies. So science does not always fit the textbook definition, with hundreds of groundbreaking discoveries not following the classic scientific method. The evidence thus challenges this long-standing concept of science.

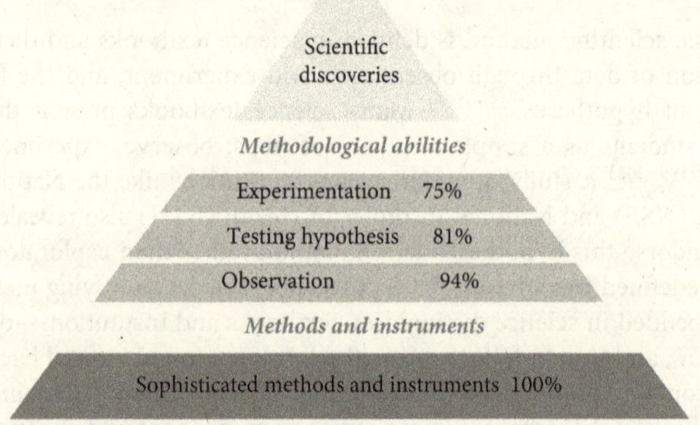

Figure 3.3 Methods of science pyramid: *Breakdown of methodological approaches we use in making discoveries*

The data cover science's 761 major discoveries.

What about differences across fields? Surprisingly, the classic scientific method was not applied in making about half of all nobel-prize discoveries in astronomy, economics and social sciences. Why? Experiments are not always possible—astronomers cannot manipulate stars in a lab and economists cannot run controlled experiments on entire economies. Some discoveries are made through open-ended exploration, without testing a pre-established hypothesis, while others are more theoretical. Even in physics, about a quarter of breakthroughs broke from tradition. This reveals a key insight: the traditional scientific method does not capture the full reality of discoveries and, more importantly, it overlooks the fact that science's major discoveries depend on applying sophisticated methods (like statistics and randomisation techniques) or tools (like centrifuges and advanced computers) (Figure 3.4). Science is about getting our hands on the right tools to break new ground.

When we evaluate science's major discoveries, what is the fundamental method we apply to be able to do science and spark discoveries? We identify a common feature we can reduce the method of science to: science's major discoveries have relied on using *sophisticated methods and instruments*. These external resources—from lasers to chromatography—extend our mind and senses, allowing us to see, measure and analyse the otherwise unobservable that makes up most of science today. Unlike observing, hypothesising and experimenting—largely internal cognitive abilities—scientific tools are material artifacts that can be shared, refined and built on by others (Figure 3.4). Applying sophisticated methods or tools is a necessary condition for discovery. Without them, discovery and scientific progress is not possible—this reveals a universal principle of scientific methodology.

The sophisticated scientific method is actually more unique to science than the classic method; after all, the most used scientific methods and tools—like particle accelerators, electrophoresis methods and x-ray crystallography—we largely only use in science. But we often make observations, test hypotheses and experiment in business, industry, public policy and even in our everyday lives and they are not just prototypical or distinct to science. Recognising the enormous importance of complex tools adds an essential element to understanding science and how science has evolved.

In its early origins, science was often grounded only on directly observing, hypothesising or experimenting—while today, these activities are only possible for making discoveries by applying and refining complex tools (a topic we explore later in Chapter 7). The *classic scientific method*—or one or two of its three features—was more likely to be applied in its traditional form, without complex scientific tools, when early scholars like Bacon described it in 1620.[(204)] At the time, the first two major tools that shaped 17th-century science, the microscope and telescope, were recently invented, and tools did not yet dominate science in the same way. Bacon's vision of science aligned with the prevailing understanding of science and was limited by the tools of his time. But since then, *sophisticated scientific methods* have transformed how we observe, experiment, test hypotheses and solve problems—in much more diverse, complex and efficient ways than ever before. *Just as science itself has evolved and expanded, so too should the classic scientific method.* Its broad and general description is better understood as a *basic method of reasoning* used for human activities (non-scientific and scientific alike).

Let us look at some of the nobel-prize discoverers who did not directly apply or generally could not apply the classic scientific method in making their groundbreaking discovery. Einstein did not himself run traditional experiments when he developed the law of the photoelectric effect in 1905, nor did Franklin, Crick and Watson when they uncovered the double-helix structure of DNA in 1953 using observational images generated by Franklin. The British Roger Penrose did not himself make direct observations when he formulated the mathematical proof for black holes in 1965, nor did the Russian-Belgian Ilya Prigogine when he created the theory of dissipative structures in thermodynamics in 1969. The Danish Niels Jerne did not directly test a hypothesis when he developed the natural selection theory of antibody

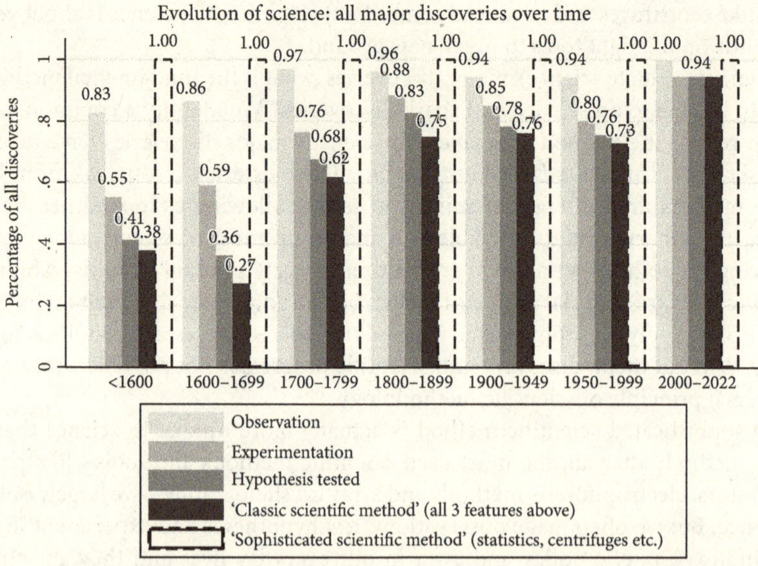

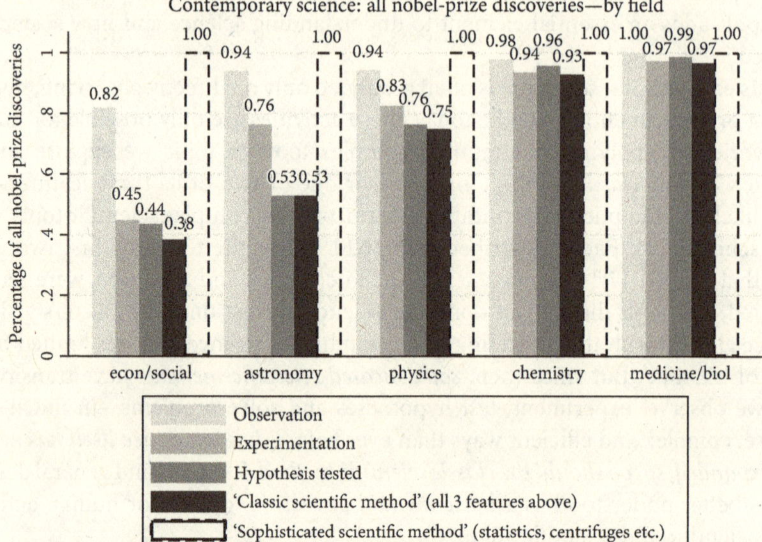

Figure 3.4 *Breakdown of discoveries we make using the classic and the sophisticated scientific method—across time and fields*

The data reflect science's 761 major discoveries (including all nobel-prize discoveries) (Figure a)—and all 533 nobel-prize discoveries (Figure b). Each of these discovery-making publications is categorised as using observation if the study describes collecting observational data (using eyesight) (bar 1 in the figure), as using experimentation if the study ran an experiment (bar 2), as testing a hypothesis if the study formulated and examined a proposed explanation (rather than doing exploratory research) (bar 3) and as using the classic scientific method *if the study applied the three features (bar 4). Finally, the publication is categorised as using the* sophisticated scientific method *if the study applied a complex scientific method or instrument (bar 5). When we combine all nobel-prize and other major discoveries over the same time period across these five fields, we find that for example the share of discoveries made applying the classic scientific method is 40%, 35%, 75%, 93% and 89%.*

formation in 1955, nor did the Canadian James Peebles when he created the theoretical framework of physical cosmology in 1965. If we were to abide by the classic scientific method, Copernicus, Einstein, Franklin, Crick and Watson and many others would not be seen as applying it, as they did not directly conduct experiments to trigger their seminal breakthroughs. Yet these scientists became iconic figures of science.

Many nobel-prize discoveries have been awarded for breakthroughs that lacked theoretical underpinning, from radioactivity and x-rays to viruses and scientific tools. While individual scientists and breakthroughs at times bypass steps of the traditional scientific method, the method can be seen as often eventually applied at the collective level by the scientific community over time. Scientific progress, in this sense, is distributed across scientists over generations, making it a cumulative endeavour where one picks up where another left off. Yet the sophisticated scientific method is in fact implemented across science's major discoveries. By reframing the scientific method not as a rigid, linear sequence of steps a scientist applies but as a more flexible, tool-driven process that scientists actually apply, we gain a critical advantage. We shift our focus from formulating hypotheses to identifying the most effective tools and methods to solve complex problems—with or without a pre-established hypothesis, as many of the greatest discoveries highlight.

So we need to reframe the (classic) scientific method from mainly inside our heads—observing, hypothesising and experimenting—to the (sophisticated) scientific method mainly in the world outside that we explore and measure with our cutting-edge scientific methods and tools. For tools extend the very limits of our mind, generate completely new kinds of data and uncover insights about the world far beyond what we could have imagined. Scientific tools, like scientists themselves, come in many shapes, sizes and levels of sophistication. To do science, we do not just observe; we combine mathematics with precise measurement instruments, merge statistics and AI with controlled experiments, link x-ray crystallography, spectrometers and particle detectors to systematic observation, and make hundreds of other powerful combinations. Think of the incredible diversity of methods used in immunology, oceanography, neuroscience and astrophysics, or chemistry, agronomy and behavioural economics.

Our tools make it possible, for most phenomena in science, to observe, experiment and test hypotheses in the first place and do exploratory research—and in new and innovative ways otherwise out of our reach. The *sophisticated scientific method* integrates these features into our central methods and tools, creating an adaptable system of discovery (Figure 3.5). Even replication, a main feature of science, is deeply tied to sophisticated methods (not the classic method). Using advanced tools like x-ray devices and statistical methods, researchers can replicate discoveries (while simply observing, experimenting and testing hypotheses is too broad and too susceptible to each researcher using them differently). But with sophisticated methods, science becomes more accurate and reliable and less prone to human error, while enabling us to much better evaluate the quality of research.

Ultimately, with the classic scientific method, we would not be able to label many of our greatest scientific discoveries as scientific—despite their profound impact on our lives. This traditional view, seen as a golden principle connecting the scientific

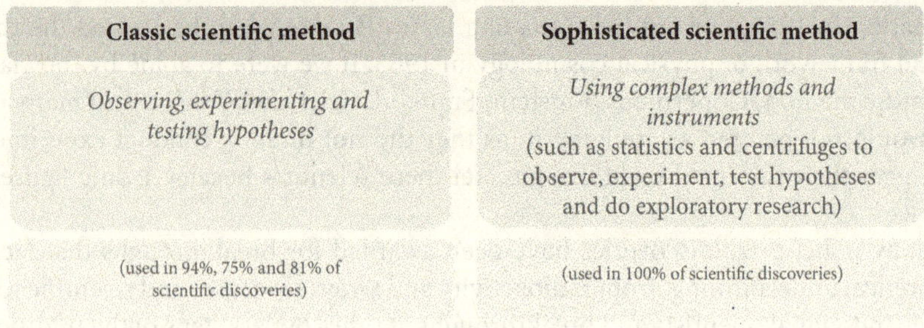

Figure 3.5

community together, can be misunderstood as universal. It is an idealisation that can at times be confusing and misleading, especially for students and less-experienced researchers, when learning about science and realising that it does not always apply or focusing their attention at times in the wrong direction. In fact, adhering to it as a guiding principle can stifle innovation and constrain us from developing new ideas and breakthroughs. How we do science and spark major breakthroughs thrives on our diverse methods and tools, *so we can best view the method of science as leveraging our sophisticated toolbox*. We need to reform the *classic scientific method*, integrating and redefining it as the *sophisticated scientific method*. Since methods and instruments—from mathematics to microscopes—extend our mind and senses and are essential for doing science today, we lay out a definition that better reflects actual scientific practice:

> Scientific methodology is the use of sophisticated methods and tools that enable us to observe, experiment, test hypotheses, solve problems and do exploratory research.

Sophisticated methods are general-purpose, meaning we can apply them to different questions and domains. This definition offers a more accurate understanding of scientific methodology. It shifts our attention to upgrading our sophisticated toolbox—the very engine that enables us to push science forward. Ultimately, the best path to discovery is not the classic scientific method but its extension: the sophisticated scientific method.

At this point, we have gotten through enough evidence to return to the related but broader question: what is science? While science is generally defined as the study of the 'world through observation, experimentation, and the testing of theories' (Oxford English Dictionary),[(74)] this view misses something crucial. At its core, science is powered by methods and tools we create—from telescopes to statistical simulations—that enable observation, experimentation and testing theories. Without them, science today is not possible; so science needs a better definition based on actual evidence:

> Science is the study of the natural and social world using methods and tools to observe, experiment, solve problems and develop theories—in order to describe, explain and predict the world.

As a final thought, to develop new tools and ideas, we use not only observation and measurement but also imagination and abstraction that go beyond the periphery of the measurable scientific method. Take Emmanuelle Charpentier at Umeå University and Jennifer Doudna at Berkeley for example, who imagined and devised a unique new genome-editing technique CRISPR. Their method functions like genetic scissors that precisely cut a DNA molecule at a specific location. This method has transformed the life sciences and our understanding of how to change the DNA of animals, plants and even humans.(114) Take Allan Cormack at Tufts University and Godfrey Hounsfield at EMI Laboratories, who conceived computer-assisted tomography (CT scans) by imagining the different pieces and devising the necessary computing system. How a CT scan works is that clusters of x-ray beams pass through our body from multiple angles, generating detailed internal images of our body with computer calculations. The instrument marks a vital advance in the medical world and some of us who have had injuries have taken a CT scan. Tool discoveries rely on imagination—from Wilson conceiving the first particle detector that visualised particle tracks and Ruska inventing the first electron microscope, to Gilbert and Sanger designing DNA sequencing.(95)

Conclusion

It is tempting to think of science as revolutionary, as sudden paradigm shifts make for appealing and fascinating stories. But science—its scientific methods and the discoveries and fields they enable—is best described by (cumulative) evolution rather than (paradigm-changing) revolution. Our methods and fields would not even be conceivable if they were not deeply cumulative. New and expanded tools better explain scientific advances than new paradigm-changing theories that arise from those expanded tools.

By recognising this, we can accelerate the pace of cumulative progress by reconceptualising scientific change: rather than thinking of elusive revolutions, we need to systematically expand our cumulative toolbox as the engine of science and discovery. Ultimately, because science evolves does not mean that it is not cumulative—it is precisely what makes science cumulative, by building on and refining our collective methods and knowledge of the world. By standing on the shoulders and cumulative methods of giants, we keep seeing further and developing new discoveries, life-saving medicines and transformative technologies. To reflect how science actually works, we need to also extend the *classic scientific method* (which we do not always apply) with the *sophisticated scientific method* (which we do always apply).

So far, we have explored the two most influential and cited explanations of science—and then sketched out the method maker's new perspective. Examining science's major discoveries, we reveal that science and discovery actually follow a logical pattern, unlike Karl Popper's explanation of science that rejected this possibility (Chapter 2).(17) We also find that the history of science is highly cumulative, unlike Thomas Kuhn's explanation of science that is shaped by non-cumulative paradigm shifts (Chapter 3).(15) Both the logical and cumulative nature of science are deeply embedded in the ever more powerful tools we use to spark new advances.

4
Discovery-makers

How younger age, interdisciplinary education and resources can support discoverers and our new cutting-edge tools

Summary

With our expanding toolbox driving major discoveries and cumulative science, this opens an important question: how do our powerful tools connect to our broader environment? How do they interact with the demographics, institutions and resources that help support the scientific process—and develop new tools? And more intriguingly, who are the scientists behind our biggest discoveries? By studying science's biggest discoverers and their broader traits, we uncover a unique portrait—a general profile of the greatest scientists in history. This is key for us to tackle our central puzzle: uncovering what consistently drives discoveries across science—and what is only relevant in certain discoveries. Strikingly, we find that interdisciplinary scientists, those with broader method training, are behind about half of all nobel-prize and major non-nobel discoveries over the same period. This highlights the importance of crossing scientific borders and combining methods across disciplines. We also unexpectedly find that younger, more motivated scientists are the ones behind our greatest advances: scientists over 50 made only 7% of all nobel-prize discoveries and 15% of non-nobel discoveries. Surprisingly, those over 60 made only 1% and 3%. And where do discoverers work? Most work at less-known institutions at the time of their discovery: scientists at top 50 ranked universities made up only one third of both nobel-prize and other major discoveries. This is important—it debunks the assumption that only top-tier institutions are where breakthroughs happen. We also dig deeper: exploring the geographic location, gender and wider country conditions of science's greatest discoverers.

Overview

So far, we have been unpacking *how* and *when* discoveries emerge. In this chapter, we dig deeper and fill in the gap: *who* makes them and *where* do they happen? Despite science profoundly shaping the course of human history, we know surprisingly little about who makes science's breakthrough advances. Understanding who

The Engine of Scientific Discovery. Alexander Krauss, Oxford University Press. © Oxford University Press (2026).
DOI: 10.1093/oso/9780197829790.003.0005

science's great discoverers are—the personal side of science—intrigues both scientists and the general public. Classic work on scientific discoverers goes back at least to Boleslaw Prus in 1873,[169] Florian Znaniecki in 1923,[205] Francis Galton's *English men of science: their nature and nurture* in 1874,[206] and especially influential has been Harriet Zuckerman's *Scientific elite: Nobel laureates in the United States* in 1977 (Box 4.1).[44] Researchers often explore a limited sample of discoverers, covering one time period, scientific field or country like the US or UK.[41,43,85,207–209] While valuable, these snapshots do not allow us to draw broad, general insights about discoverers. A more comprehensive approach is needed—one that spans major discoverers across science, time periods and discoverers' broad traits to uncover the fuller picture.

Typically, researchers studying the traits of scientists often also focus their lens on one factor—reflecting the broader trend in today's specialised scientific landscape that adopts a more narrow focus. Here we widen the lens to all nobel-prize and major non-nobel discoverers, exploring the broad range of demographic traits: their age,[12–14] their education level,[44,85] their interdisciplinary training,[83,84] their gender,[210,211] their country of residence,[212–214] and their religious affiliation[44]—all at the time of their discoveries. We also investigate their institutions—university ranking[208,215,216]—and their broader social and economic features, including income per capita and population size of the country they lived in.[213] By exploring this broader set of supporting factors, we find unique patterns in the background traits of science's eminent discoverers across fields and history—including the wider context they work in. We reveal how science's prominent discoverers have changed over time: they have shifted towards greater interdisciplinary education and method training—going against the common trend to specialise—and they are still making breakthroughs at a young age, though they are not as young as before. We also dig deeper to analyse what shapes discoveries by controlling for the common influencing factors and comparing how important they are (for interested readers, more details on the data are in the online Methods Appendix).

Box 4.1 The sociologists of science Harriet Zuckerman and Robert Merton

The American sociologist of science Harriet Zuckerman, while working at Columbia University, packed her bags and travelled throughout the US to interview Nobel laureates. She pioneered studying the traits of prominent scientists, asking them about their background, family and research. In her seminal book *Scientific elite: Nobel laureates in the United States* in 1977, she tells the story of how being 'a Nobel laureate is, for better or for worse, to be firmly placed in the scientific elite' and the history of science.[44] She offers insight into the lives of Nobel laureates, 'as we follow the laureates from their social origins through their formal education and apprenticeships in the craft of science to their major contributions at the research front'.[44] One of Zuckerman's core arguments is that discovery is more of a collective effort than an individual enterprise—it is embedded in a social context, with the scientific elite made up of talented individuals influenced by institutions, education, social stratification and mobility.

By exploring the experiences of scientists winning the Nobel prize, described as 'the ultimate scientific award', she observes power, authority and influence in science's award system.[44] Zuckerman argues that the path to these ultra-elite of science can be highly competitive and hierarchical, where only a few can achieve high success. Some Nobel laureates, she found, dedicated their lives to a single breakthrough, often sacrificing personal time and other ambitions along the way—with Nobel laureates achieving a kind of scientific immortality after they pass away.[44] Yet Zuckerman's study only focused on a small sample of American Nobel laureates. In this chapter, we explore the broader spectrum of demographic traits for all nobel-prize and other major discoverers.

Robert Merton—Zuckerman's main inspiration and later her collaborator—was another pioneering sociologist of science.[44] He argues that discoveries take place within a scientific community with shared norms, values and institutions. Perhaps his most famous insight is the influential finding of the 'Matthew effect' in science—the idea that prominent scientists receive disproportionately more credit for their research than less prominent scientists with equally important contributions that often go unnoticed.[178] The social structure of science generates a self-perpetuating cycle, where recognition breeds more recognition, solidifying the status of elite scientists. At 83, Merton married Zuckerman, with the duo leaving behind a large mark in the sociology of science. In short, sociologists of science focus on the interesting social conditions in which scientists work. They make valuable insights about science just like historians, scientists, psychologists and philosophers (Chapters 1–3). But while they explore the environment and traits surrounding discovery, they leave the key question unanswered: what actually drives science's breakthroughs—the common force we are unpacking as we progress through the book.

Most discoverers have an interdisciplinary education with training in broader methods to make breakthroughs

With new tools driving new discoveries, what factors and traits can support scientists in getting their hands on and creating new tools to advance science? We begin by exploring education—a natural starting point in scientists' training. We find that 88% of science's major discoveries since 1600—when doctoral awards began to spread—have been made by researchers with a PhD at the time of their breakthrough. The figure rises to 96% for all nobel-prize discoveries (awarded since 1901). In other words, most discoverers are highly trained. As science has expanded, a paradox has emerged: the more complex the world we study, the more sophisticated our tools must be to study the ever greater complexity. The two move closely in parallel. Today, to trigger new discoveries demands training in ever more advanced tools, from cutting-edge statistical methods to high-resolution electron microscopes, along with acquiring extensive bodies of knowledge. Truly mastering our best tools takes time, but is key.

Over the history of science, dozens of great discoverers—including Faraday, Tesla and Dalton—completed at most only secondary schooling. Yet by mastering newly developed tools on their own, such as the galvanometer, electric generator and eudiometer, including mathematical methods, these scientists were able to spark groundbreaking discoveries. While universities offer structured training and greater access to cutting-edge methods, they have not always been necessary for achieving discoveries. Yet since the 18th century, discoverers have been more and more likely to hold a PhD and work as professors (as we see in Figure 4.1a).

But what about interdisciplinarily education? Today, scientists typically specialise and work in one field—a common norm of science. Yet we find a unique pattern here: most nobel-prize discoveries—54%—have come from scientists with two or more degrees in different fields by the time of their breakthrough. The figure is 42% for major non-nobel discoveries over the same period—serving as an independent benchmark or control. Some fields rely on interdisciplinary training more than others: in medicine and biology, discoverers are most likely (69%) to receive degrees in multiple fields, compared to physics (39%), where specialisation is much more common (Figure 4.1d). Interdisciplinary education is on the rise, with over 70% of major discoveries since 2000 made by scientists who completed (at least) two different degrees. Simply put, discoverers are trained more *broadly* than their peers, and not necessarily more *deeply*.

The National Research Council in the US argues that challenges at the frontiers of science can often require the intersection of disciplines. Deeply complex problems in both science and society can demand crossing scientific borders.[58] So advances can arise at the edges and boundaries between fields.

Scanning science's biggest discoveries, we find that innovation at times comes from combining methods across fields or harnessing a method from one field in another. Interdisciplinary training expands our toolkit: equipping us with skills in tools from different fields. It enables us to overcome rigid disciplinary constraints, merge seemingly unrelated methods in innovative ways and develop entirely new integrated methods. In short: it allows us to bridge gaps between disciplines by adopting a completely new approach to a complex problem that would otherwise remain isolated. Take the German Max Delbrück for example. Trained in physics at the University of Göttingen, he turned to biology in the 1930s, bringing with him cutting-edge tools from physics—including advanced statistical methods. With these unconventional tools, he was able to tackle unanswered questions in genetics and reveal that bacteria develop through mutations. His breakthrough 1943 paper *Mutations of bacteria from virus sensitivity to virus resistance* helped open molecular genetics. Similarly, the German Konrad Bloch's unique educational path—completing degrees in chemical engineering at the Technical University of Munich and biochemistry at Columbia University—gave him a special edge. It enabled him

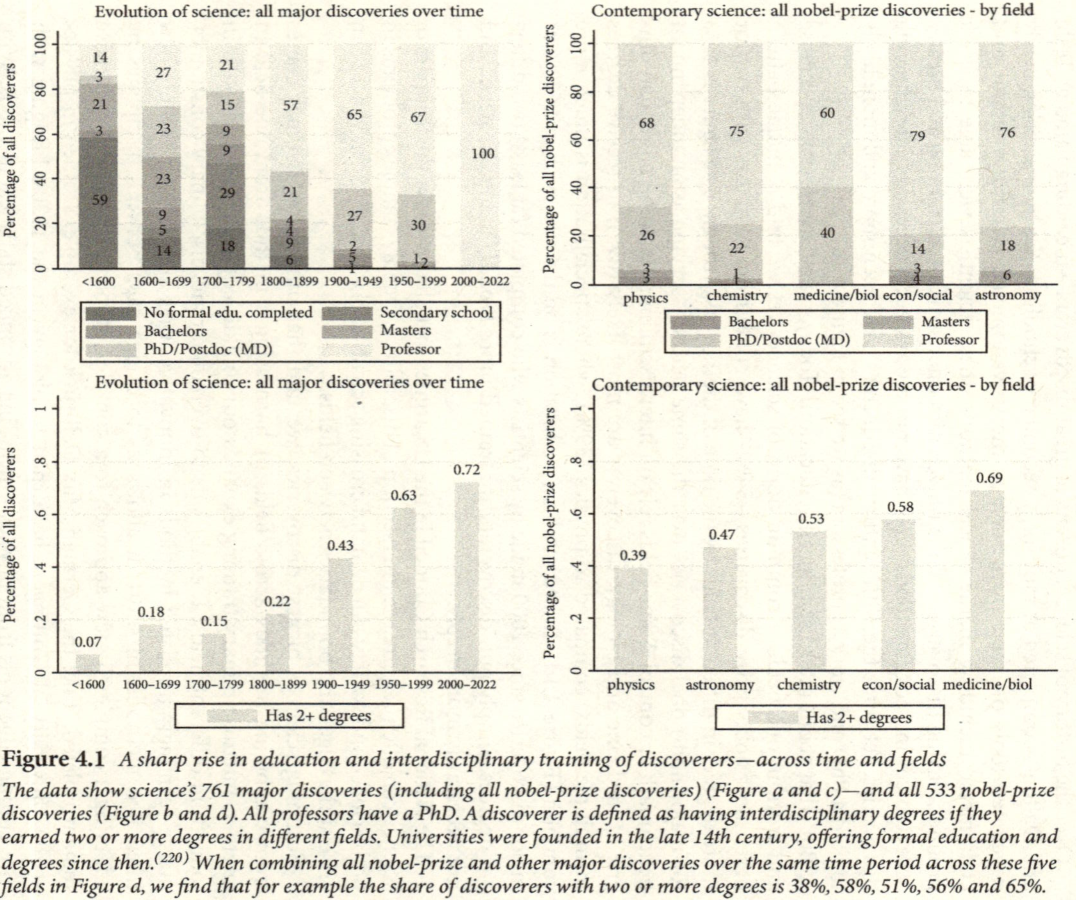

Figure 4.1 *A sharp rise in education and interdisciplinary training of discoverers—across time and fields*

The data show science's 761 major discoveries (including all nobel-prize discoveries) (Figure a and c)—and all 533 nobel-prize discoveries (Figure b and d). All professors have a PhD. A discoverer is defined as having interdisciplinary degrees if they earned two or more degrees in different fields. Universities were founded in the late 14th century, offering formal education and degrees since then.[220] *When combining all nobel-prize and other major discoveries over the same time period across these five fields in Figure d, we find that for example the share of discoverers with two or more degrees is 38%, 58%, 51%, 56% and 65%.*

to apply new isotopic labelling methods to discover the mechanism and regulation of cholesterol. He uncovered how cholesterol and fatty acids are converted in our body.

The British Frederick Sanger studied natural science, biochemistry and medicine at Cambridge, and combined his methodological training to apply new gel electrophoresis methods and pioneer one of the most influential techniques in the history of genetics: DNA sequencing.(184) Not only have these powerful sequencing methods catalysed multiple discoveries, but Sanger himself also won two Nobel prizes for his breakthroughs. The Swedish Svante Pääbo first studied humanities, including history of science and then medicine at Uppsala University, before later pursuing postdoctoral studies in molecular biology and then becoming the director of the Max Planck Institute for Evolutionary Anthropology in Leipzig.(217) His diverse background enabled leveraging the new DNA sequencing methods to explore ancient DNA, leading to the groundbreaking discovery of the Neanderthal genome and rewriting how we understand human evolution.(218) The Canadian Donna Strickland, trained in engineering physics and optics at McMaster University, worked together with the French Gérard Mourou, trained in physics at the Sorbonne. Together, they employed new laser technology to develop a new breakthrough technique: the method of ultrashort high-intensity laser pulses. Their breakthrough method paved the way for laser eye surgery.

The German Hermann von Helmholtz received a PhD in medicine and also studied physics and mathematics at the Humboldt University of Berlin. This allowed him to apply new mathematical principles and physical analysis that other physiologists did not. He could then make—at just 26—a foundational discovery that transformed part of physiology and physics: the principle of the conservation of energy.(219) Then there is the British Rosalind Franklin, who studied physical chemistry at Cambridge and then turned to biology, ultimately applying x-ray crystallography methods developed in physics to capture the first remarkable image of the double-helix structure of DNA.(1)

History makes it clear: some of our greatest discoveries happen when we step outside our field and adopt powerful tools in other fields. Yet delayed discoveries are pervasive in science—often occurring when researchers are unaware of powerful tools outside their own field to tackle the very problems they seek to solve with less effective conventional tools (Chapter 1).

There is a trade-off at play: as our knowledge and the methods we use to generate that knowledge expand over time, researchers have specialised into ever-narrower fields. Yet making unexpected connections commonly relies on harnessing unexpected methods across domains. While the universal scholar is an ideal of the past, the small share of researchers who push the frontier are more likely to defy the trend towards specialisation and venture beyond their own domain.

Although breakthrough research is often interdisciplinary, the very structure of scientific institutions—universities, academic journals and even prestigious awards like the Nobel prize—reinforces traditional disciplinary borders.(83) To foster discoveries, we need to challenge and rethink these outdated disciplinary divisions and how we can promote and reward interdisciplinary research, including the

Nobel Prize. Independent of disciplines, we need to reward the best research. Beyond education and training, there is another closely linked factor here: researchers' age.

Greater productivity and impact of younger scientists

Thomas Kuhn famously argued that 'Almost always the men who achieve these fundamental inventions of a new paradigm [or major breakthrough] have been either very young or very new to the field whose paradigm they change. ... for obviously these are the men who, being little committed by prior practice to the traditional rules of normal science, are particularly likely to see that those rules no longer define a playable game and to conceive another set that can replace them'.[(15)] Kuhn developed this hypothesis by studying a small sample of theoretical discoverers, mostly in early physics, like Einstein. But does this hypothesis hold up when we test it on science's major discoveries across fields? Indeed, Einstein was only 26 when he published his nobel-prize-winning paper on the law of the photoelectric effect in 1905, in the prestigious journal *Annalen der Physik*. He did this while working at the Swiss patent office. As Einstein himself boldly put it: 'A person who has not made his great contribution to science before the age of 30 will never do so'.[(221)] Yet do the conditions at Einstein's time reflect science today?

We uncover a striking finding: the golden age range of high productivity and impact in science is between 35 and 45 years of age, with exactly 50% of all Nobel laureates in science falling into this age range when uncovering their prize-winning discovery. The average age of discoverers is 39 years. This is the prime of most scientists' careers, with the odds drastically dropping as age increases afterwards. We find that only 7% of all nobel-prize discoveries and 15% of major non-nobel discoveries over the same period arise after 50. And more remarkably: only 1% and 3% after 60. So today, we can revise Einstein's claim and say: a person who has not made their great contribution to science before the age of 45 is much less likely to do so (as we uncover in Figure 4.2a). For the group of nobel-prize discoverers, we can partly explain this by the average life expectancy at present and the average 21-year gap between making the discovery and receiving the prize. Yet on average, Nobel laureates in science are 60 years old when they actually receive the prize.

So why does the sweet spot of discovery seem to peak in a researcher's prime years? Younger, untenured researchers are often more motivated and ambitious to make their mark and achieve a breakthrough. Older researchers with already secure, tenured positions often do not face the same degree of external pressure—publish or perish. Younger researchers, those just entering a field, can have another advantage. These researchers are trained in the latest, up-to-date methods and technologies. They bring a fresh perspective to problems, without always accepting established assumptions, and can be more open to exploring new techniques. The evidence challenges the popular belief that greater age and experience are synonymous with innovation.

In the past, some researchers uncovered groundbreaking discoveries very early in their careers. The Indian student Subrahmanyan Chandrasekhar, at just age 21,

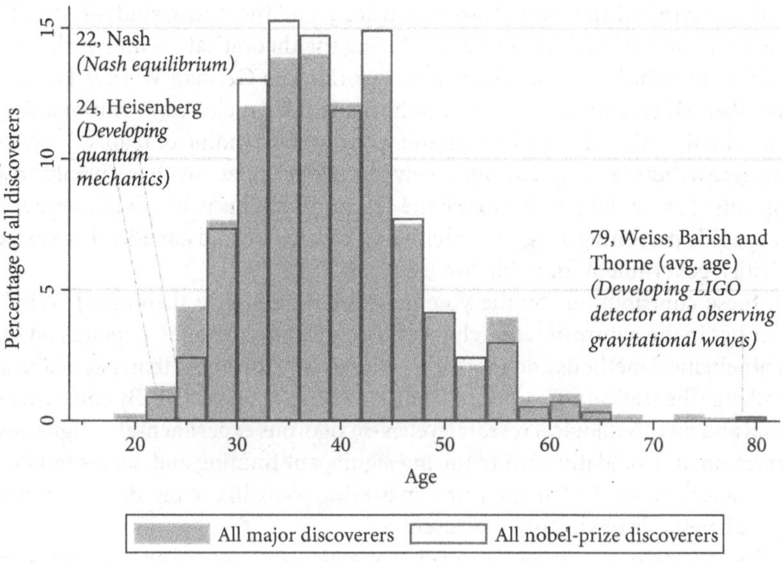

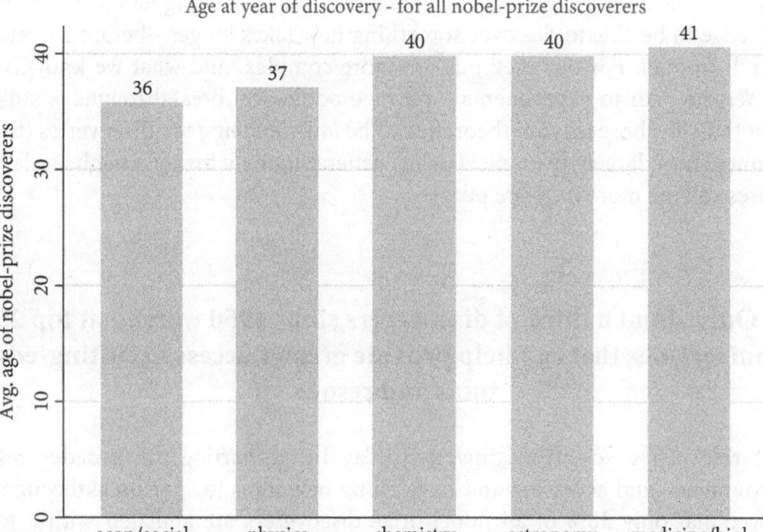

Figure 4.2 *The golden age range of high productivity and impact in science is between 35 and 45 years of age*

The data cover science's 761 major discoveries (including all nobel-prize discoveries) (Figure a)—and all 533 nobel-prize discoveries (Figure b).

described the physical processes governing the evolution of stars, making him the youngest scientist ever to make a nobel-prize-winning discovery. Close behind him was the British student Brian Josephson, who created a framework for tunnelling supercurrents at 22, and the American John Nash, who introduced the game theory concept of Nash equilibrium also at 22. The Swedish student Svante Arrhenius developed the electrolytic theory of dissociation at 24, and the British student Alan Turing devised the Turing machine also at 24, laying the theoretical groundwork for modern computing that later transformed our world. The German Werner Heisenberg, British Paul Dirac and Danish Niels Bohr made their major contributions to quantum mechanics at 24, 26 and 28, reshaping our understanding of reality.(95) All these young researchers, except Turing, received a Nobel prize for these breakthroughs. (Tragically, Turing did not live to see the full impact of his work, dying mysteriously from cyanide poisoning at age 41, after being persecuted and chemically castrated by the British government for being homosexual).(222)

All these contributions by the youngest discoverers were theoretical. Why? The answer lies in the nature of research itself. Theoretical work often depends on the latest mathematical methods and modelling, along with little more than paper and a pen, to challenge theoretical assumptions from a fresh, new perspective. By contrast, experimental and methodological research relies on rigorous experimental designs, testing, observation and validation, all requiring significant training and access to lab space and advanced tools. It demands first mastering tools like x-ray devices, advanced statistical methods or electron microscopes.

Today, it is extremely difficult to achieve a major discovery at such a young age. But why is that? Simply because acquiring the ever-expanding method training and knowledge to be able to discover something new takes longer—before we reach the research frontier. For our methods are more complex, and what we know is more vast. Yet the path to experimental and methodological breakthroughs is still often longer than for those that are theoretical. The *low-hanging-fruit* discoveries (the simpler ones) have largely been picked—so generating new breakthroughs today often requires solving more intricate puzzles.

Only about a third of discoverers since 1950 worked at top 25 universities that can help provide greater access to cutting-edge tools and resources

What role can a scientist's university play in supporting the broader research environment—and accessing and developing new tools to spur breakthroughs? We uncover that only 30% of all nobel-prize discoverers are affiliated with a top-25-ranked university, and also 30% of major non-nobel discoverers (the uncrowned laureates) over the same period—serving as an independent control group. Yet globally, less than 1% of researchers work at these elite institutions. Expanding the scope to the top 50 universities, we see that 38% of all nobel-prize discoverers and 34% of major non-nobel discoverers were at these institutions. So being at a top university is far from a prerequisite for making breakthroughs, providing hope for researchers not at top-tier institutions.

The share of discoverers at a top 25 university is higher in astronomy (41%) and economics and social sciences (49%) (Figure 4.3). Why? We would expect this because in astronomy, the most advanced radio telescopes, laser interferometers

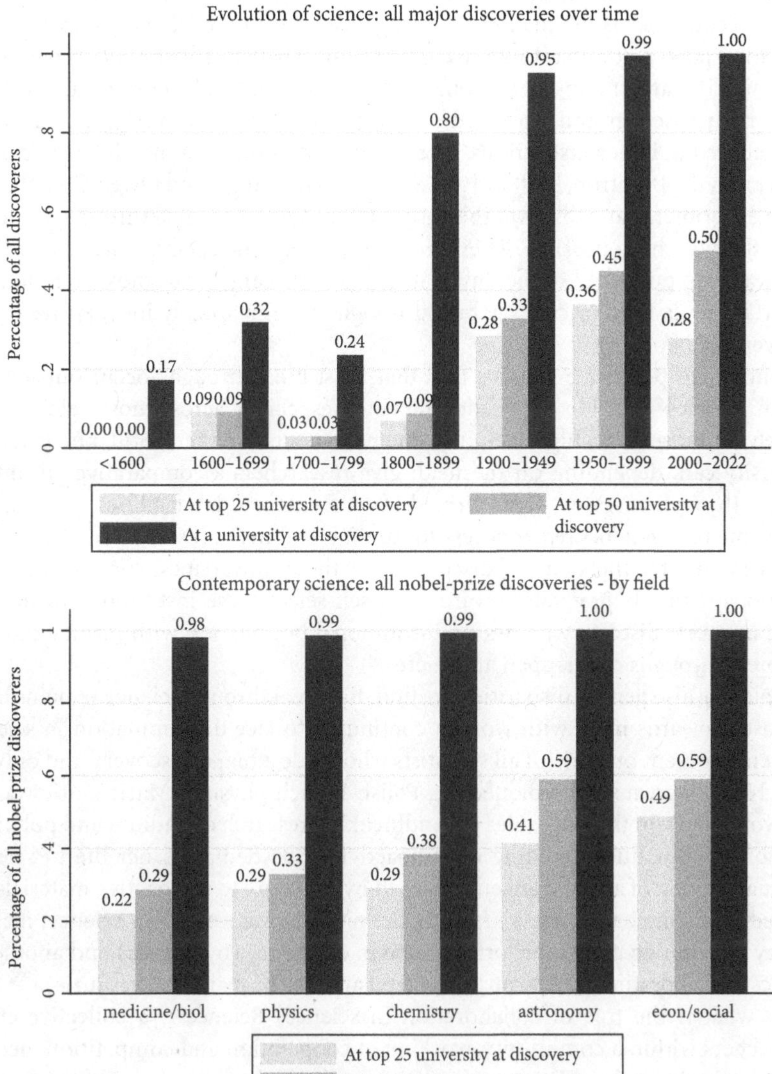

Figure 4.3 *Only about one in three discoverers since 1950 worked at a top 25 university*

The data show science's 761 major discoveries (including all nobel-prize discoveries) (Figure a)—and all 533 nobel-prize discoveries (Figure b). The data reflect the discoverers' university affiliation at the time of the discovery, using the QS World University Rankings in 2021[223] *as a common reference point. Most top universities have remained among the top over time. For earlier centuries, we should view data with caution: while most discoveries have been made while today's top 50 universities existed, some did not yet exist before the 1800s (Figure a). By combining all nobel-prize and major non-nobel discoveries over the same time period across these five fields, we find that, for example, the share of discoverers at a top 25 university is comparable: 22%, 28%, 29%, 42%, and 48%. (And beyond university location, we explore geographic location but also the religious background of discoverers in Appendix Figure 4.1.)*

and space observatories needed to make breakthroughs are concentrated there—and because economics is the field most strongly centred in the US, where institutional reputation plays a dominant role. We find a similar pattern in other fields with some of the world's largest and most sophisticated instruments, like particle accelerators, electron microscopes and advanced x-ray technology needed to trigger discoveries, concentrated at better institutions. At Berkeley for example, a massive new particle accelerator, the Bevatron, built in 1954 enabled Emilio Segrè and Owen Chamberlain to run experiments and discover the antiproton in 1955. The breakthrough would not have otherwise been possible. Their groundbreaking article, *Observation of antiprotons*, was then published in the Physical Review. Yet many of our most common and important tools used to catalyse breakthroughs are remarkably inexpensive (as we uncover in Chapter 6).

Contrary to popular belief, we find that most breakthroughs occur outside top-tier universities—outside the traditional elite—especially across most fields that do not rely on those large, high-tech instruments often based there. Still, being at a top university can, depending on the field, give researchers a comparative advantage, not just through access to sophisticated instruments and lab facilities, but can also offer some researchers greater access to funding and networks of researchers, when necessary. Yet, for the share of discoverers at these universities, some of the more ambitious naturally gravitate towards and self-select these institutions in the first place. In short: discovery knows no institutional boundaries, with history showing that breakthroughs can happen anywhere.

Exploring also gender disparities, we find that breakthrough science remains heavily biased towards men, with women continuing to face discrimination in science. Women represent only 5% of all scientists who made a major discovery and only 3% of all Nobel laureates. The pioneering Polish-French physicist Marie Curie was the first woman to win the prize. Her groundbreaking research on radium and polonium transformed our understanding of radioactivity. She dedicated her life to science, eventually dying of aplastic anaemia caused by exposure to radioactive materials she worked with. She has become arguably the most iconic woman in science, making history by winning two Nobel prizes—one in chemistry (by herself) and another in physics (collaborating with Henri Becquerel and her husband Pierre Curie).[(95)]

But what is the role of collaboration in science? Science is a collective effort. Researchers within a community, working in cooperation and competition, need to inevitably build on the existing tools and research of others that contribute towards a breakthrough. Discovering DNA's double-helix structure is an example of a deeply collaborative effort. It was not just the work of Watson and Crick at Cambridge. But the remarkable breakthrough was only possible because of the pivotal x-ray work produced by Rosalind Franklin and her student Gosling—but as we highlighted, Franklin was discriminated for being a woman and was not fully recognised during her lifetime. Her trailblazing work relied on powerful x-ray crystallography methods developed by von Laue and the Braggs, who used x-radiation identified by Röntgen.[(125)] The research also built on early work on DNA by Miescher, and was supported by parallel work on DNA by Wilkins and his group of colleagues at King's College London, along with others.[(1)] The collaboration spanned across time and geography. Some scientists create the tools needed to carry out the research (like von Laue's and the Braggs' x-ray methods), others may uncover the observational or experimental insights applying these tools (like Franklin's x-ray images), and others may then develop a theoretical explanation for the evidence (like Watson and Crick's theoretical model).

Discovering the Higgs boson at CERN and unravelling evolution's multiple mechanisms are other classic examples of great collaboration—triumphs requiring the collective effort of hundreds of researchers working together over time. Larger teams can at times better leverage different methods, combine them and integrate more expertise. For relevant fields, they can help pool resources for more cutting-edge instruments and can have greater access to different technologies and lab equipment. While we focus on scientific superstars, these pioneers build on a foundation that was made before them and makes the last step towards discovery possible.

Funding and the scientific community

Ask scientists what drives discoveries, and the two most common factors stated are research funding[(6–8)] and teams and the scientific community.[(9–11)] Given the vast breakthroughs achieved in science, the role of governments after the Second World War changed: for the first time, public funding for science was spent strategically. With this shift, a global awareness emerged of the importance of science: the Chinese Academy of Sciences was founded in 1949, the National Science Foundation (NSF) in the US in 1950 and the German Research Foundation (DFG) in 1951. Before the war, science spending was negligible, making up only 0.1 billion US$ in 1930 in the US. After the war, the US arose as the most important player in science funding. Research and development (R&D) spending in the US surged nine-fold, increasing from 1.5 billion US$ in 1945 to 13.6 billion US$ in 1965. Strikingly, in the 1940s and early 1950s, over 80% of this spending was targeted to national defence alone. At the time, the number of PhDs awarded in the US also jumped 10-fold, from 1634 awarded in 1945 to 16,340 in 1965 (Figure 4.4).

Yet science's biggest discoveries throughout history cannot simply be attributed to strategic science funding and a large, trained scientific community—because these factors largely only began after the Second World War. *By 1945, more than half of science's major discoveries—almost 400 breakthroughs—had already been made, including the majority of major non-nobel discoveries since 1875.* If strategic science funding had played a very strong role, we would see a very strong sustained spike in discoveries following the establishment of major science agencies and large-scale funding. Yet we do not see a sustained spike (Figure 4.4). Instead, we uncover a common pattern throughout the history of science: major discoveries consistently follow after the newly invented methods needed to make them.

Remarkably, before the mid-1930s, scientists already developed many of the central methods and tools of science—and the discoveries they enabled. These include modern microscopes since 1873, particle detectors since 1911, x-ray crystallography methods since 1913, the centrifuge in 1924, modern statistics in 1925, particle accelerators since 1929, electrophoresis in 1930 and chromatography in 1931. To create these powerful tools, large-scale research funding and a large scientific community were not the most important factors—or even an important factor—as they did not yet exist. These breakthrough tools were made at relatively low costs, and each then triggered multiple major discoveries. So while research funding fosters our system of science, if we trace the path of discovery up to the breakthrough itself, one key driver consistently emerges: the development of new methods and tools that directly spark new discoveries. We dig deeper and unpack this topic in Chapter 6.

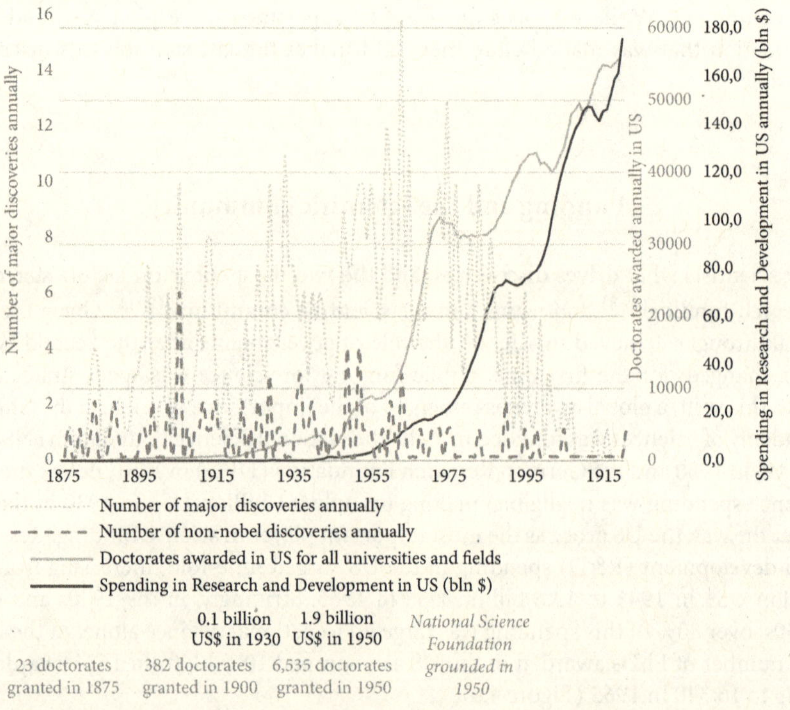

Figure 4.4 *Trends in science funding and the growing size of the scientific community in the US, in relation to major scientific discoveries*

The data reflect science's 620 major discoveries made since 1875—and the 90 major non-nobel discoveries since 1875. Other major discoveries before 1875 are not included. Once again, all data—throughout the book—reflect the year discoveries are made—not when the Nobel prizes are awarded. The data on R&D spending are from the Federal Reserve Economic Data,[224] *and total PhDs awarded are from the NSF.*[225]

Science's major discoveries reveal that expensive tools are not always needed for most breakthroughs, beyond exceptional discoveries using instruments like large particle accelerators at CERN and space telescopes at NASA. Several of the ten most used central methods and tools of discovery are remarkably low-cost, including statistical and mathematical methods, light microscopes, electrophoresis, chromatography methods, centrifuges and thermometers (and later the PCR method and other assay techniques). We can buy these new for less than a thousand or even few hundred dollars (Chapter 6). Statistical methods including fast-growing statistical programmes like R can be downloaded for free and some AI programmes have low costs, and play a critical role in analysing vast datasets.

We see that nobel-prize discoverers in countries within the bottom two quintiles (the poorest 40%) have access and use common tools at similar rates as those in wealthier countries (Figure 4.5). Discoveries using common sophisticated instruments are not just concentrated in the richest countries. Many fields rely on affordable and easily accessible methods like advanced statistics and electrophoresis—and these, as expected, show little variation across income levels. Once a minimal income threshold is met, researchers in less affluent countries do not seem to face significantly greater constraints to triggering breakthroughs—though historically, discoveries have been concentrated in Europe and North America. Yet some countries' researchers live in—like their university—can influence access to more advanced instruments and greater funding, infrastructure and government support. Wealthier countries often house some of the more cutting-edge technological and computing facilities and specialised labs needed for some discoveries. Greater income per capita (more resources) and a larger research team (more researchers) can foster the basic conditions of science—but for most discoveries, they have not been needed (Chapter 6).

Researchers are also not in science for the money, otherwise they would take their PhDs and go into industry with higher salaries. An exceptional and tragic case is

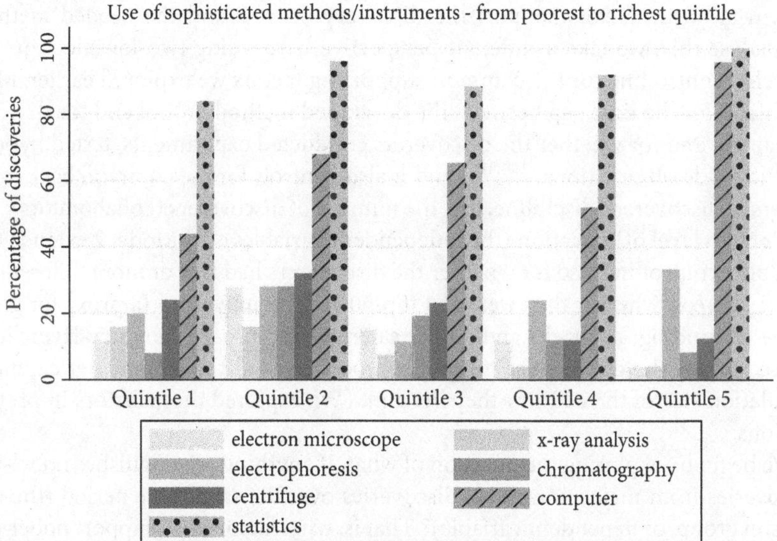

Figure 4.5 *Scientists in poorer and wealthier countries have comparable access and use of various sophisticated methods and tools, for nobel-prize discoveries since 1975*

The data reflect the 125 nobel-prize discoveries since 1975. These methods and tools were first developed by 1950 and in wide use by 1975; to reflect this, these income quintiles are calculated using the broad set of countries in which nobel-prize discoverers lived from 1975 to 2022. The income data is adjusted for inflation, and we analyse the period from 1975 to 2022 to control for larger variations in income over time.

Nikola Tesla, the visionary discoverer who did pioneering work in electricity and wireless transmission, but died impoverished and living alone in a New York City hotel room despite his enormous contribution to science.[95]

Simply investing more money and people into science is not a focused strategy. Yet researchers' salaries and recurring overhead costs account for the vast majority of total science spending—it sustains the day-to-day running costs. The large increase in public science funding and the expansion of the scientific community since the 1950s have fuelled an extraordinary surge in publications—a surge that seems almost natural in an ever-growing academic community shaped by a publish-or-perish culture where frequent publishing is often prioritised over high-impact breakthroughs. Yet the surge has lacked a guiding framework—an evidence-based understanding of what triggers groundbreaking discoveries. Despite large funding bodies in place since the mid-20th century (like NSF and DFG) and early 21st century (like the European Research Council), no funding agencies or large research communities yet have large strategic programmes targeted to the development of transformative new methods and tools across science. Surprisingly, science's best tools happen to emerge as ad-hoc projects as a specific demand arises, like constructing a cutting-edge spectrometer or supercomputer (Chapter 6).

Predicting new discoveries and new methods (regression results)

After exploring how new methods drive discoveries, we turn to the broader question: what other factors support the discovery process and the needed methods? To analyse this, we take a different perspective here, using two logistic regression models—controlling for the common supporting factors we explored earlier. Model 1 controls for the time gap between the developed method or tool and the discovery it enabled, and for whether the discoverers conducted experiments, tested hypotheses and made observations;[196,198] and it also controls for the common background factors of discoverers' discipline, age, the number of discoverers (collaborators), their gender and level of education (the independent variables).[85] Model 2 extends these variables, controlling also for whether the discoverers had two or more different academic degrees, whether they were at a top-50-ranked university (a proxy for greater access to funding, networks and higher salaries) and broader country-level factors. These include the discoverers' geographic location, and the income per capita and population size of the country they lived in. (We explored these factors in previous sections.)

We begin by exploring the question of what, if anything, distinguishes nobel-prize discoveries from major non-nobel discoveries over the same time period (the comparison group, or dependent variable)? That is, what factors may support nobel-prize discoveries? We find that nobel-prize discoveries are more likely to be made faster—within a shorter time span between the developed method or tool and the resulting breakthrough. They are also more likely to be made by highly educated researchers (PhDs or professors), by two or more researchers and through experimentation—while controlling for the set of supporting factors including discipline, age, gender, university ranking and geographic location. Being at a top university is not an

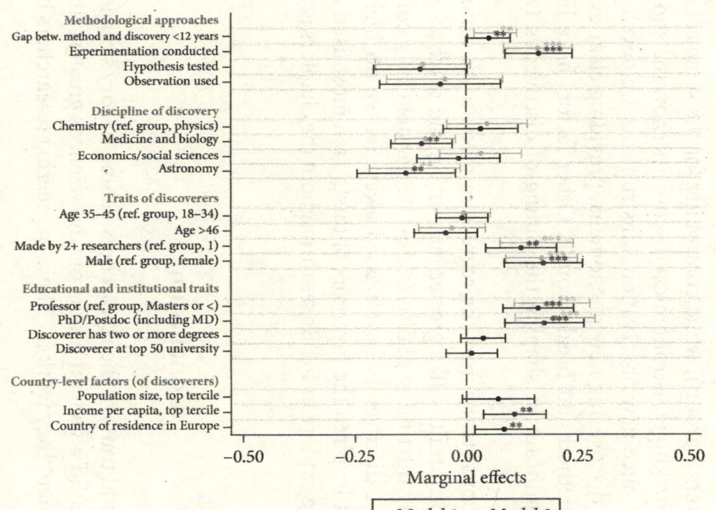

Figure 4.6 *The enabling factors of nobel-prize discoveries*

The data reflect 616 major discoveries that include all nobel-prize discoveries compared to major non-nobel discoveries made over the same time period.

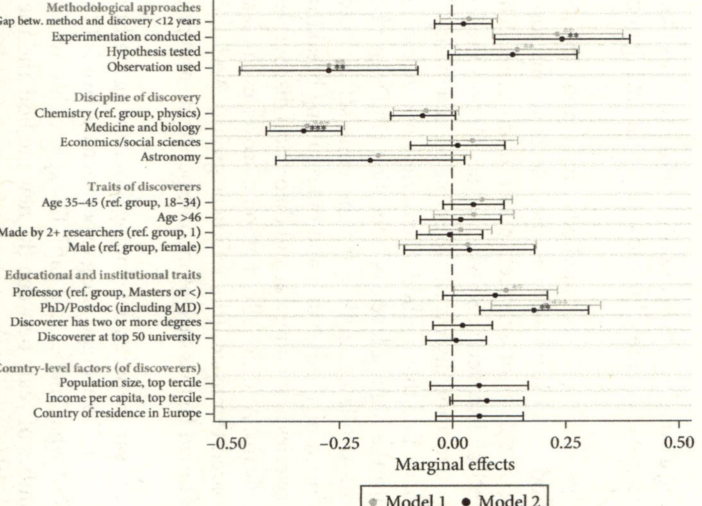

Figure 4.7 *The enabling factors of method discoveries (major new methods and tools)*

The data reflect 697 major discoveries since 1575, including all nobel-prize discoveries. For interested readers, we provide the technical details and model specifications below Appendix Figure 4.2.

important predictor of making a nobel-winning discovery. Broad factors like income per capita and population size also play a more limited role in influencing nobel-prize breakthroughs (Figure 4.6).

Next, we explore the essential role of methods and tools in triggering breakthroughs, with method discoveries—from the electron microscope to electrophoresis—responsible for 25% of science's major discoveries. The other 75% are empirical and theoretical discoveries. So what factors support these method discoveries? When we compare method discoveries to empirical and theoretical ones (the new dependent variable), we find that method discoveries, too, are most strongly supported by researchers with the highest levels of education (PhDs or professors) and through experimentation—while controlling for the same set of factors. The time span between the developed method or tool and the resulting breakthrough is very similar for method discoveries and empirical and theoretical discoveries, so the effect is not statistically significant. Additionally, method discoveries are least likely to emerge in medicine compared to physics (the reference group). And once again, we see a more limited role of broad factors such as income per capita and population size (Figure 4.7).

Finally, we turn to the gap in years between developing the enabling method or tool and making the discovery using it. On average, the gap between the two is 14 years for all nobel-prize discoveries. But what shapes this gap? To explore this, we use linear regression to analyse the number of years between the new method and the discovery (the dependent variable). We find that the strongest factors influencing this gap are the specific discipline and time period in which the breakthrough occurred. Discoveries in medicine and biology tend to have a gap of about 7 years longer than those in physics (the reference group), while controlling for the same set of factors earlier and time periods. Strikingly, astronomical discoveries show the shortest time gaps, as we would expect: once new observational instruments are developed, they more immediately enable discovering new phenomena, providing clear evidence of the direct link between method and breakthrough. Other factors have little effect or are not statistically significant in shaping the timing of discoveries (Appendix Figure 4.2).

The overall trends are consistent across the regression models. We unfortunately cannot explore psychological traits—like greater motivation and drive among some discoverers—in influencing the timing of discoveries. We cannot easily collect these traits simply because most discoverers have passed away. Yet we do highlight the remarkable stories of what inspired the inventors of the top ten most used and powerful tools in science (as we explore in the Boxes throughout the chapters and further in Chapter 6).

Conclusion

We explored the features and traits of the researchers behind history's biggest breakthroughs, enabling us to build a general profile of who science's greatest discoverers are and the basic environment they work in. These pioneering researchers have been evolving, shifting towards greater interdisciplinary education and method training, and they are not as young as before, but still overall young. One background trait stands out remarkably among the rest: about half of all discoveries are made by scientists with at least two degrees in different fields. This underscores a critical insight. Researchers trained across disciplines and methods are more likely to make new method connections across fields, sparking breakthroughs at the intersection of fields. Not only interdisciplinary but also early career researchers are central in the process. We found that the golden age of innovation in science falls between 35 and 45—with very few major breakthroughs occurring after 50 and almost none after 60. For younger, more motivated researchers generally bring more recent training in cutting-edge tools and fresh perspectives. Science, at its core, needs to reward those willing to take risks and challenge the status quo. Strikingly, scientific progress is not confined to elite institutions—with seven in ten discoverers not at top-tier universities. Groundbreaking research can emerge anywhere, as long as researchers have the right tools and the basic conditions (basic facilities and resources) and freedom to do research.

So what do we learn by exploring the background traits of discoverers? There are critical opportunities here: we could achieve more breakthroughs if research institutions and agencies better incentivise researchers to push beyond the common trend of narrow specialisation. By fostering interdisciplinary innovation that blends novel methods across fields, we open up entirely new ways to tackle problems and spark innovative solutions. These unexpected solutions would not have been possible in a single discipline—from MRI and the gene-editing method CRISPR to AI methods. It is crucial for us to break down the artificial boundaries between fields to enable researchers to explore connections they could otherwise not make. To unlock untapped areas of innovation, we need to remove disciplinary barriers to sharing powerful tools across fields and prioritise developing new integrated tools. Also providing incentives for underrepresented groups—including female researchers—is important to ensure a more inclusive scientific community that taps into the full range of human potential and fosters discovery.

Here we traced the evolution of science's greatest discoverers and the scientific system they do research in. We later dig deeper into how the tools these discoverers used are commonly the central driver in powering our biggest breakthroughs (Chapter 6). But before that, we explore in the next chapter how innovations in tools also trigger new scientific fields—not just discoveries. By understanding the methods-and tools-powered nature of science, and how we can accelerate it, we set the stage for more strategic and rapid scientific progress.

5
The birth of fields

How new methods and tools launch new disciplines

Summary

Pioneering discoveries, such as the structure of DNA, the periodic table of elements and quantum theory, are embodied in our scientific fields, such as biology, chemistry and physics. A discovery captures an immediate advance, while a field captures our discoveries and methods—our ever-expanding body of knowledge over time. Scientific progress unfolds—and is measured—in three fundamental ways: through new discoveries, methods and fields. But how do new fields actually emerge? What is the key driver enabling them? Despite fields encompassing our immense scientific understanding, no comprehensive answer to this fundamental puzzle yet exists. Here we systematically trace the origins of science's major fields including hundreds of fields spanning across science, offering a unique opportunity to uncover what triggers new fields. Remarkably, we find that fields are kick-started by a new method or tool—from advanced telescopes to electrophoresis—as they enable a completely new perspective to the world and without them, the fields would not be possible. About a quarter of fields are the new method or tool themselves—from laser physics and computer science to x-ray crystallography and econometrics—forming entire disciplines around new techniques. Our extraordinary development of new statistical techniques, x-ray devices, microscopes and spectroscopes has fuelled a wave of new disciplines: each enabling over ten new fields to emerge. The common link uniting these diverse fields is not specific theories, large teams, more funding or even serendipity—it is that each relied on the same kind of powerful tool, used in remarkably different domains. The speed at which science expands is not random. The pace of opening new research domains is mainly determined by how quickly we create new tools: particle detectors launched high-energy physics, microscopy techniques triggered neuroscience and randomised controlled trials kick-started experimental economics. This simple yet powerful principle—if we begin to deliberately develop transformative methods—holds the key to accelerating scientific progress and enabling us to spark a *tool revolution* in science.

The Engine of Scientific Discovery. Alexander Krauss, Oxford University Press. © Oxford University Press (2026).
DOI: 10.1093/oso/9780197829790.003.0006

Overview

Scientific fields embody our vast, ever-expanding scientific, medical and technological advances and bodies of knowledge over time. But despite their tremendous impact, we still do not understand how new fields emerge. This is a fascinating and critical question because unravelling the mystery would enable us to explore entirely new domains.

New fields can seem to arise from discoveries so groundbreaking that they create a new scientific community and body of knowledge. When we look closer, some new fields seem to emerge as a novel specialised branch of established disciplines, like molecular biology and developmental psychology. Others seem to grow out of the integration of two or more fields, like biophysics and cognitive science. Some seem to arise as contrasting approaches within the same domain, like theoretical and experimental physics and economics. Still others seem to develop due to shifting societal or environmental forces, like telecommunications and climate science. And these different paths can overlap. But what is the force enabling new fields in the first place, which then reflect greater specialisation, integration, diverging approaches and adapting to new external conditions? Given the diverse paths, can we identify a unifying explanation of how we give rise to all fields? If we could explain and even help predict their emergence, we would not just be passive observers of scientific growth—we could actively develop new research domains and speed up advances.

The most common explanations for how fields emerge are through paradigm shifts in theories,[15,16,73] new research programmes[182] or splitting or merging scientific communities.[226] But do they actually explain how new fields come about? A highly influential explanation, proposed by Thomas Kuhn, describes fields arising and evolving through fundamental changes in scientific theories—paradigm shifts.[15,16,73] But many applied fields do not involve a theory. In fact, no new field emerged even from the classic paradigm shift from the Ptolemaic earth-centred theory to the Copernican sun-centred theory, or the shift from the theory of continental drift to the theory of plate tectonics to explain large-scale geologic changes. Another influential explanation proposes that major new research programmes lead to new fields.[182] But many research communities—such as those studying new viruses, astronomical objects and global warming mitigation—have expanded our knowledge without spurring distinct new fields. Another influential explanation argues that scientific communities split or merge through new collaboration networks that can bring about new fields.[226] But these shifts follow after a new field arises as an offshoot, rather than its driving force. Sociologists also offer explanations for how social movements and conditions can influence the scientific landscape and some disciplines.[227] Yet, each of the proposed explanations is in fact mainly a consequence of new fields rather than their cause—they generally follow rather than precede their emergence.

While researchers have long sought to explain new fields, efforts have mainly centred on research outputs, such as new theories[15,73] and citation patterns,[3,4,14,23,24] and aspects of fields like collaboration networks and productivity.[10,11,21–23] But are these really the most relevant metrics to focus on? In a thought-provoking article published in *Science*, researchers challenge this conventional approach: 'Citations, publication counts, career movements, scholarly prizes, and other generic measures are crude quantities at best ... and their ability to predict the emergence of a new field or the possibility of a major discovery may be low'.[3,4,14,21–23,39,40,61] Yet explaining and better predicting new fields is a key goal of the *science of science*.[3,4,14] So if these standard metrics fall short in capturing the birth of new disciplines, what does capture them?

Achieving this goal requires us to dig deeper and take a completely new approach. Here we shift our attention from commonly exploring outputs (especially article citations and theories) to investigating the inputs—especially new methods and tools that enable entirely new perspectives to studying the world. The idea that we can study the drivers of new fields to identify successful ways of advancing science, but that scientific fields have not yet been systematically analysed and linked to their underlying methods and tools, may seem ambitious. By tracing the origins of science's major fields including hundreds of different fields and the innovations in tools that enabled them, we show that this alternative approach is not only attainable—it may be the best strategy we have to understand how we develop new fields and how we can do so faster. We also explore the broader conditions behind the scientists kick-starting the fields.

What do we find? A striking pattern emerges: about one hundred new methods and tools that won a Nobel prize have opened new fields, such as x-ray crystallography, electron microscopy, quantum computing, mass spectrometry, climate modelling, laser spectroscopy, phase-contrast microscopy and econometrics. Each of these is a powerful new method or tool, each earned a Nobel and each is the foundation of the new field it triggered—fundamentally reshaping the way we explore the world.

Take the electron microscope, our prime example of a key instrument leveraged in multiple fields. Developed in 1933, this pioneering instrument does not just vastly magnify objects, it helped unlock an entirely new domain of biology: modern cell biology. By uncovering the intricate architecture of cells, it allows scientists to link cellular structure with function. It has transformed how we understand the causes of disease and enabled developing targeted therapies, paving the way for advances in regenerative medicine. The maser, invented in 1954 as the precursor to the laser, laid the foundation for new fields: laser spectroscopy and quantum electronics. It enables us to probe the dynamic behaviour of chemical reactions and the structure of molecules at an unprecedented level of precision. X-ray crystallography methods, created in 1913, gave birth to molecular biology. By unravelling the molecular structure of DNA and proteins, it reshaped our understanding of life itself—and remains a cornerstone for life-saving research in drug design and disease treatment.[95]

Exploring the birth of fields, we uncover a common pattern across them: whenever a scientific community undergoes a major change in the way it understands the world, it is preceded by a major methodological change in the way we study, measure and theorise about the world that was not possible before.

Measuring and tracking the birth of scientific fields

To understand how fields emerge, we need to trace their origins and the pioneering scientists who sparked them. Here we analyse science's major fields including hundreds of fields spanning the physical, life and social sciences—from physics, chemistry and biology to psychology, economics, anthropology and computer science. Those less interested in the details of how we collected the data can skip this paragraph. We verified that all fields return at least 25 results on Google Scholar by searching 'field of x' (such as 'field of chemistry' or 'field of genetics'), with most generating hundreds or even thousands of results. To ensure all fields are recognised fields, we took several steps. First, we included all fields that emerged directly from nobel-prize-winning research in science—fields recognised in the prize motivation or opened by scientists who earned the prize.[95] These make up 257 fields.

Next, to broaden the scope, we also included all fields since 1500 pioneered by scientists featured in the seven science textbooks listing the world's 100 greatest scientists across science.[96–102] These make up 116 other fields. This adds up to 373 fields and we then verified that the scientists who opened the fields are described in scientific publications as the 'founder', 'father' or 'mother' of the discipline. Finally, we focus on fields established at universities, confirming that a total of 213 of the 373 fields—such as cognitive science, astrophysics and microbiology—exist as departments, institutes, centres or schools within universities. Each discipline is defined as a separate field or subfield—for example, the emergence of modern physics, then particle physics and later laser physics. (For emerging and smaller fields, additional analyses are available in the Online Appendix.)

So what defines a scientific field? A field is commonly characterised by shared research topics, methods and tools, a scientific community and at times theoretical models.[228] When a field emerged is established by the *first* article that kick-started its development—starting a growing research community and body of knowledge. The first publication that triggered each field is the main source for compiling the data; it includes the central method or tool applied in that publication, is commonly highlighted by the authors and made the field possible—and also includes the year they were published.

The origins of new fields: the power of new methods and tools

To begin, we examine the emergence of fields since 1500, tracing the year a new method or tool was created by the year the field arose applying it—unlocking an entirely new way to explore the world. Each groundbreaking new method or tool appears as a vertical line (|) and each field that emerged using it appears as a dot (●)—seen in Figure 5.1. Think of for example the development of femtosecond (ultrafast) spectroscopy in 1985 (shown as a vertical line |) that led the Egyptian Ahmed Zewail to open the field of femtochemistry in 1988 (shown as a dot ●). This pioneering instrument enables us to much better understand chemical reactions on very short timescales—mere quadrillionths of a second—using lasers. Suddenly, we could witness molecules in motion, reshaping our understanding of fundamental chemistry. Think of the invention of the microscope with the silver staining technique in 1873 that made it possible for the Italian Camillo Golgi and Spaniard Santiago Ramón y Cajal to launch the field of neuroscience that same year. This trailblazing technique revealed the first images of our nerve cells and enabled understanding the structure of our nervous system. It opened a new era in studying the intricate architecture of the brain and how we think, feel and remember.

Think of the creation of a new 270-power microscope in 1673—capable of observing objects just one-millionth of a meter in length—that enabled the Dutch Antonie van Leeuwenhoek to give birth to the field of microbiology in 1674. For the first time in history, we could study the invisible world of bacteria and it even redefined what it means to be human. History reveals a key insight: new microscopy techniques—like new spectroscopes and controlled experimental methods—are representative of how new fields are born after inventing the right tools; and without them, they would not have been possible.

We uncover a strong and direct link between new tools and the birth of new fields: examining the publications that kick-started science's new fields shows that each new field—one that studies and measures an unexplored part of the world—consistently begins by adopting a groundbreaking new method that enables studying and measuring the world in a new way. New method innovations precede the disciplines they trigger—with *new* meaning we use it for the first time ever for a specific problem.

Some tools spark new fields almost instantly, while others lie dormant for decades before scientists pick up on the tool's potential. This relationship is mapped out here, where the length of each horizontal line reflects the time between the method's invention and the field it enabled (Figure 5.1). Two striking trends emerge: the lag between the two is shrinking, and the rise of new fields and bodies of knowledge is not slowing down (Appendix Figure 5.1).

We uncover a very strong pattern here. The top ten methods and tools most used in nobel-prize discoveries have been so powerful that they each kick-started not one but five or more fields—in ways their inventors never anticipated: new statistical and mathematical methods gave rise to fields such as experimental economics and empirical finance. Spectrometers launched fields such as molecular spectroscopy and exoplanetary science. Electron microscopes triggered fields such as modern cell biology and electron microscopy. X-ray methods gave birth to fields such as molecular biology and x-ray crystallography. Chromatography enabled fields such as organocatalysis and signal transduction. Centrifuges kick-started fields such as enzymology and centrifugation. Electrophoresis enabled fields such as DNA sequencing and proteomics. Lasers sparked fields such as laser physics and laser spectroscopy. And particle accelerators and detectors paved the way for fields such as high-energy physics and solid-state physics. Beyond these top ten tools, others—such as geiger-müller counters and game theory methods—have also given birth to at least five fields. These powerful, general-purpose tools explain most scientific progress because they each enable us to access a part of the world otherwise out of our scope. *A powerful insight emerges from the fact that each of these tools kick-started multiple fields: what links these diverse fields is not a particular theory, research team, more funding or*

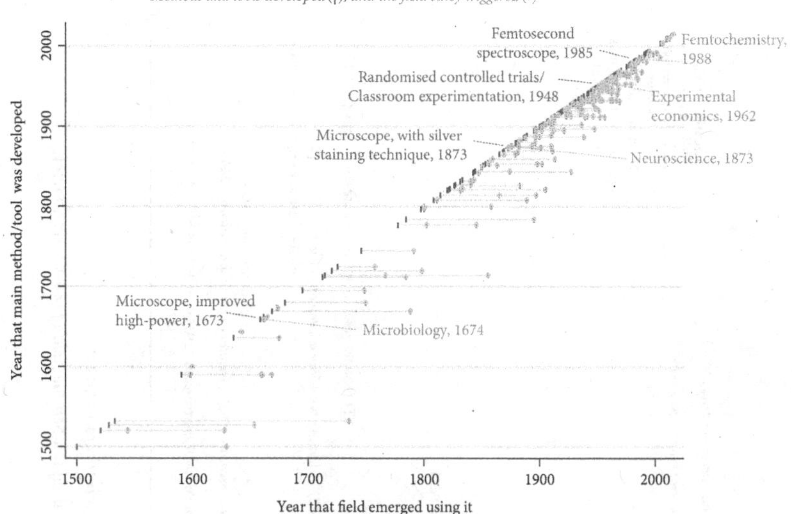

Figure 5.1 *New scientific fields emerge through new central methods and tools we create*

The data reflect 213 established fields developed since 1500 and the methods and tools that enabled them—with four examples illustrated. A new tool at times triggers multiple fields, appearing as multiple dots on the same horizontal line. We find that the year we create methods is closely correlated with the year the resulting fields emerge—94% of the variation.

even serendipity—it is that each relied on the same powerful tool, used in remarkably different domains.

What makes this connection astonishing is that none of these tools were specifically developed to open these fields, yet they still consistently sparked multiple fields across different disciplinary domains. Through quasi-experimental reasoning, we see that the new tools were not intentionally designed to create the new fields, but still did with their transformative power. The fact that entirely new domains of research emerged from tools designed for other purposes highlights a fundamental insight about scientific progress: *new fields commonly come from the unexpected power of new tools.* Without them, the new fields would not be possible. This unforeseen link between the two is key to understanding how science unfolds. Establishing that the new method (the independent variable) originally had no direct connection to the new field it ultimately enabled (the outcome) helps us isolate this causal link. And it reduces other explanations—such as funding, teams and institutions as the key triggers—as the necessary tools were already invented.

So if we focus on outcomes in studying how fields emerge—like major new paradigm shifts in theories, research programmes, splitting or merging scientific communities, or even accumulated citations—we fall short in explaining the birth of disciplines. For these outcomes generally only *arise at or after their emergence.* But new tools, in contrast, can in fact explain and help predict new fields because they must *arise before their emergence* and be applied to kick-start them. This gives us a different way to think about this relationship: we need to track both the baseline (when tools are born) and the endline (when fields are born). Think of how randomised

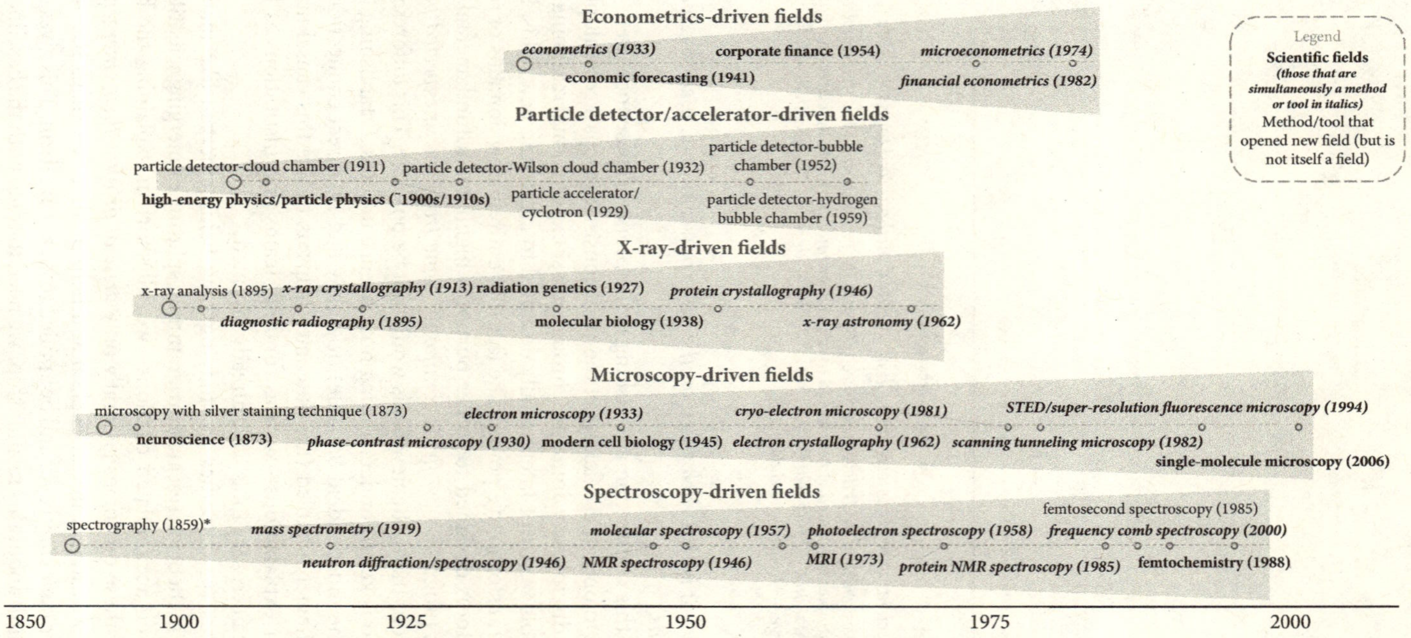

Figure 5.2 *New central methods and tools kick-start new scientific fields (illustrated with nobel-prize-winning methods)*

**The first spectrograph, developed in 1859, is the only tool included that did not receive a Nobel prize, but was used to open the field of mass spectrometry.*

controlled trials designed in 1948 for clinical medicine [133] foreshadowed the eventual (or inevitable) rise of experimental economics. This new branch of economics took shape when Vernon Smith adopted the transformative experimental method in economics in 1962 with his paper *An experimental study of competitive market behavior.*[229] With the creation of modern statistics and biostatistics in 1925,[137] we could predict that the field of econometrics (statistical analysis in economics) would eventually develop—which took off in 1933 (Box 5.1).[230] With the invention of the digital electronic computer in the 1940s and 1950s, we could predict that computer science would eventually arise—which did in the 1950s.

The first particle detector that visualised particle tracks, built in 1911, gave birth to high-energy physics. The groundbreaking 1911 paper *On a method of making visible the paths of ionising particles through a gas* was published in the Proceedings of the Royal Society. The field then rapidly expanded with the creation of more advanced detectors, and also particle accelerators since 1929. These enable us to understand the nature of the particles that make up the very building blocks of matter and our physical world. Creating NMR spectroscopy in 1946 launched the fields of MRI and protein NMR spectroscopy. This helps us decode the structure and dynamics of proteins and develop life-saving medical drugs (see Figure 5.2).[95]

An improved cathode-ray oscillograph, invented in 1922, marked the birth of neurophysiology, helping unravel how our nervous system functions and diagnose and treat neurological diseases. Game theory methods, first conceived in 1928 and expanded in 1950 with Nash equilibrium, reshaped economics by giving rise to information economics—a field that explains how information impacts our decisions and economy. The transistor, a tiny semiconductor device built in 1947, made the field of microelectronics possible, bringing about mobile phones and personal computers and laying a key foundation of modern society. It was created at Bell Labs by John Bardeen and Walter Brattain and later refined by William Shockley. At the time, they had no idea that the transistor would forever change the world and kick-start what has become a massive field that shapes our everyday lives through the technologies we use. They set off a technological transformation that has changed the way we communicate, work and live.[95] Each of these groundbreaking tools earned a Nobel prize. Without innovating new methods, these new research domains—and the world-changing technologies they enabled—would not have been possible.

Box 5.1 How Ronald Fisher developed modern statistics—enabling powerful analysis across science and opening new fields

We now zoom in on statistical methods—one of the ten central methods and tools most used to make nobel-prize discoveries and trigger new fields—before zooming back out to the broad patterns across science. Statistics transforms how we study the world by allowing us to collect and analyse vast amounts of data and run large experiments on almost any phenomenon—from tracking diseases in populations and decoding cellular processes to mapping astronomical bodies, modelling complex economies and even uncovering what shapes discoveries and fields. The foundation of modern statistics developed in stages around the early 20th century, culminating in 1925 with the now classic book *Statistical methods for research workers.*[(137)] This groundbreaking book, written by the London-born Ronald Fisher at the age of 35, marked the first full-length book on statistical methods, playing a pivotal role in establishing and spreading modern statistics.

Fisher's path to transforming statistics was anything but conventional. After earning a bachelor's degree in astronomy and teaching several years in public schools, he received an offer in 1919 to work with the renowned statistician Karl Pearson. But due to a long-standing feud between the two, Fisher declined and instead joined Rothamsted, an agricultural research institution.[(231)] Here he ran plant-breeding experiments using and refining statistical techniques. This hands-on research environment provided the inspiration for his 1925 book that offered solutions to many real-world problems his colleagues faced with handling data.[(232)] In his book, Fisher writes: 'Daily contact with statistical problems as they presented themselves to laboratory workers stimulated the purely mathematical researches upon which the new methods were based'.[(137)] And after Peason retired in 1933, Fisher took over his position as chair at University College London.[(231)]

In this landmark book, Fisher developed some of today's most widely used statistical methods across nearly all fields. Among his most influential contributions, he created the analysis of variance (ANOVA) technique that enables us to assess the average differences among sub-groups or experimental conditions within a study. Whether comparing the effects of different medical treatments, agricultural practices or economic policies, ANOVA became an essential tool for detecting meaningful patterns in data. Fisher also redefined how we determine statistical significance by popularising the p-value, or probability value. He proposed using a 1 in 20 probability threshold as a standard to evaluate whether a result arose by chance in an experiment. He wrote: 'The value for which P = .05, or 1 in 20 [chance] ... it is convenient to take this point as a limit in judging whether a deviation is to be considered significant or not'.[(137)] This threshold remains widely used in scientific research, though debates continue over whether 0.05 is an optimal cutoff (Picture 5.1). Fisher also pioneered using the important technique of randomisation in experiments—a technique to reduce bias by selecting samples randomly rather than arbitrarily. He applied the principle first in his agricultural experiments.[(232)] Remarkably, he developed these pioneering methods through his everyday work and using little more than a calculator, without a team or additional resources.

His statistical breakthroughs transformed fields like agriculture and biology and paved the way for others, like medical statistics, econometrics and complexity science—demonstrating how powerful new methods can ripple across disciplines in ways its creator never anticipated. He also played an essential role in founding population genetics and shaping the modern evolutionary synthesis—merging Darwin and Wallace's theory of evolution with Mendel's heredity. Fisher is seen as the father of modern statistics and even as 'the greatest of Darwin's successors'.[(233)]

For his contributions, he was knighted in 1952.[232] Yet he remains one of the greatest unsung heroes of science: most scientists have likely never heard of Fisher, even though nearly all scientists have applied his methods. His work also set the stage for randomised controlled trials, a method pioneered by Bradford Hill. The method is one of the most important statistical innovations in fields like public health, economics and psychology—essential for testing the effectiveness of medicine we take and policies we adopt.[234] Fisher did not just refine statistics, he redefined how we think about and understand the world.

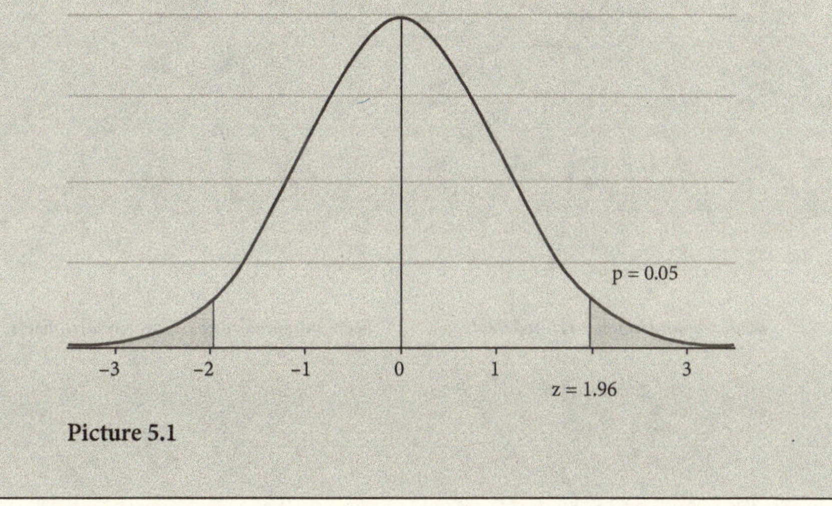

Picture 5.1

The most powerful tools we leverage to drive new fields across disciplinary areas

What are the most transformative tools across different disciplinary areas? New research domains cluster around key methods that trigger them. The methods and tools that acted as catalysts in opening most fields within physics-related disciplines are new mathematical techniques, lasers, spectroscopes and cathode-ray oscillographs. Within chemistry-related fields, these groundbreaking methods and tools are new spectroscopic methods, x-ray methods and mathematical techniques. Within medicine- and biology-related fields, these are new optical microscopes and other vision-enhancing tools like x-ray devices, spectroscopes and electron microscopes. Within economics- and social science-related fields, these are new statistical methods and economic modelling methods (as mapped out in Figure 5.3). In biology for example, without the microscope—our classic instrument—several fields could not have emerged, from microanatomy that studies the structure of tissues and organs, to bacteriology that investigates bacteria and their links to disease. The historical pattern is striking: before we change the way we explore and view the world, we first change the tools we design to study the world. Because we can observe these changes in the rise of new disciplinary branches before and after the invention of the key tools, new disciplines do not emerge randomly (Appendix Figure 5.2).

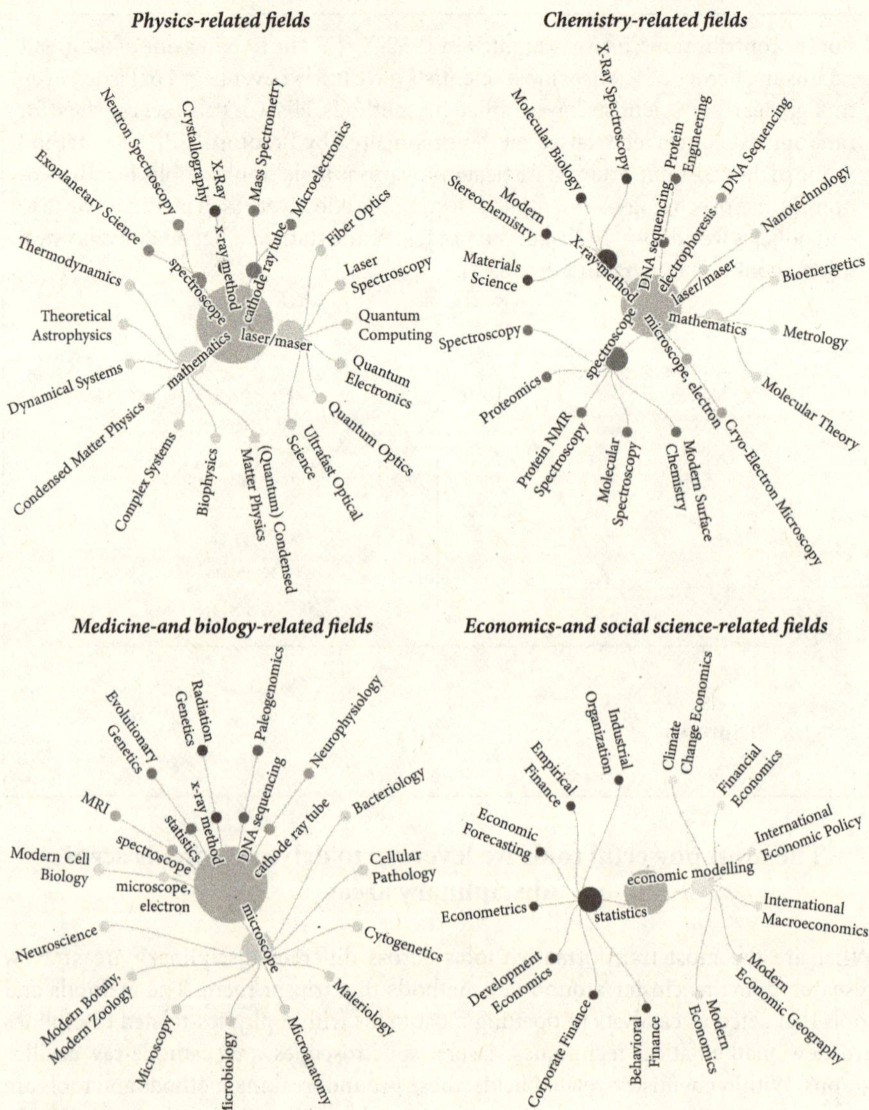

Figure 5.3 *Mapping the central methods and tools that open new fields—a network analysis*

The data reflect the central methods and tools we used in kick-starting established fields since 1600—reflecting 18, 16, 15 and 13 fields (from the top, left to right). Each method or tool was applied in developing at least three fields within any disciplinary areas.

Take the fascinating field of exoplanetary science. It launched in 1995 with the groundbreaking study published in Nature—*A Jupiter-mass companion to a solar-type star*. The field was made possible by a new echelle spectrograph, invented in 1993. New tools expose entirely new realms of exploration, they are indispensable for shedding new light on the known and revealing the unknown. It is those who conceive and refine these tools who amplify our research scope and fields by tackling the bottlenecks of our methods and mind.

But new tools do not just give birth to fields; for over a quarter of them, they are the defining feature of the discipline itself—and not what we study using it. Fields like electron microscopy (1933), x-ray crystallography (1913), mass spectrometry (1919) and neutron spectroscopy (1955) emerged with the tool's invention. Such methodological fields are often inherently interdisciplinary as we leverage these tools across the broad domains of chemistry, biology, medicine and physics—with each tool earning a Nobel prize. Reinvention and fusion are also important: dozens of fields have emerged when we merge cutting-edge tools from different domains together, like computational chemistry, quantum interferometry and statistical mechanics.

Most major method discoveries—supported at times by interdisciplinary work—establish new fields

While new fields are consistently driven by method innovations, not every method innovation fuels a new field. This raises a crucial question: why do some major method discoveries establish new fields while others do not? To answer this, we first examine the extent to which major method discoveries (all nobel-prize-winning and major non-nobel method discoveries) trigger new fields. An extraordinary pattern emerges: these major new methods and tools reflect about one in four major discoveries in science, but among them, a remarkable 82% opened a new field. These include field-triggering tools like laser cooling launched in 1985 by Steven Chu at Bell Labs; DNA amplification pioneered in 1985 by Kary Mullis at Cetus Corporation; and neutron spectroscopy created in 1955 by Bertram Brockhouse at the Atomic Energy of Canada. This striking finding—that major method discoveries are much more likely to lead to new fields than not—holds true across time and disciplinary areas (Appendix Figure 5.4). But can broader demographics, institutions and geographic location support new fields arising? To test this, we compare a control group of major method discoveries that did not establish new fields with those that did. This enables us to examine the differences, between the two groups, in factors that can support disciplines arising.

Among method discoverers, a striking 65% of those who have triggered new fields have worked interdisciplinarily, while this figure drops to just 39% for those who have not sparked a field (Figure 5.4a). But it is not just about combining two scientific communities through new collaboration networks;[(226)] rather, new fields are more likely to emerge when we fuse methods across disciplines—either integrating methodological approaches from different domains or applying methods in completely new domains.

Other factors, like discoverers' level of education, gender and age, show little to no difference between establishing new fields and not establishing them. These factors seem less important behind fields emerging. Method discoverers who launched new fields were more likely to work at a top 50 university and be based in North America, but these factors are not statistically significant when controlling for the range of demographic factors.

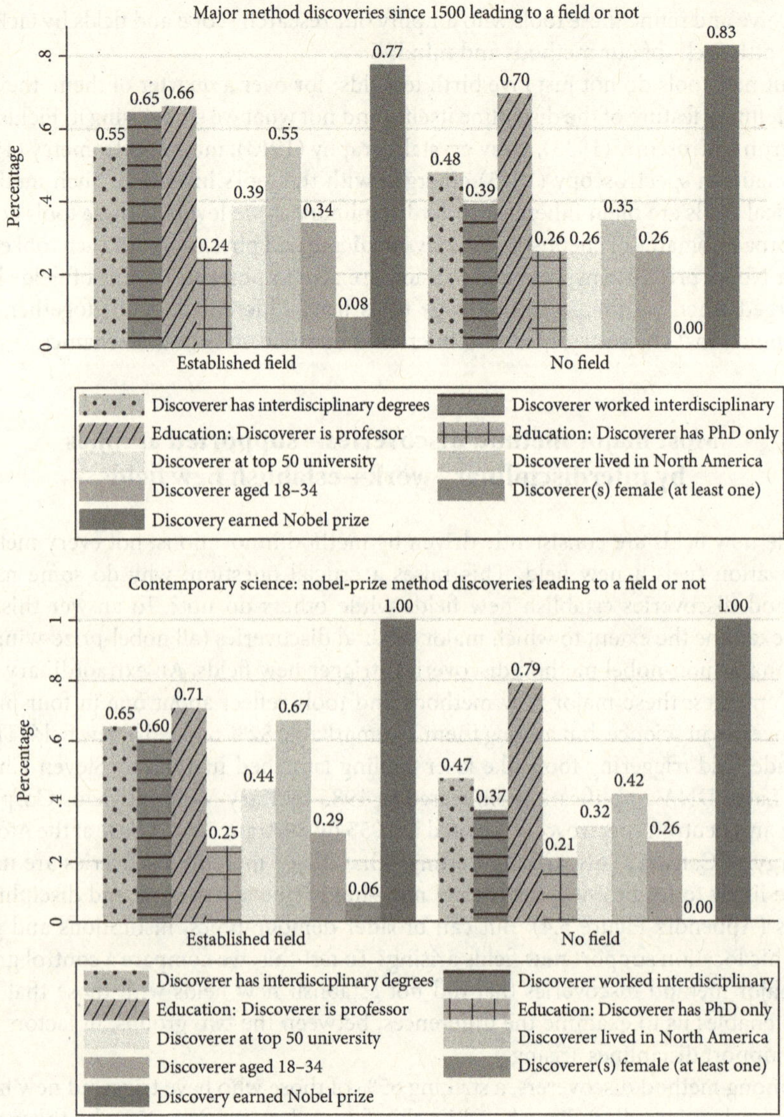

Figure 5.4 *Major method discoveries leading to new established fields compared to those that did not, by features at the time the discovery/field emerged*

The data reflect a total of 85 major method discoveries since 1500, with 62 leading to established fields and 23 not leading to a new field (Figure a). And the data represent a total of 67 nobel-prize-winning method discoveries, with 48 and 19 discoveries in the two groups (Figure b).

So we next explore what predicts whether major method discoveries establish new fields or do not. To do this, we use logistic regression to analyse the demographics, institutions and geographic location of these discoverers (as independent variables). A method discoverer working interdisciplinarily is the only significant predictor of a new field emerging, while controlling for these other factors—and considering the small sample of less than a hundred major method discoveries (Appendix Figure 5.3). In other words, when we step beyond the boundaries of our own field and blend methods, we increase the odds of breaking new ground and domains.

Box 5.2 How Archer Martin and Richard Synge developed simple chromatography methods that transformed chemistry and kick-started new fields within chemistry and biology

In chemistry, the first step is to isolate a substance from natural materials like plants or animals. The next step is to identify the substance and determine its composition. To achieve this, the Russian Mikhail Tsvet pioneered the method of chromatography in 1906, yet it was not until the Austrian Richard Kuhn refined the method in 1931 that it began spreading through the scientific community.(235) Kuhn, who received his PhD in chemistry at just 22, used chromatography to discover a new type of carotene (a vital component of vitamin A) and provide new insights into vitamins B2 and B6—research that earned him the Nobel prize. At this time, Archer Martin was studying biochemistry at Cambridge.(49) In 1938, Martin began working at Wool Industries Research Association and Richard Synge joined as a research student the following year. The two British biochemists soon began working together on wool felt and its amino acid composition.

Synge's studentship received funds from the wool industry thanks to Hedley Marston, who advised him to study the amino acid makeup of wool and to start by improving the methods for analysing amino acids: 'If you work steadily at that for five years, you will revolutionize the whole of protein chemistry'.(236) This challenge was a source of inspiration for developing a new method, along with a technique he came across in a key study. He described this technique as 'countercurrent fractional extraction [and] until then it had not occurred to me that an extraction column could be used to separate two substances of rather similar partition coefficient'.(49) His realisation underscores a crucial lesson: scientific progress often hinges on being aware of available methods to tap their potential. Synge did not need five years—he achieved his goal in just three. In 1941, he completed his PhD and published the landmark study with Martin on the powerful new method of partition chromatography, paving the way for the explosive surge of chromatography that transformed chemical analysis.(237)

Creating partition (paper) chromatography—with Synge at just 27 and Martin at 31—earned them the Nobel.(49) The method is straightforward: substances—like amino acids—are separated in a mixture to establish what they are made of. The process itself is surprisingly simple: a single drop of the chemical substance is placed on a strip of paper. The paper is then soaked in a solvent (like chloroform or alcohol) and the different components begin to spread out, forming distinct coloured marks that reveal the mixture's components. That is it. This simple method requires just water, filter paper and a solvent—all very cheap lab supplies (Picture 5.2).[237] As Martin himself remarked in his Nobel speech: 'All of the ideas are simple and had peoples' minds been directed that way the method would have flourished perhaps a century earlier'.[49] Any chemist can use the method—no expensive equipment or research teams required. The powerful method enables us to separate proteins, amino acids, carbohydrates and sugars. It has led to many new medical treatments and biological advances that have benefited our lives, laying also the foundation for new fields like bioenergetics and signal transduction.[235]

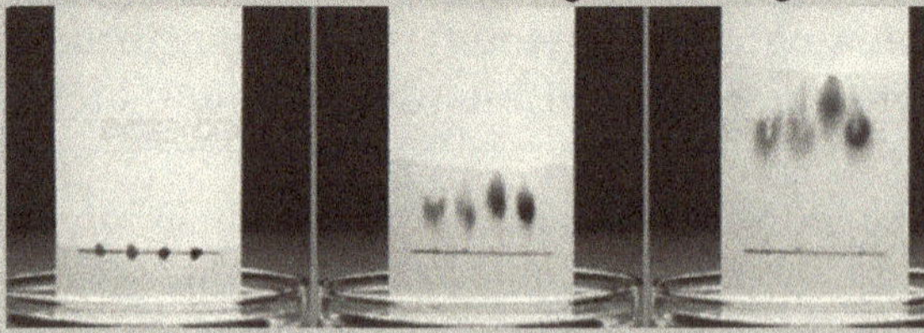

Picture 5.2 *Paper chromatography method. Reproduced from light the minds.*

Before closing, think of major fields that have stagnated and those that have recently grown rapidly. Now think of how tool development relates to these differences. It is applied fields across science—like experimental physics, economics and biology—that are largely thriving. In contrast, theoretical fields—like theoretical physics, economics and biology—have largely stagnated, at times locked in long-standing debates (Chapter 10). Applied research thrives because of new, frontier-opening tools, methods and data they produce. Think of the James Webb space telescope exploring the early universe, the LIGO interferometer detecting gravitational waves, and sequencing and high-throughput tools decoding the entire human genome.

Can method innovation explain this vast divide? Indeed, the main driver of a field's growth or stagnation is commonly the power and novelty of its tools and methods. The fields that expand most rapidly commonly apply the most powerful new tools—think of the field of AI that is driven by new machine learning methods, to genomics powered by DNA sequencing methods, and genetic engineering driven by the CRISPR gene-editing method. So when do fields actually grow fastest? We find here four pathways: one, through such method inventions that allow asking questions not possible before. Two, through upgrading—or three, integrating—computational, statistical and experimental tools that enable researchers to analyse such previously intractable problems with massive, complex data. Four, through cross-disciplinary borrowing that can involve fields also combining tools—like neuroscience merging tools from biology, computing and cognitive science, and behavioural economics mixing methods from psychology and economics. And when do fields stagnate? Not because we run out of ideas or papers, but because we run out of ways to explore, test and generate new findings—using new methods. The key insight is simple: fields that begin treating tool-building and method-design as central—not auxiliary—will be the ones generally unlocking the new frontiers of research.

Conclusion

Scientists like Galileo, Newton, Mendeleev, Hooke and Mendel were pioneers in testing new methods and evidence. Yet they could not foresee whether and how their individual contributions would fit the construction of an immense system of knowledge. What they were contributing—eventually leading to the fields of physics, chemistry and biology—became clearer over the centuries. Today, we have a far clearer view of the evolving edifice of science and its ever-expanding structure and complexity. While we are not fully aware of the immensity of what is beyond our planet, we have amassed vast bodies of complex knowledge, from laser physics and genetics to climate science and AI, that were incomprehensible just a few generations ago. A key goal of the science of science is to understand how these bodies of knowledge grow—from the 17th century to the cutting-edge developments of today.

Because each publication that opened a new field used a new tool—and the study, including often experiments, could only be conducted with that tool—we find that new fields consistently emerge through new tools. Because they enable novel insights and testing those insights that would not have been possible before. Through the new tools we design, we can trace what new fields fundamentally rely on: new microscopes leading to microbiology, computers launching computer science and radio telescopes giving rise to radio astronomy. For these necessary tools allow us to observe what is otherwise too small, too vast, too fast, too far beyond our mind's capacities to imagine. The history of science is ultimately a history of expanding our human senses—crafting new ways to detect, measure and understand our world that open new domains of knowledge. Here we identify the fundamental principle that fields share in common: we consistently kick-start new fields through newly invented methods and instruments that reflect a new way to perceive the world not previously feasible (Figure 5.1). Yet nearly all scientists commonly just use conventional methods they are trained in to study a problem.

We also uncovered that most major method discoveries do not just launch a new field but often multiple new fields. Science's most powerful tools are rarely confined to their discipline of origin; instead, most spill over into other disciplinary areas to unexpectedly trigger new fields that their inventors never anticipated. This underscores the causal link of new fields driven by new tools that would not have been possible without them. This new *methods-to-fields* principle holds across history and disciplinary areas and helps redefine the predictability of new fields emerging after and where we make new method advances. Unlike traditional explanations—including paradigm shifts in theories, evolving research programmes or splitting or merging scientific communities—this principle does not focus on just outputs but on what precedes and causes new fields in the first place. Beyond new tools, there are supporting factors that can help influence when a field emerges—like funding,[(6–8)] collaborations,[(9–11)] and developments in other fields—but they cannot directly start a field on their own.

This tool-driven principle highlights the need for us to redirect much greater attention to upgrading our toolbox. By prioritising the development of new methods and technologies, we can create an environment for accelerating new breakthroughs and fields. Ultimately, the origin of a major breakthrough is the central event, and whether the breakthrough takes place in an established field or creates a new field is often less important.

Expanding our tools is where the frontier of research lies, yet we do not give enough attention to this frontier research. A guiding principle for researchers seeking to break new ground is: when we hit upon an interesting problem facing a method or tool we are using or come across an idea of how to tackle that problem, we should drop everything and pursue it—because history shows that this is generally how new scientific advances and fields are born. The most promising thing to hear in science is often not just 'I have a new idea' but rather 'I have a new method that I can apply'. In the final chapter, we sketch out the constraints to our best tools—the bottlenecks that, if we overcome, would vastly advance science.

Shifting our research focus to this powerful principle of tool innovation would speed up how we spur new advances and the pace at which we can enter unmapped terrain at the borders of science. Such a shift would mark a tool revolution in science itself. But this raises key questions: if we become better at developing new methods, could we better anticipate and predict tomorrow's new scientific advances and domains? If we learn to recognise the signs of methodological bottlenecks and methodological leaps before their impact unfolds, could we actively create and shape the future of science? And crucially, what concrete steps do we need to take to design our next big methods and tools? We tackle these key questions in the next and foundational chapter. But one thing is clear: tomorrow's great scientists—those best equipped to tackle society's pressing challenges—are generally those who are best at developing new tools or who take advantage of the new tools.

6

The discovery engine

How we invent the powerful methods and tools of discovery—and a new field: The *Methodology of Science*

Summary

The first step to understanding scientific progress is uncovering that new methods and tools consistently spark new discoveries and fields. The second step is now tackling the essential question: how do we actually develop our best tools of discovery? To answer this, we begin by tracing the extraordinary stories behind the top ten most influential toolmakers—from Ernest Lawrence's particle accelerator to Theodor Svedberg's centrifuge. Yet remarkably, no theory explaining how we make major innovations in tools across fields exists. Even for the first step, where researchers have proposed factors that can support breakthroughs, we still have no general theory of scientific discoveries. Here we introduce a general theory of science—*the new methods-driven discovery theory*—that explains how we trigger science's new discoveries and fields by creating new methods and tools. These are placed at the centre of understanding scientific progress. The evidence here challenges conventional belief: we find that hundreds of major discoveries have emerged without additional funding or larger teams, but by leveraging recently invented tools that provide the completely new perspective. We illustrate that—without effective tools to detect or measure—more funding and collaborations are not enough, our scientific theories cannot be meaningfully developed or tested, and unexpected or serendipitous observations cannot be made. New tools are commonly the most central factor because they directly trigger new breakthroughs and enable otherwise entirely unattainable insights. So what if we no longer wait for new discovery tools to emerge by chance but begin deliberately prioritising their development? How many big breakthroughs are we missing because we have not yet strategically focused on designing the needed tools? In this foundational chapter, we map out the crucial pathways to create new tools—the discovery engine. We introduce a taxonomy of scientific methods and the idea of setting up methods labs and hubs—as incubators of innovation—that catalyse tool creation. This theory and these pathways can provide a foundation for a new field targeted to developing methods that can accelerate new advances: the *Methodology of Science.*

The Engine of Scientific Discovery. Alexander Krauss, Oxford University Press.
DOI: 10.1093/oso/9780197829790.003.0007

Overview

'The important thing in science is not so much to obtain new facts as to discover new ways of thinking about them', the physicist Lawrence Bragg said anecdotally. The novelist Marcel Proust noted that 'The real voyage of discovery consists not in seeking new landscapes, but in having new eyes'. And the biochemist Albert Szent-Gyorgyi mentioned that 'Discovery consists of seeing what everybody has seen and thinking what nobody has thought'. Yet Bragg's nobel-winning discovery of the structure of crystals in 1913 was only possible by developing a new x-ray spectrometer the previous year.[(125)] Szent-Gyorgyi's nobel-winning discovery of isolating vitamin C in 1932 was made possible by applying recently created assay techniques and the ultracentrifuge. Here we move from personal anecdotes about an individual discovery to systematically analysing the greatest discoveries across science. This broader lens enables us to draw a broader insight: *Major discoveries consist of creating new tools that see and measure the world in ways that nobody has before.*

By constructing microscopes and telescopes, we have uncovered a world of microorganisms in our intestines, nanoparticles and molecules, and have peered into the depths of the universe to rewrite our understanding of its origins. By creating advanced statistics and high-speed computers, we have reduced the limits of human cognition, allowing us to process vast data about basically any phenomenon in science and reveal previously hidden relationships. A general principle of discovery emerges across science: every scientific breakthrough is, at its core, a breakthrough in how we observe, measure and understand the world using a new method. New tools are key to unlocking the mysteries of life and the universe by fundamentally reshaping our perception. But despite their impact and completely transforming our world, we still have not yet answered crucial questions: how do we actually develop these powerful tools of discovery? How can we upgrade them faster? And how can we better use them to their fullest potential?

Extending our scientific toolbox is about tackling the very constraints of our mind, senses and current methods to perceiving the world—opening realms of knowledge otherwise hidden from view. *Our scientific methods and tools not only reduce human constraints, errors and biases but also chance and luck in how we make sense of the world* (Chapters 1–2). They are what enable us to experiment, measure and control our environment, human biology and the world around us in ways that we never imagined. Without them, discoveries—whether the invisible forces of quantum mechanics to the neural circuits that give rise to human thought—would remain out of reach. This is why the best researchers at sparking new advances are generally those who are best at recognising and seizing the power of new tools. In fact, about half of science's major discoveries are triggered by researchers who develop a novel method themselves, while the other half rely on cutting-edge methods pioneered by others (Figure 1.4b).

Why is it so crucial for researchers to be aware of the methodological limits, challenges and opportunities of their research? Because these researchers are better able to push these limits, tackle these challenges and take advantage of these opportunities. They are better able to design, experiment with and leverage new tools that often redefine what is possible in science.

In this foundational chapter, we cover a lot of ground and provide many of the central explanations in the book. We reveal how the leading, extraordinary tools of discovery were developed. We then examine how these powerful tools stack up against supporting factors like funding and larger teams—and often make science's major theories and much of scientific imagination possible in the first place. This sets the stage for introducing the new methods-driven discovery theory. From this perspective, we map out a taxonomy of discovery tools—and we propose the creation of global methods labs and hubs for catalysing tool innovation. We then lay out the practical steps we can take to design and innovate tools—whether within such labs or through individuals. The opportunities are huge, and understanding this process is key to accelerating scientific progress. What emerges is a foundation—grounded in evidence and theory—for a new field we introduce here: the *Methodology of Science*—a field that fundamentally shifts the focus from *what* we discover to *how* we discover and how we can discover faster. Together, the insights offer a roadmap for speeding up progress.

With new tools driving groundbreaking discoveries, what then drives how we develop our remarkable tools of discovery?

Establishing new methods and tools as the engine of science and discovery (Chapters 1–5) leads us to the next and equally important question of what the engine of new methods and tools is. But researchers studying scientific methods focus on a specific method—like statistical techniques or microscopy.[137,238] (Or philosophers investigate scientific methodology abstractly from a conceptual perspective, exploring principles like falsifiability of theories or the classic scientific method of hypothesis testing).[180,204,239] Yet remarkably, there are no systematic studies, applied or theoretical, that explain how science's diverse new methods and tools—from cutting-edge telescopes to supercomputers—emerge across fields and how we can speed up their development.

Here we tackle this key question from different perspectives and angles. We begin by examining the ten central methods and instruments most used to spark discoveries and the inventors behind them—our great method-makers of science. These innovators, from Ernst Ruska's electron microscope to Max von Laue's x-ray diffraction, did not just solve pressing method challenges—they reshaped the future of discovery. Each of the top ten transformative tools and methods triggered multiple nobel-prize discoveries across fields, far beyond their original domains. But what drove these researchers to shift gears from traditional scientific research within their fields to creating these new method innovations—to developing entirely new ways of doing scientific research? What compelled them to rethink the methods they used to explore the world and how can we rethink the methods we use? By tracing their career paths and the key turning points in their work, we uncover six key factors behind science's greatest tool breakthroughs.

One, we find that the pioneering inventors of the top ten general tools of discovery generally ran into methodological limits and were then motivated by the need to create a new tool by pragmatically *overcoming a constraint to an existing tool.* Working within conventional scientific fields, they encountered barriers that hindered them in studying key phenomena using available tools. So they turned the problem around and set out to fill these method gaps—with for example DNA analysis being very slow and inefficient until Kary Mullis developed a simple way to rapidly amplify DNA that ultimately transformed genetics: the PCR method.

Two, these researchers each *shifted their research focus* from addressing scientific questions to tackling method and tool questions. They studied and worked in established fields like physics and chemistry; but transitioned from traditional science to an unconventional pursuit of extending our tools to explore the world in new ways—with for example Theodor Svedberg studying macromolecules and colloids but then turning to how to measure their properties more accurately that led to inventing the ultracentrifuge.

Three, they took a very practical *experimental approach*, not simply theorising about the limitations but tackling specific shortcomings of our tools that held back progress—with for example Ernst Ruska refining electron lenses, step by step, until he had a working electron microscope with vastly improved resolution like nothing before.

Four, these innovators commonly produced the new tool *at low or modest costs* and not in large research teams. Motivated, resourceful researchers can make foundational breakthroughs with little resources and simple materials—from Mullis' PCR method and Tiselius' electrophoresis to Fisher's modern statistics.

Five, they were *primarily interdisciplinary* scientists, with eight out of ten trained in different fields. This enabled them to generate new connections by often drawing on methods and evidence from different domains—with for example Ernest Lawrence's degrees in physics and chemistry enabling him to better understand the nature of atomic interactions and design the first practical particle accelerator.

Six, these inventors were *not prominent scientists before designing the groundbreaking tool.* Most were relatively unknown and did not make a major discovery before—with for example Mullis an unknown biotech researcher and Ruska, Tiselius and Synge PhD students at the time (Table 6.1).

A remarkable insight emerges: science's most important toolmakers are not elite researchers at the peak of their careers but rather typical scientists when looking at their background. This is important: it highlights that applied and resourceful researchers—regardless of reputation and institution—can turn their attention towards tackling practical method problems that can transform science.

So what specifically motivates us to develop new discovery tools and methods? Let us take a closer look at science's most transformative tools. Svedberg created the ultracentrifuge to overcome the limits of the ultramicroscope in studying particles.[143] Tiselius then invented electrophoresis that addressed constraints of the ultracentrifuge and the struggles of biochemists to separate proteins effectively.[142] Martin and Synge designed partition chromatography to resolve the shortcomings of earlier chromatography methods.[237] Each of these inventions unlocked new ways of separating and studying substances. Von Laue generated x-ray diffraction that enabled going beyond traditional x-ray analysis and probing the atomic and molecular structure of materials.[120] Ruska pioneered the electron microscope that broke the optical barrier of light microscopes by bypassing the diffraction limit of visible light that prevented visualising tiny structures.(112) Bloch and Purcell developed NMR spectroscopy that advanced earlier techniques, especially Rabi's molecular beam method, to uncover a powerful new way of probing atomic nuclei with unprecedented precision.(240) Fisher invented modern statistical methods to tackle the inadequacies of earlier data analysis techniques, providing rigorous tools to design experiments and draw reliable conclusions.(137) Townes created the maser that overcame limitations in generating and controlling electromagnetic waves and soon led to the laser—though he did not have a specific application for the tool in mind.(141) Mullis invented the PCR method to make the process of DNA amplification and analysis much more efficient and rapid.(129) Lawrence constructed the particle accelerator that enabled vastly expanding existing models of linear accelerators.(241)

This trend is striking across science: it is often applied researchers—not generally theorists—who hit upon and spot a tool's defects, but rather than just commonly accepting them they seek to tackle them head-on. It is generally technical bottlenecks—when methods fall short and progress hits a wall—where the birth of breakthroughs begins. *The origins of science's best tools challenge a long-standing assumption—that theories play the central role in driving science—yet science's most transformative tools generally arise not from abstract theorising but from practical deficits in earlier tools.*

In fact, Lawrence's inspiration for example came in 1929 when reading articles in the university library at Berkeley one evening. He wrote that he 'came across an article in a German electrical engineering journal by Wideröe on the multiple acceleration of positive ions. Not being able to read German easily, I merely looked at the diagrams and photographs of Wideröe's apparatus ... This new idea immediately impressed me as the real answer which I had been looking for to the technical problem of accelerating positive ions, and without looking at the article further I then and there made estimates of the general features of a linear accelerator'.(241) That year, he published his landmark paper, *The production of high speed light ions without the use of high voltages*, in the journal Physical Review. At just 28, the youngest professor at Berkeley at the time, Lawrence had designed the first particle accelerator, giving an entirely new impulse to high-energy physics.(241) We explore the inspiration of these great toolmakers deeper within the boxes (in Chapters 1, 5 and 6).

Surprisingly, these stories behind the ten most influential tools in sparking discoveries in scientific history challenge the popular narrative about scientific progress. The stories reveal a deeper insight: the overlooked flaws of tools are key in driving scientific progress. Great breakthroughs commonly begin not with just a grand question but with a technical obstacle impeding investigation. What does this mean? It means we need to pay much closer attention to the gaps in our instruments rather than just the gaps in our knowledge. Because tools shape the questions and theories we can even formulate and the discoveries we can make (we explore this in greater detail later).

After zooming in on science's top tools, we now zoom back out and broaden the focus. We compare the features of the top ten nobel-prize-winning methods and tools—those sparking the greatest number of later nobel-prize discoveries—with the features of all other 523 nobel-prize discoveries across science. We find that these top ten discovery tools were all developed through direct observation and experimenting, while other nobel-prize discoveries were less likely to rely on these approaches (as we see in Figure 6.1a). Science's pioneering method-makers shared several unique and surprising traits: not only were eight out of ten interdisciplinary—each earning two or more degrees in different fields before their breakthrough; but half of them were not professors and only completed a PhD or less, and were under 35 (see Figure 6.1b for a comparison with other nobel-prize discoverers).

We next take an even broader approach by examining all 149 nobel-prize-winning method discoveries—including the top 10—and compare them to all other 384 nobel-prize discoveries (experimental and theoretical breakthroughs). What stands out is that these groundbreaking methods and instruments were more likely to require experimenting and lead to a new field, than other discoveries (Figure 6.1c).

Exploring science's most influential toolmakers deeper, we uncover a unique pattern—one that challenges a common assumption about where innovation happens and who drives it. We find that among these ten great toolmakers, one group consists of *low-profile researchers* not at top universities, developing four of the ten most powerful tools. A second group is made up of *academic outsiders*—researchers in tech and biotech companies that are very applied research environments, inventing three of the ten tools. This is surprising since 99% of all other nobel-prize discoverers were at universities or research institutions. And a third group involves *higher-profile researchers* at top 50 academic institutions, creating three of the ten (Table 6.1). Stepping back and exploring all 149 nobel-prize-winning method breakthroughs we find a similar trend: almost two-thirds of these toolmakers were not based at the world's top 50 universities (Figure 6.1d). Innovation usually begins not with privilege or prestige—but with a technological gap and someone eager to try a different approach.

These ten pioneering methodologists were also younger: just 34 years old on average at the time of their innovation, compared to 39 for all other nobel-prize discoverers. The most transformative tools that change the course of science can emerge before a career has even properly started—with three publishing their tool breakthrough as part of their PhD thesis. Another unique finding is that most of these ten general tools were generated at low cost, requiring less than 1000 US$ (in 2025 prices) (Table 6.1).

Since major discoveries begin with *new* methods that provide a new perspective to a problem—not with *existing* methods that represent conventional research—how do we develop those new methods? Four common ways exist, as we introduced in Chapter 1 and explored more deeply in this chapter. One powerful way is *creating entirely new methods* not yet conceived—like chromatography for separating chemical substances. A second transformative way is *extending methods in novel ways* that represent a new method not ever used before—like gas chromatography that vastly upgraded chromatography. A third innovative way is *combining methods in novel ways* that introduces a new method not yet leveraged before—like x-ray crystallography that joins x-ray methods with crystal analysis.[120] A fourth foundational way is

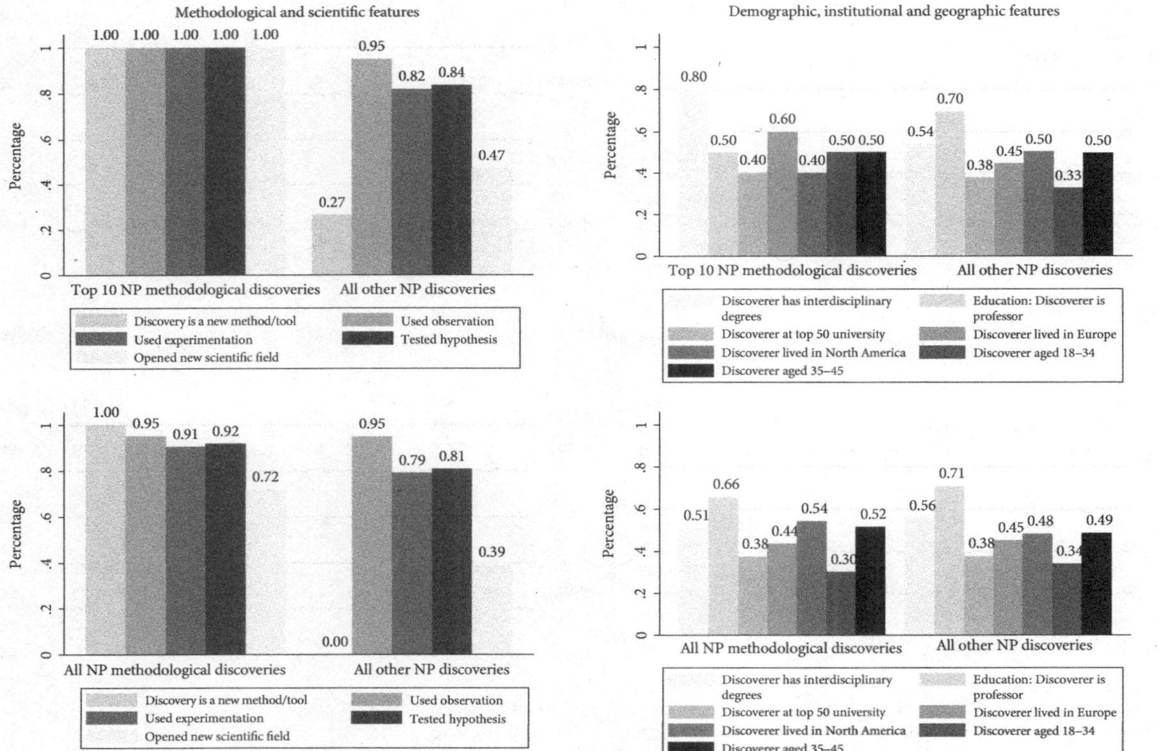

Figure 6.1 *Nobel-prize-winning methods and tools compared to all other nobel-prize-winning discoveries, by features at the time of the discovery*

The data reflect all 533 nobel-prize discoveries: with the top 10 nobel-prize-winning methods and tools compared to all other 523 nobel-prize discoveries (Figure a and b), and with all 149 nobel-prize-winning methods and tools compared to all other 384 nobel-prize discoveries (Figure c and d). NP stands for Nobel Prize. We outline the top 10 nobel-prize-winning methods and tools in Table 6.1, with the particle detector included here instead of modern statistics as it did not win a Nobel—each of these tools has been applied to develop at least four later nobel-prize discoveries.

Table 6.1 How science's ten central methods and tools were developed that triggered most nobel-prize discoveries

			Traits at the time of the discovery:										
Method or tool	**Year developed**	**Inventor's name**	**Education level**	**Age**	**# of people making tool**	**University/ Top 50 ranked**	**Field of research**	**Field of degree(s)**	**Won Nobel prize for the tool**	**Cost to develop tool***	**Country of birth**	**Fields using the tool**	**How the method or tool works—and its general use**
PCR method	1985	Kary Mullis	PhD	41	1	Cetus (biotech company)/no	Chemistry	Biochemistry; Chemistry	Yes, 1993	Low	USA	Medicine, genetics, biochem-istry	Amplifies a small DNA sample (e.g. a drop of blood) quickly into millions of copies—and enables medical diagnostics (like detecting HIV) and palaeontologic findings (identifying species using fossils) (Box 1.3)
Maser/laser	1954	Charles Townes	Professor	39	1	Columbia University/ yes	Physics	Physics; Modern language	Yes, 1964	Medium	USA	Physics, computer science, astron-omy, medicine	Produces concentrated beams of light or microwaves—and used to probe molecular structures and deep space, and applied in DNA sequencing, fibre optics, laser surgery and computers
Spectrometer, NMR*	1946	Felix Bloch, and Edward Purcell	Professor	38	2	Stanford University/ yes	Physics	Physics; Electrical engineering	Yes, 1952	Medium	Switzerland /USA	Chemistry, physics, biology, medicine	Visualises magnetic fields around an atomic nucleus—and used to decode the structure of proteins, produce life-saving drugs and observe tissue in our body
Chromatography, partition	1941	Archer Martin, and Richard Synge	PhD (research was part of PhD thesis in 1941)	29	2	Wool Industries Research Association/ no	Chemistry	Physics, chemistry, physiology, biochemistry	Yes, 1952	Low	UK	Chemistry, biology	Separates substances in mixtures and determines their composition—and enables identifying and understanding e.g. amino acids and sugars (Box 5.2)
Electron microscope	1933	Ernst Ruska	Engineering degree (research was part of PhD thesis in 1934)	26	1	Technical University of Berlin/ no	Physics	Electrical engineering; Physical sciences	Yes, 1986	Medium	Germany	Biology, medicine, physics	Magnifies miniscule objects using the wavelength of an electron—and enables studying microorganisms, nanoparticles, electrochemical reactions, crystals and molecules (Box 1.1)
Electrophoresis	1930	Arne Tiselius	PhD (research was part of PhD thesis in 1930)	28	1	Uppsala University/ no	Chemistry	Chemistry	Yes, 1948	Low	Sweden	Chemistry, biology	Separates substances by moving charged particles in a fluid—and applied to analyse e.g. DNA and proteins, paternity tests and DNA fingerprints (Box 6.1)

Particle accelerator	1929	Ernest Lawrence	Professor	28	1	Berkeley, UC/yes	Physics	Physics; Chemistry	Yes, 1939	Low (to high)	USA	Particle physics, nuclear medicine	Propels charged particles to high speeds using electromagnetic fields—and used to study matter and proteins, develop new drugs and cancer therapies
Modern statistics	1925	Ronald Fisher	Bachelors	35	1	Rothamsted (agricultural research institute)/no	Statistics and genetics	Astronomy (w/specialisation in mathe-matics)	No	Low	UK	All scientific fields	Used to collect and analyse large datasets and run experiments—and applied to examine almost any phenomenon, from diseases in populations to planets, societies and even science itself (Box 5.1)
Centrifuge	1924	Theodor Svedberg	Professor	40	1	Uppsala University/no	Chemistry	Chemistry	Yes, 1926	Medium	Sweden	Chemistry, biology, clinical medicine	Spins samples at ultrahigh speeds at over 20,000 revolutions a minute—and used to separate small particles in fluids, gases and liquids, like cells and viruses (Box 6.1)
X-ray diffraction	1912	Max von Laue	PhD	33	1	University of Munich/no	Physics	Mathematics; Physics	Yes, 1914	Low	Germany	Chemistry, physics, material science, biomedicine	Scatters x-rays using crystals—and applied to uncover crystal structures, atomic structure of matter including proteins, and identify cancers (Box 1.2)
Average of 10:	...	...	*40% professors*	*34*	*1.2*	*30% at a top 50 university*	...	*80% inter-disciplinary*	*90%*	*Low to medium*	*70% in Europe*	...	...
Comparison group—all other nobel-prize discoveries:	...	...	*69% professors*	*39*	*1.4*	*38% at a top 50 university*	...	*54% inter-disciplinary*	*100%*	...	*45% in Europe*	...	...

The 10 central methods and tools are the most used to make the 533 nobel-prize discoveries. The comparison group (the last row) reflects all other 524 nobel-prize discoveries—with modern statistics not included as it did not win a Nobel prize. While the particle detector (1911) is at the threshold of making it into the top ten, the PCR method (1985) is included as it illustrates a powerful, more recent tool discovery that also received a Nobel prize and was used in triggering 10 major discoveries. The year reflects when the tool was first created, while all have been vastly improved. *The cost to develop the tool is based on three categories: *low* under $1000 US, *medium* between $1000 and $10,000 US and *high* above $10,000 US, in 2025 prices. We describe the cost for seven of the ten tools in the Boxes in Chapters 1, 5 and 6. Here we illustrate the other 3: it cost Townes about $500 US to create the maser in 1954 ($5944 US in 2025 prices).(242) Developing the first NMR spectrometer in 1946 cost Bloch $450 US ($7379 US in 2025 prices).(243) Creating the first operational particle accelerator prototype in 1929 cost Lawrence only about $25 US ($467 US in 2025 prices),(244) and later scaled up accelerators about $2500 US ($46,750 US in 2025 prices).(245) **While the first spectrograph was invented in 1859, an entirely new form of spectroscopy based on magnetic resonance—namely NMR spectroscopy—has since become far more important.

adopting new methods developed in other fields—like NMR spectroscopy created in physics and then used for the first time *ever* in chemistry, medicine and biology.[240] In each of these cases, we triggered major discoveries with new methods *never* employed before to a given question. When we make a novel methodological leap—whether merging two tools from different domains or repurposing tools in new ways—innovation often soon follows.

Digging deeper we uncover two broader underlying strategies to advance new methods: tackling a bottleneck facing a tool or opportunistically experimenting with tool designs. Which strategy is more common and successful? Scanning the method innovations sparking science's over 750 major discoveries reveals a clear trend: the most common strategy is *constraint-driven innovation*. Most new tools arise when researchers hit a wall—a method obstacle when doing research—and then shift their focus to extend, combine and reinvent tools to overcome the obstacle. The less common strategy is *exploratory innovation*—where researchers invent new tools not to solve a pre-established problem but simply to probe into the unknown and possibly discover something new. Yet some breakthroughs can be based on elements of both, a strong exploratory component and a known bottleneck that requires experimenting with a new method to find out if an assumed phenomenon exists.

Consider a striking example of how constraint-driven and exploratory innovation can merge: the invention of the maser by Townes in 1954. He had no specific application in mind—only to amplify beams of microwaves. He famously described the maser as 'a solution looking for a problem'.[141] But over time, researchers began experimenting with this peculiar invention in more and more areas—from medical clinics and physics labs to telecommunications and computer industries.[141,242] We use the maser's offspring, the laser, widely in different technologies in our daily lives, from barcode scanners, smartphones and laser printers to high-precision surgeries.[246] So does it pay off to experiment with new methods without a clearly defined use? The payoff is often high—precisely because there are very few full-time researchers dedicated to inventing methods themselves, and because what begins as tinkering with a new possible tool, like the maser, can turn into an unexpected cornerstone of science and technology. This insight opens a vast space of untapped potential. Science and technology are often intertwined. Yet science can often be more submissive to new technology.

So beyond internal scientific pressures to expand our toolbox, there are also external pressures—technological and societal. Who invented for example the six most influential tools that drove 17th-century science? Surprisingly, three of the tools—the barometer, vacuum pump and calculus—were created by scientists, while the other three—the microscope, telescope and statistics—were developed outside of academia for non-research purposes. But scientists recognised their potential, refining and extending them over centuries. The microscope and telescope emerged far from the lab: at the turn of the 17th century, non-scientists developed them by tinkering with and expanding eyeglasses and magnifying glasses.[46] Descriptive statistics were born in the 1600s out of the practical need to collect and track demographic trends and mortality in rapidly growing European states.[247] Interestingly, the centrifuge—typically associated with lab science—was also invented outside the walls of academia in 1864 in Germany, originally used for separating liquids from paint and later cream from milk.[248] But the device was eventually picked up on 60 years later, in 1924, when Svedberg vastly expanded it with the ultracentrifuge, enabling extraordinary findings in fields like biochemistry and molecular biology.

Scientific progress, then and now, is driven by practically minded toolmakers. New methods in artificial intelligence, machine learning and big data are increasingly pioneered by tech companies for commercial use and then applied by scientists. These not only make science more efficient but also highlight an often symbiotic relationship between industry and science.

Our new tools commonly as the central driver of discovery—beyond supporting factors, from additional funding to larger teams

What are seen as the most important factors driving scientific discoveries? Reviewing the literature we find a consistent pattern. Most frequently, researchers highlight two key factors for how discoveries emerge: greater *research funding* (for salaries and sustained long-term programmes)[6–8] and *team collaborations and size* (for sharing expertise and research tasks).[9–11] These two factors compete for the top spot and are especially emphasised by economists and scientometricians (Chapters 4 and 6). Other commonly discussed factors driving breakthroughs are *greater education* and human capital, providing a trained and skilled workforce in the long term.[44,85] At times *interdisciplinary research* used to integrate perspectives and tackle complex problems.[83,84] And *younger researchers* who can be more productive and motivated (Chapter 4).[12,13,86] Beyond these, the more elusive role of *serendipity*, sudden insights and chance are at times stressed especially by psychologists, historians and philosophers of science (Chapter 2).[17–20,30,146] Then come broader basic forces: *societal and policy challenges*—such as pandemics like COVID, climate change and clean energy research, and government priorities—can shape some research areas.[227] And *computers, internet and AI* make science more efficient and automate processes, especially in data-intensive fields (Chapters 6 and 11).[90,249]

At a deeper level, some examine *revolutionary theories*—especially highlighted by theoretical physicists and philosophers (Chapter 3).[15,16,73] Others explore the role of *problem-solving*.[15,87,88] Broad *institutional structures* such as universities, journals and research networks also facilitate knowledge exchange,[208,215,216] while *scientific awards* can help incentivise innovation (Chapter 4).[41,95] Some researchers also highlight the growing role of *open access publishing*, speeding up the flow of knowledge.[250,251] Finally, *academic freedom* is needed—the space for exploring new methods and insights, risky projects and accepting 'failure' and changing paths as new opportunities arise.[252] Taken together, researchers exploring scientific progress largely focus on these broader factors and basic conditions.

These factors vary largely by discovery, by scientific field and by scientist, and alone are not enough to make discoveries. We lay out, throughout the book, how these factors and background conditions can foster the basic environment for developing methods and science. But it is new tools that are the very spark of major discoveries across the sciences.

Some broad factors, like a scientist's age or a serendipitous moment, are often beyond our direct control. Since 1900, 80% of discoverers were under 45—and about 15% of discoveries involved an element of serendipity. But even those unexpected moments are typically sparked by new tools—ones that visualise and detect the unexpected (Chapter 2). Other supporting factors can be more within reach like large funding (for certain instruments and research) and a large scientific community (for doing research using instruments). These factors mainly emerged after the second world war, with the establishment of national science foundations in the 1950s. Yet by then, most of science's leading methods and tools were already invented (Table 6.1). And more than half of science's major discoveries were already made by 1945, including the majority of the major non-nobel discoveries since 1900 (Figure 4.4). Since 1900, 60% of discoverers did not work at top 50 universities—institutions with generally greater funding and resources. And since 1900, 33% of discoverers were not professors and 5% did not have a PhD—and even 36% worked in low-resource contexts with average annual incomes below 10,000 US$ (Chapter 4). These facts show that discovery does not depend on elite institutions, elite scientists and elite funding.

Let us first explore deeper the factor many assume is the lifeblood of scientific progress: *research funding*.[6–8] It is easy to assume that bigger budgets can lead to bigger breakthroughs. But we can test that assumption by examining science's over 750 discovery-making studies, including publications such as the Nobel prize lectures describing these discoveries. We find a striking insight: hundreds of high-impact breakthroughs are sparked by low-cost methods and tools. These include those among the ten most used in science: statistical and mathematical methods, light microscopes, electrophoresis, chromatography, thermometers and centrifuges—and later the PCR method and other assay techniques. Most of these tools of discovery were developed with small budgets—and today we can buy them new for less than a thousand or a few hundred dollars or are even free, like most statistical methods (Table 6.1). Tools, once acquired, can also often be shared, multiplying their impact.

What is remarkable is the range and depth of nobel-prize discoveries powered by these inexpensive tools: from identifying viruses and developing life-saving vaccines, to inventing transformative statistical and mathematical methods and isolating essential chemical and biological substances. The German Robert Koch for example discovered the deadly bacteria causing tuberculosis using a light microscope together with simple and cheap staining techniques, a thermometer, syringe and guinea pigs. The Canadian Frederick Banting and Scottish John Macleod uncovered insulin using a colorimeter, scalpel, syringe and extracted pancreases from dogs—with the groundbreaking findings leading to the treatment of diabetes. The Israeli Daniel Kahneman revealed how we think and make decisions under uncertainty by using innovative but simple survey methods, statistical methods and decision-making experiments and deviated from expected utility techniques. The American Michael Kremer developed experimental development economics by adopting randomised controlled trials—created in clinical medicine—and applying the method to conduct the first RCT on an education intervention.(133,135) We highlight many other low-cost but transformative discoveries—including method discoveries like partition chromatography and the PCR method—throughout the book.

These resourceful nobel-prize discoverers illustrate a recurring pattern across hundreds of breakthroughs achieved with minimal resources: they leverage innovative yet simple methods like statistical techniques or light microscopes, borrow recent cutting-edge methods from other fields and repurpose or upgrade basic lab instruments. Strikingly, these inexpensive discoveries span across major fields—from microbiology and chemistry to epidemiology, genetics and cognitive science. This reveals an important insight about the discovery process: *large-scale funding is not always the norm and—aside from researchers' salaries and standard overhead costs—many discoveries require no additional funding.*

When we look at the hundreds of fields spanning the biological, chemical, physical, cognitive and social sciences, we find that the most expensive instruments are concentrated in a few subfields, especially within branches of physics like particle physics and astronomy. They include the Hubble space telescope constructed in 1990 by NASA and the European Space Agency which enabled discovering the accelerating expansion of the universe in 1998. The massive radio telescope built at Cambridge in 1967 led to discovering pulsars later that same year. The laser interferometer (LIGO) upgraded in 2015 in the United States made it possible to detect gravitational waves that same year.(115,138) The large hadron collider (particle accelerator) developed by CERN in 2008 enabled revealing the Higgs particle in 2012.(117) The Manhattan Project—a monumental effort requiring nuclear reactors and advanced isotope separation techniques—paved the way for the first atomic bomb in 1945. Beyond physics, some exceptionally expensive projects in genetics and medicine have also invested vast funding in major cancer research and the human genome project.(113) High-performance supercomputers, space programmes and other particle accelerators also demand immense resources. These are among the most popular examples of the most costly instruments and initiatives in scientific history that relied on much government funding.

These exceptional cases are often mentioned as proof that science can be very expensive because these instruments are very expensive—yet they only make up a small fraction of major discoveries. While people's attention is naturally captured by these extraordinary initiatives in a few fields, they overlook the over 750 major discoveries that make up the very foundation of science. The price of discovery is often lower than usually assumed. During the centuries between Newton and Einstein, we find that science's discoveries even flourished on low budgets and small scales. Scientists funded themselves or received only modest support from universities or private foundations. The key is that it is resourceful researchers—challenging conventional thinking—who often drive transformative breakthroughs.

Take also the often discussed role of *large research teams or communities* in advancing science. More researchers working together can cumulatively build on complex methods and findings, pooling expertise to tackle large-scale problems. Fields like biomedicine, climate science and high-energy physics are thriving examples of this collaborative spirit. As Ernest Lawrence—the inventor of the particle accelerator—said, 'No individual is alone responsible for a single stepping stone along the path of progress'[241] since we work within a broader scientific community.

Yet a share of transformative discoveries has been achieved by researchers working alone. These range from the discovery of superconductivity by Kamerlingh Onnes, the first exoplanet by Michel Mayor and Didier Queloz and x-rays by Wilhelm Röntgen, to the discovery of cells by Robert Hooke, bacteria by Antonie van Leeuwenhoek and the moons of Jupiter by Galileo. The history of these groundbreaking discoveries began with these researchers using a new tool powerful enough to penetrate the boundaries of the known—without an existing research programme among a scientific community or any existing theory. These discoveries were sparked not because of team size, funding or institutional prestige, but once researchers had these extraordinary tools at their disposal (Chapter 2 on serendipity). A significant share of breakthroughs is also triggered by individual discoverers using a tool already developed in one field within another field to open an entirely new domain without yet a research community. An influential study published in *Nature* found that individual researchers and small teams are most likely to generate breakthrough research, not large-scale collaborations—which attract our attention.[10] Nobel-prize discoveries have also commonly been awarded to a single researcher.[95] In *Scientific Elite: Nobel Laureates in the United States*, Zuckerman highlights that 'laureates were especially concerned to have the record clear for their more significant work, and particularly in their prize-winning research papers'.[44]

When discoveries happen in new teams and collaborations, or new interdisciplinary settings—not just established ones—it is generally precisely because new methods or tools from different researchers are turned on a problem for the first time. They bring the new lens. It is not just about who is on a team—it is generally more about what methods they use that reveal what others could not yet see. From this perspective, it is at times about forming the kinds of collaborations that make new methods possible.

Whether scientists work alone or collaborate with others, science on the whole naturally progresses collectively over the long run. Scientists are rarely isolated from the broader scientific community, with 99% of all nobel-prize discoverers working at or affiliated with a university or research institution at the time of their discovery (Appendix Table 4.1). We can think of scientific progress like an Olympic race. Some compete by training in coordinated relay teams, pushing each other and improving their performance, but when the finish line nears they break away from the pack to try and take the gold (like Crick and Watson pushing their way to the front and others to the side in depicting DNA's structure). Others train in a large pack without much competition (like the thousands of scientists working together to detect the Higgs particle). Still others take a more solitary route in unknown terrain (like Onnes uncovering superconductivity)—in each case guided by cutting-edge tools.

A key insight emerges from examining the over 750 biggest discoveries: we find that significant shares of discoveries are made by scientists who can be in small or large teams, low or high funded, young or old, at lower or top ranked universities, interdisciplinary or not, guided by theory or not, or involve a serendipitous moment or not (Chapters 2, 4 and 6). But the key factor common across all discoverers is that they directly leverage a new method to be able to break new ground—by enabling us to explore the world in a new light. *No major discovery has been made without first applying a novel method or tool; and major extensions to our tools, from microscopes to spectrometers, from electrophoresis to particle accelerators, have generally always led to new discoveries—this is as causal as we can get* (Chapter 1). This insight leads us to a distinct conclusion: we commonly over-focus on *who* is doing the research, when we should pay closer attention to *what* they are using to do it and *how* they got their hands on it.

Broader cognitive, economic, societal and geographic factors do not change just before we make each discovery. These supporting factors generally remain stable over time, often lying beyond our control. They serve as the soil. But the seed that sparks the discovery itself is a new tool, lying in our direct control that we actually shape, design and refine—with most created just a few years before major discoveries (Appendix Figure 1.9). There is also at times not too much more we can say about factors like money and the scientific community in fostering science, except that they do.

Common responses from researchers, on what can fuel groundbreaking research, generally circle around money and collaborations but also curiosity and serendipity. These factors support but are hardly unique to science; they are found widely in business, industry and public policy. For both scientists and non-scientists, a basic salary is generally needed to pursue work; teamwork can help merge perspectives and methods; psychological traits of curiosity and drive can lead to exploring new questions; and then there are serendipitous moments that can also arise. *Far from being distinct to science, these common factors apply to most human activities. Yet what truly sets science and discovery apart are scientific tools and methods—by enabling us to explore and uncover entirely new types of complex knowledge.* Scientific tools like advanced particle accelerators, x-ray crystallography, space telescopes and electrophoresis are largely exclusive to science and not used in other human domains. Looking at the ten most used methods and tools in science, we generally just use statistical methods in business, industry and public policy—and even then, they generally do not employ the complex techniques involving deep scientific modelling and rigorous controlled experiments (Table 6.1). Scientific tools uniquely define how major discoveries emerge. Machine learning methods are developed by researchers with advanced academic backgrounds also at tech companies and are gaining much importance in science. Pressing global challenges—whether in the race for new vaccines, cancer treatments and responses to aging populations, or technological solutions to climate change and renewable energies—can also influence science. They can shape some of our research priorities and the demand for discovering new solutions with new techniques.

The most widely followed strategy for researchers aiming to spur new advances seems straightforward: identify an open research question and search for funding support and collaborations. This model guides projects funded by science's largest institutions, such as the European Commission and the National Science

Foundation. But is it the best model for innovation? Asking original questions not yet explored, thinking outside the box, challenging long-held assumptions and filling blind spots in what we know are central aspects of science. And new tools—by their very nature—enable us to do this more than other factors. Following new evidence and data wherever it takes us is critical, especially when it contradicts what we currently know. Experimenting in new ways and taking risks, even if the outcomes are uncertain, is key to advancing science—as unexpected data can often lead to unexpected findings. And once again, it is new tools—from x-ray methods to machine learning—that empower us to do this and explore uncharted territory.

Factors like funding, teamwork, researchers' age and background can support research, but without innovative tools they alone cannot lead to breakthroughs. Hundreds of major discoveries have emerged without additional funding or collaborations, but from the moment a new tool was first applied that enables the new insights—as evidenced by discoverers in the discovery-making studies and Nobel prize lectures. New tools are commonly the central factor because they directly trigger new ideas and innovations and open entirely unimagined frontiers—more so than other factors can across science. This is highlighted in the conceptual framework in Figure 6.2. There are several reasons why tools are commonly the most direct factor we can influence in sparking discoveries:

- we develop most tools of discovery within a few years—and often the same year—as the breakthroughs they enable that are otherwise not possible (Chapter 1).
- we need to generally invest the greatest effort and time into tools, from design and refinement to application.
- we directly modify tools as we actively develop or transfer them across fields, giving scientists unparalleled control to trigger innovation.

At the end of a chain of questions into what shapes and guides new advances, we get to new tools. These are not aids but the very levers of discovery. They visualise the invisible: microscopes and telescopes reveal what is too small or too distant for the naked eye. They quantify the non-quantifiable: statistics, thermometers and measurement scales measure quantities with great precision. They sense the unsensible: spectrometers, radar devices and radio telescopes detect and measure light, radio waves and signals far beyond human sensory limits. They isolate the non-isolated: centrifuges, chromatography and electrophoresis separate substances like proteins, DNA and viruses. They simulate the unsimulated: computer technology models complex systems—from neural networks to the Earth's climate—that are too vast to study directly. They automate the unautomated: machine learning methods analyse massive datasets, run experiments and generate predictions by spotting patterns at exceptional speed. They image the non-imageable: x-ray devices, MRI, ultrasound and CT scans create extraordinarily precise images inside the human body. More than just standing on the shoulders of giants, these powerful tools enable the giants of science. These remarkable tools are the direct force uncovering what we did not even know existed—they are the discovery engine.

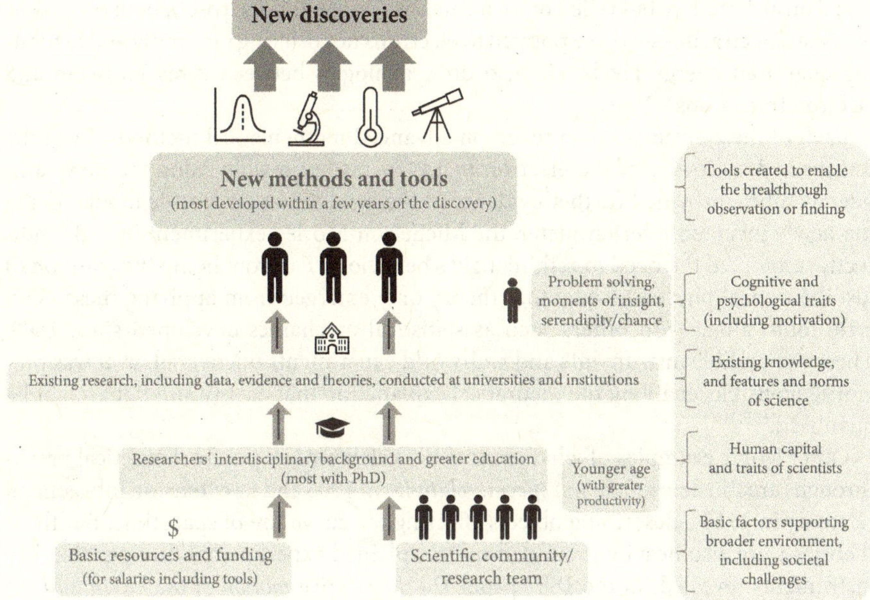

Figure 6.2 *The foundation of discovery: new methods and tools—our most direct levers of progress—and the factors that support them*

Our powerful tools also trigger our major theories of science by enabling us to see and describe what was hidden before. But how exactly do our greatest scientific theories fit into this tool-driven picture?

How our powerful tools drive scientific progress and make our best scientific theories possible—beyond the role of theories

Some of science's most celebrated discoveries are at times seen as purely theoretical masterpieces of the mind, developed by brilliant thinkers without using tools and experiments. Take the classic example of *quantum theory*—describing the behaviour of subatomic particles at the smallest scales. At times, the theory is seen as uncovered by thought. But taking a closer look, the story shifts. Max Planck's quantum hypothesis in 1900, the starting point of quantum theory, did not emerge from just a flash of inspiration. He directly built on findings made possible by precision instruments like spectrometers and bolometers that measured emitted light across wavelengths with exceptional detail—namely the famous blackbody radiation experiments. And he applied the model of linear oscillators to explain energy exchange.(53) These instrumental experiments did not just *inspire* a theory; it *enabled and demanded* one. The tools revealed the deficits that required a new interpretation. Einstein's work on the photoelectric effect in 1905 was not just a product of thought, but was firmly rooted in unexplained experimental findings using electroscopes and cathode ray tubes to observe how light liberated electrons from metal surfaces.(54) He then interpreted what the tools had shown. Bohr's quantum model in 1913 was not just based on abstraction, but relied on rich data gathered with spectroscopes using prisms to reveal spectral lines—these pointed to electrons not orbiting randomly and exhibiting quantised energy levels. He also drew analogies between x-ray emission and electron transitions.(55)

Each of these scientists also relied on advanced mathematical methods. Even the unexpected discovery of the electron in 1897—made possible using the new cathode ray tube—provided further evidence of the need for a new atomic model. Using the newly invented interferometer, the Michelson-Morley experiment in 1887 indirectly reinforced the need to rethink light's behaviour. Far from being the triumph of just theoretical physicists, quantum theory only emerged from applying these powerful tools, along with others such as statistical mechanics developed since 1868. These remarkable instruments and real-world experiments uncovered what was previously invisible, enabling the theoretical explanation that energy and light might be quantised.(95,103)

Other iconic examples of what are at times celebrated as purely theoretical breakthroughs are Einstein's *theory of special relativity* in 1905 and its extension into general relativity in 1915—describing objects affecting the curvature of spacetime. But these theories were also heavily grounded in unexplained experimental findings that key instruments revealed. In the 1887 paper *On the relative motion of the Earth and the luminiferous ether*, Michelson and Morley measured the motion of the Earth relative to the luminiferous aether that was thought to carry light waves.(50) Surprisingly, the experiment demonstrated, by relying on the new interferometer, that the speed of light remained constant regardless of the Earth's motion through space. This extraordinary finding challenged the existing notions of absolute space and time. It laid the evidence that enabled developing Einstein's theoretical interpretation—that space and time could be relative. This landmark experiment was widely debated by leading physicists at the time, including Einstein who discussed the 'luminiferous ether' in his 1905 paper.(253) And in Einstein's words: 'the Michelson and Morley experiment had made it clear that phenomena obey the principle of relativity'.(51,52) Yet Einstein is said to have later downplayed the influence of these experimental findings on his thinking.(52)

Maxwell's equations in 1865 also laid the mathematical foundation for understanding electromagnetic waves, and led Einstein to reconsider how light behaved in different reference frames.(99) Tensor calculus—a new mathematical method—was also indispensable for capturing the experimental findings and expressing gravitational fields in Einstein's equations. He did not invent these methods but adopted them. Discovering the electron with the cathode ray tube not only informed quantum theory but also highlighted inconsistencies in classical physics that relativity later resolved.

Einstein's key role was in interpreting and restructuring these experimental findings that revealed what was not conceivable before the experimental findings.(82) These extraordinary tools created the very possibility of seeing the world differently in the first place—and enabled Einstein to make the conceptual leap. Far from just confirming predictions, tools make new kinds of predictions possible. Technological advancements, including particle accelerators, telescopes, atomic clocks and gravitational wave detectors, then enabled precise experimental tests that validated and extended Einstein's predictions with remarkable precision. These again made new theoretical explanations possible that were not imagined before.(99) Likewise today, theoretical physicists are fundamentally guided by experimental and methodological discoveries—for example by the particle physics experiments at CERN that inform their theories.

Taking a step back, Einstein is famously credited with the most iconic equation in science that is associated with special relativity: $E = mc^2$. The simple formula states that the energy E of an object at rest (such as an atom) is equal to its mass m multiplied by the speed of light c squared. Similar mass-energy formulas, such as $E = kmc^2$, were already developed by Umov in 1873 and Preston in 1875, and later also by Thomson in 1881, Heaviside in 1889 and Poincaré in 1900.(254) In 1904, Hasenöhrl derived $E = 3/8mc^2$ while studying blackbody radiation—and his results were consistent with special relativity.(255) In the 1903 paper published within the Proceedings of the Royal Veneto Institute of Science, De Pretto developed the very same formula $E = mc^2$,(256) two years before Einstein's same version appeared in the German-based *Annalen der Physik* in 1905—the same journal where Hasenöhrl had published his work the year before.(255) In 1906, Einstein then recognised Poincaré's mass-energy equivalence.(257) And in 1907, Einstein acknowledged that others may have arrived at similar conclusions: 'It seems to me to be in the nature of things that other authors might have already elucidated part of what I am going to say. However, bearing in mind that the problems under consideration are being treated here from a new standpoint, I felt that I should be permitted to forgo a survey of the literature (which would have been very troublesome for me)'.(258) Einstein critically connected the equation with the theory of relativity. But without Einstein we would still likely have a theory of relativity which other leading scientists, like Hendrik Lorentz, were also working on—and special relativity was originally called the Lorentz-Einstein theory.(255,256)

Darwin's *theory of evolution* from 1859 is often seen as a product of just reasoning and observation—as drawing sweeping conclusions from observing Galápagos finches and a long voyage. But in reality, Darwin relied on vast amounts of experimental data and scientific tools, both his own and those of others.(57) His advanced microscopes—our prime example of a key scientific tool—enabled him to examine barnacles, corals and plant structures in great detail over the years. These shaped his understanding of variation, specialisation and adaptation across species.(56) Darwin also conducted systematic experiments: he bred pigeons and plants under controlled conditions. These provided him with direct evidence of how artificial (human) selection could cause distinct traits to emerge across generations. These experiments did not just enable drawing the analogy for natural selection—they were also living laboratories for understanding how selection pressures shape species. The findings led Darwin to the conclusion: if humans could shape animals over time, so could nature.(57) He also applied comparative anatomy methods to study finches, dissected animals and drew on the latest fossil discoveries—providing him with further evolutionary evidence of how extinct species relate to living species. And he was also influenced by Thomas Malthus' views on population growth—that we can compete for limited resources as a force of survival. By synthesising vast data across biology, geology and palaeontology and using natural experiments, Darwin was able to pull the pieces together into a coherent theory with unprecedented depth—one that transformed how we understand life and its origin.[57,82]

So why do the most celebrated scientific ideas and theoretical advances—quantum theory, relativity, evolution, DNA's double helix—commonly follow newly applied tools? It is because conceptual breakthroughs do not just deeply depend on advanced tools and unexplained experimental findings and puzzles, but are fundamentally made possible by them in the first place. Cutting-edge tools laid the essential foundation of evidence on which scientists could then create the theoretical explanation for what they could not have seen or conceived before. Here lies a key insight—tools do not just support the scientific process and make it more efficient, as commonly believed, they largely define what problems we can conceive and make our biggest scientific ideas and experiments possible.

Without the microscope, no cell theory could have emerged; without x-ray crystallography, no DNA double helix model could have arose, and so on. Science is often seen and taught as theory coming first, with observations then merely testing them. But science's groundbreaking theories are commonly born from unexpected experimental observations. *New tool → unexpected finding → new theoretical explanation of that finding.* Theoretical physicists like Einstein claim that 'it is the theory which decides what we can observe'. Here, the broader evidence of discoveries across science leads to a deeper insight: *it is the tools that define what we can observe, measure and experiment and ultimately what theories we can even conceive in the first place to explain surprising findings those tools uncover.* So while in theoretical fields—like theoretical physics—observations often tend to be thought of as theory-laden,[15] observations and theories are actually always tool-laden across science.

Screening science's over 750 biggest discoveries, we find that only very few exceptional cases are popularly described as stemming from just 'thought experiments'. The idea that some theoretical breakthroughs can arise from just abstract thought is a somewhat romanticised view in the public imagination. The perception is especially common about theoretical physics, where scientists study at the smallest and largest scales (subatomic particles and the vast universe). That requires at times greater interpretation and abstraction from available evidence than is typical in most other fields across the natural and social sciences. In reality, such popular narratives of the lone genius are constructed only after new instrumental tools produced the unexpected findings and enabled interpreting those findings in a new light. Even Einstein's most famous and iconic description—imagining chasing a beam of light—was only part of his retrospective narrative decades later, not in earlier publications, interviews or communications.[52] But we find that each of the few later descriptions actually fundamentally relied on pivotal enabling tools. The narrative came later; the tools came first—by revealing the puzzling problems that needed a theoretical explanation.

Scientists—from Einstein to Darwin—are endowed with the same senses and motor skills as others, but they excelled at synthesising existing experimental findings and tools *before* developing their theory. And *afterwards*, major theories generally only become recognised as theoretical breakthroughs once they survive rigorous experimental testing with the tools that validate them—as we see scanning nobel-prize discoveries. *Even our best theories can only ever be confirmed using tools and methods.* After all, Einstein received the Nobel prize in 1921 not for his relativity theory, but for his work on the photoelectric effect that did have clear experimental confirmation at the time.[54]

Next, *when we explore how science's central tools emerged, we find that many were born largely as practical inventions using trial and error, experimenting and solving real-world constraints to our tools*: from Kirchhoff and Bunsen's spectrometer, Martin and Synge's chromatography and Tiselius' electrophoresis, to Mullis' PCR method and Svedberg's centrifuge. From the first microscope and telescope born from lens crafting, to the thermometer based on the expansion of fluids; from Wilson's particle detector inspired by observations of cloud formation, to the Geiger counter built on observing ionised gases, a number of tools first began as hands-on applied breakthroughs rather than by building on an existing theoretical framework. For Röntgen's discovery of x-rays—an entirely unexpected observation—no theory was possible. Such tools were developed *before* theory even knew what questions to ask to develop them. But applying them then led to over a hundred major applied discoveries into invisible worlds, without yet theoretical explanations. *So why has theory at times not been crucial for building some tools, especially first-generation tools? Because the goal is not explanation, but functionality—demanding testing and prototyping.* But other central methods did rely in part on a theoretical foundation. Fisher's statistical methods solved real-world problems but also used practical principles of probability, and Ruska's electron microscope tackled real limits in resolution but also applied practical insights from quantum mechanics. When used to help optimise tools, applied theory is typically used—focused on solving real problems—rather than abstract theory, traditionally used to describe nature.

Why are methods and tools so fundamentally important in propelling discoveries, often playing a more crucial role than scientific theories in science's over 750 major discoveries? Why are tools commonly the central driver in breakthroughs? There are several reasons:

- *Tools enable us to not only directly see, measure and experiment but also think in ways unimaginable without them.* Remarkably, using them we can perceive previously unknown and inaccessible phenomena across all fields—from microorganisms and nerve cells to radio waves, subatomic particles and distant galaxies. By making the invisible visible, we have sparked hundreds of major discoveries through exploratory research and experimenting using new tools and with limited theoretical understanding. Tools—microscopes, x-ray devices, machine learning—observe and measure the unknown in ways that theories simply cannot, while we develop and revise theories (explanations of the world) from what tools reveal (Chapter 1).
- *Tools make discovery more strategic and less driven by chance.* They shift discovery from chance to a deliberate and effective strategy—by enabling systematic, repeatable ways to explore unknown terrain. Uniquely, serendipitous discoveries, from x-rays to pulsars, generally reduce the role of theory as they are sparked by new tools of discovery before anyone knew how to explain them—turning observation into theory (Chapter 2). So tools often *precede* and inform the theories we develop.

- *Tools accelerate the pace and scale of discovery.* Better methods—computational, experimental, statistical—generally evolve faster than the theories meant to explain what they uncover. New technologies like high-powered computing, machine learning and data analytics can speed up the pace of science. By collecting, analysing and visualising enormous volumes of data faster and with unprecedented precision, they allow us to test, refine and update discoveries including theories quicker and open new research areas (Chapter 1).
- *Tools push us past the boundaries of current theory, and without them our best theories could not have generally been imagined, much less tested.* Conceptual breakthroughs occur after new tools are invented—from radio telescopes and particle accelerators to x-ray detectors. Because these tools unveiled invisible realms no one had conceived and enabled entirely new types of questions. Powerful new instruments reveal unexpected phenomena that current theories cannot explain. And theories do not explain themselves—they rely on the evidence that instruments provide. New evidence in turn forces us to reevaluate established theories. Detecting for example exoplanets using new spectrographs upended existing models of how planets form and sparked new theoretical explanations. And when it comes to the origin of our groundbreaking tools, we found that they generally emerge not from abstract theorising but from practical deficits in earlier tools and experimenting.
- *Tools spark breakthroughs across unrelated fields.* Numerous transformative tools each triggered over a dozen discoveries—from electron microscopes and x-ray devices to spectrometers and chromatography—and often in unexpected fields far from where they emerged (Figure 1.7). What connects the distinct discoveries that each of these tools made possible is not a common theory, research teams, more funding or serendipity, it is using the common powerful tool—each unlocking groundbreaking findings across different fields. Versatile methods (like the gene-editing method CRISPR) are initially created in one field but then imported in others (with CRISPR transforming genetics, medicine, agriculture and environmental science).
- *Tools make science scalable, precise and reproducible.* High-precision instruments enhance efficiency and allow researchers around the world to replicate experiments using standardised measurement.

So the reason why tools often prove to be the more critical driver of discovery than theories is straightforward: the history of science shows they often not only resolve more of science's deepest questions but also spark more new fundamental questions and answers unimaginable before. Our tools foundationally shape not only what we know but how we know it.

Across science, most fields are driven by experimental and observational methods and have a limited theoretical footing. Think of medical researchers and epidemiologists, ecologists and environmental scientists, archaeologists and palaeontologists, psychologists and the wide range of other applied scientists. When asked how discoveries are made, they highlight collecting data and running experiments using tools as the very cornerstone of science. *A look across the fields of science shows that many disciplines thrive with little or no established theoretical foundation. Yet what unites all scientific fields is not theory—but methods and tools that ground knowledge in empirical reality. Without these tools, fields never get off the ground. No field can function without evidence derived from methods and tools.* In some fields, researchers can have stronger synergies between technological advancements and the theoretical insights they spark, like with biotechnologists, chemists and materials scientists. At times, this synergy can help motivate tool design—with for example limitations in microscopy combined with quantum mechanics contributing to the scanning tunnelling microscope.

In contrast, in a few subfields researchers place greater emphasis on theory, like theoretical physicists, theoretical economists and philosophers of science, often relying on mathematical methods with limited (direct) evidence. For these researchers, it is not surprising that the strongest emphasis on theory tends to come from fields where tools and experiments are used least as part of daily work. After all, they deal with phenomena far removed from everyday experience: like subatomic particles, cosmic scales and global financial markets—often expressed using mathematical equations. Yet even the most abstract theory must ultimately be based on and pass the test of reality. We require instruments before and after creating a scientific theory: *before, tools provide the surprising observations and anomalies that spark theoretical ideas and explanations; and after, they offer the only means to test and refine them.* A theory is always developed, directly or indirectly, by building on existing instrumental findings. No theory comes out of nowhere. And without the tools to create and verify theories with real-world evidence, they remain speculative—like string theory and the multiverse.

Ultimately, both popular narratives—of a lone genius striking a fortunate flash of serendipitous insight or deducing fundamental truths through abstract thought—make science feel exciting and almost magical. By studying discoverers systematically, a different and more grounded story emerges: science's major discoveries are catalysed by new tools of observation and analysis and building on experimental work that enable the surprising insights. Surprise and insight can often be engineered. Here we provide a breakdown of the different types of discoveries and people's often simplified view of how science progresses (seen in Figure 6.3). *Ultimately, we find that most of science's over 750 major discoveries are experimental breakthroughs (60%), followed by method innovations (25%), and finally theoretical advances (15%).* Theoretical advances awarded a Nobel prize are largely concentrated just in theoretical physics and theoretical economics—and are extremely rare, or non-existent, in fields like medicine and biology.

Ultimately, without robust methods and tools to detect or measure, theories cannot generally be developed or tested but also more funding and collaboration are not enough, and serendipitous observations are not possible. Nothing can substitute tools in detecting, measuring and revealing surprising new discoveries. While the role of factors varies by discovery, tools are commonly the decisive factor—the common driver we must directly apply to be able to spark the breakthrough. Major advances often do not come from where we commonly look, but from where we do not commonly look—the methodological setup. Methods generate questions, data and evidence, and theories help shape the questions and interpret the data and evidence we generate.

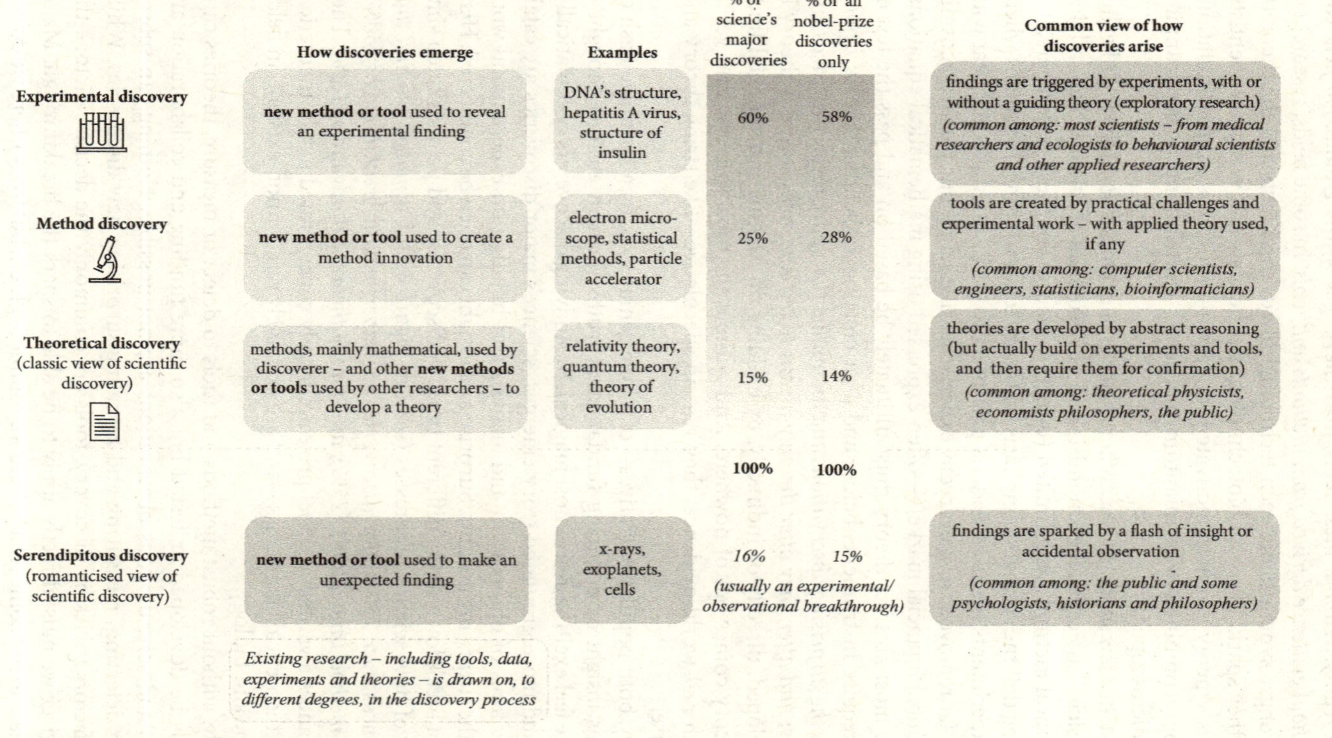

Figure 6.3 *Different types of scientific discovery*

The description is based on science's over 750 major discoveries. Method discoveries are a new major method or tool; experimental discoveries uncover a new experimental finding (and are not a method breakthrough); and theoretical discoveries introduce a new theory. Some discoveries can be technological or computational and fall into the category of method discoveries—using for example AI and machine learning.

Remarkably, many think methods and tools are stepping stones—serving their purpose until we create a discovery or theory, and are then forgotten. Yet we must learn to see our major methods as the core of scientific progress itself, and our discoveries and theories not as the final goal but as provisional outputs—providing provisional explanation and understanding. But they are always subject to refinement as we keep improving our methods that enable new evidence. The hope here is that we no longer just value science's major discoveries and theories as the pinnacle of science, but recognise science's great methods and tools of discovery as highly, if not more, as they make them possible in the first place.

New tools stretch our imagination and provide an intentional strategy for innovation—beyond the more unpredictable role of creative intuition

How we conceive reality—cells, molecules, galaxies, cosmic radiation—is largely shaped by how our tools let us see them. Scientific creativity and imagination expand not just by closing our eyes, but especially by opening them and looking through powerful instruments. These tools reveal entirely new structures and patterns in the world we could never have conceived with the naked eye and just thought: x-ray crystallography uncovered the elegant double-helix spiral structure of DNA that Watson and Crick modelled *after* observing the diffraction pattern.[1] The Hooker Telescope at Mount Wilson Observatory enabled Edwin Hubble to observe galaxies beyond the Milky Way and the universe expanding—unimaginable concepts *before* telescope resolution advanced.[2] Functional MRI scanners exposed real-time mapping of brain activity via blood flow, revealing the brain areas involved in memory, vision, emotion or decision-making. In other words, we could not imagine how the brain works in such ways but we directly *saw* it in real time. Particle accelerators uncovered the traces of particles observed in high-energy collisions. Automated DNA sequencing and sequence-tagged site maps revealed the human genome—an unconceivable genetic code consisting of three billion letters.[113] These discoveries make up central pillars of science—how we understand life and the universe—and they did not form from just deep introspection, but from what our extraordinary tools see and measure. This is a key insight that has been overlooked.

Imagination is limited to what the mind can conceive—but tools vastly expand what is observable and thus conceivable. If we cannot measure something, it is hard even to *conceive* of it scientifically—and our tools are what enable us to measure. Tools do not just confirm ideas, they generate them. Scientific progress has commonly stalled when tools stagnate, regardless of imagination: fields like theoretical physics and theoretical economics plateau when they lack new applied methods to explore and test models.

Today, the greatest leaps in science have emerged largely from our *external interactions* with tools, not just from *internal introspection*. Scientific tools are often thought of as just answering our questions, but they actually generate more new questions we did not even know to ask. Tools turn the unknown into the visible and imaginable.

Scientific creativity emerges largely outside the mind: from what our tools see, detect, measure, simulate and model. They have vastly expanded the very space of creativity itself. This is the key difference, because outside of science, creativity can often spring from within—the imagination of the artist, poet or composer expressing an inner vision through pens, paintbrushes or musical instruments.

With the lens of instruments driving what is imaginable in science today, let us look at how we develop our remarkable instruments. Science's powerful tools do not emerge from sudden creative sparks but rather inventors describe how they run into the limits of a current tool and work systematically to overcome the methodological bottleneck (as we highlighted earlier in this chapter and in the Boxes throughout the book). What inspires developing science's top tools is not just creative intuition. It is instead practical needs: increasing the power of microscopes, boosting the efficiency of particle accelerators, optimising the sensitivity of spectrometers, strengthening the robustness of statistical methods, sharpening the resolution of chromatography techniques and so on. Innovation and curiosity start when hitting a methodological wall.

When screening science's major discoveries, we find that recognising a constraint to a current tool commonly involves systematic steps: we design and experiment with prototypes of the tool, gather resources, leverage emerging technologies, test how effective the new tool is or at times bring together methods and collaborations from different fields. Systematically, we can trace these observable steps in the discovery process—but not always an individual creative idea. Tool generation is a practical strategy—requiring problem solving and often trial and error—that has proven very successful by upgrading, integrating and reorganising existing tools into something more powerful. What is called an individual's *creative innovation* can be better understood here as *strategic development* or recombination that different researchers can adopt; these strategic steps can be thought of as intentional or systematic creativity. Hundreds of discoveries emerge by applying a new lens, sensor or technique or by importing a new tool from one field into another or merging two methods from different fields to solve a problem in an entirely new way (Chapter 1). It does not matter if we label this systematic or imaginative thinking. These are powerful strategies we have in our hands to *intentionally* drive innovation—while creativity can play a role in forming hypotheses, concepts and interpretations.

Tool creation underscores a deliberate design: we find that, first, science's transformative tools are born from necessity, not creative intuition; and secondly, science's major discoveries depend more on intentionally designing and applying new tools than the creative insights that the tools can enable. Unlike creativity that is often seen as elusive, generating and applying tools embodies the most deliberate, strategic and measurable part of science. Tools make science cumulative, reliable and replicable because they allow other researchers to arrive at the same findings using the same techniques. Even the most celebrated creative insights of Einstein, Planck, Darwin and others were only possible because of the tools and methods they directly or indirectly relied on, as we highlighted earlier. Here we reframe scientific progress as an active pursuit of building and using new lenses to the world, outlining a more practical strategy than the notion of creative genius. We find that discovery is far less about *stumbling upon ingenious, unpredictable insights* and far more about *engineering insights* by constructing better ways to see, sense and think about the world. Scientific progress is less about intangible brilliant minds and more about tangible brilliant methods that make those surprising insights possible in the first place. Building the sharpest lenses enables the brightest insights and theories to arise. The more intentionally we build, the more imaginatively we can think.

With this tool-driven understanding of discovery, researchers can become aware and begin solving method limitations we face. With each new tool we use, our ability to see, generate ideas and our imaginative potential grow. *Here we shift the traditional idea of creativity—often associated with an individual mind, as in art, music or literature—to recognise that in science today, creativity is instead largely driven by our tools that transform how we see, think and imagine—and do so for multiple researchers.* In many ways, tools have become the central engine of creativity and imagination in science, fuelling our capacity to ask completely new questions and solve entirely new problems like nothing else can. The best strategy we have to foster creative insights is commonly upgrading the tools with which they can arise—just as with serendipitous insights (Chapter 2). So if one thinks of creativity as the spark for some discoveries, then tools are the fuel that enable the creative insight. There are some exceptional cases—most notably in theoretical physics—where concepts have been developed like gravitational waves and black holes, by applying sophisticated methods like tensor calculus and building on existing findings using interferometers and telescopes.

While we cannot force new breakthroughs, we can play an enormous role in actively spurring them by directing much more attention to improving our tools. This brings a powerful shift in how we think about innovation: we should not wait for it to happen—we need to design and build our way to it. This moves us from just celebrating lone geniuses—a concept especially of past centuries—to empowering the method-makers who drive progress. The concept of scientific genius in earlier eras centred on theoretical insight—as exemplified by Einstein as the most iconic scientist in history. But science has become increasingly tool-driven so that science in the future will arguably shift the focus to our extraordinary toolmakers—the great and brilliant researchers who make new theories possible. We need to replace the traditional view of discovery as a product of solitary genius—imaginative and theoretical minds—with the tool-driven view of discovery. Even if the *romantic mind* view feels more inspiring than the *mechanical, replicable tool* view, the core power of science shifted over time from our minds to the sophisticated tools of science. For tools like electron microscopes and x-ray crystallography vastly surpass the human mind in perceiving, testing and sparking theories about the world.

The new methods-driven discovery theory

'No theory exists that can reliably predict which research activities are most likely to lead to scientific advances', as the National Research Council in the United States confirms.[58] Yet by drawing here on the evidence from science's major discoveries, we introduce a new framework: the new methods-driven discovery theory. What the theory explains is simple but powerful: *new methods and tools unlock new discoveries by enabling us to see, measure and understand the world in ways that are impossible without them.* The theory explains how new breakthroughs are preceded by a leap in the way we access and understand the world through new methods. Earlier, we highlighted diverse strands of evidence illustrating this causal *tool-to-discovery* link across science's greatest breakthroughs that were otherwise not possible (in Chapter 1's section on causality).

A core insight in society and an economy is that major advances in human well-being and development are commonly fuelled by technological changes that stem from deliberate human choices—from light bulbs to phones and channelling solar energy.[259–261] Here, we uncover in science that major discoveries and fields are powered by method innovation that arises from the intentional decisions of researchers to craft and deploy new tools. New advances emerge when we do not just produce research with the same *existing* tools but shift our very means of producing research by adopting *new* tools to create knowledge from an entirely new perspective. It is no coincidence that the driving force of major advances in society and an economy (new technologies) is so similar to the driving force of major advances in science (new scientific methods and tools).

Before we go further, let us return briefly to the definitions of the central terms here. *Major discoveries* are defined as new breakthroughs that mark entirely new ways to understand the world, open new paths of inquiry and later have a proven, lasting impact on science—such as uncovering the structure of DNA and exoplanets. *Scientific methods* are systematic techniques, and scientific tools are systematic instruments that we use to study the world and extend our ability to observe, measure and analyse, and are general-purpose. The *new methods-driven discovery theory* is thus straightforward: not only does it explain our past breakthroughs, it improves predicting future ones. When we invent, upgrade, combine or adopt a new method or tool, we can predict *where* discoveries can come from next, and *when. It is about tracking the speed and direction of method innovations: this enables us to spot strong signals that discovery is soon to follow. The signal can come from a recently invented computational method, an upgraded particle accelerator, an advanced spectrometer combined with AI-tools, or a new experimental method adopted from another field. Each helps us anticipate where the frontiers of science can soon move.*

So the theory enables prediction: if a major new method or tool is developed, then the likelihood of new discoveries rises sharply (Figure 1.8). The theory predicts that making a major tool innovation sharply increases the odds of sparking major scientific advances—and more than other factors across science. We directly tested the theory across fields and history, finding strong and consistent evidence (Chapters 1–5). It is assumed in the discovery process that supporting factors—including our cognitive abilities and a minimal level of funding and collaboration—are in place to generate needed tools and research.

New transformative method innovations unlock new discoveries and fields through five key stages:

- *Method constraint*: generally, we run into a practical problem we cannot solve when using a current method or tool to study a phenomenon.

- *Method scanning*: next, we need to scan our own field and related disciplines for a method or tool that can tackle that constraint (and if none exist, we go to the next stage).
- *Method conceptualising*: at this point, we require conceiving a new method or tool designed to solve the constraint, expanding our cognitive or sensory scope.
- *Method development*: we need to then test, retest and scale up prototypes of the new method or tool, enabling us to explore the world in fundamentally new ways.
- *New methods-driven discoveries and fields*: finally, we require applying the new method or tool to trigger new findings and advances.

Take for example the extraordinary discovery of detecting gravitational waves in 2015. No tool was yet sensitive enough to reveal them. Researchers set out to develop an expanded massive laser interferometer, the LIGO detector. Uncovering the discovery that same year came down to an incredible methodological feat: they designed detectors thousands of kilometres apart that operate together with almost unimaginable precision—measuring a change in distance between the detectors' mirrors 1/10,000th the width of a proton (Figure 6.4). The high-precision instrument gave us a new window to the universe. Just two years later, the discovery earned a Nobel prize—one of the fastest recognitions in Nobel history.[115,138]

Understanding new tools as the key causal driver of breakthroughs offers a new way to understand scientific progress. Drawing from the *general analysis* of science's over 750 major discoveries, the evidence lays the foundation for this *general theory* of discovery. While we described the practical features of this new method-to-discovery view earlier (in the section on the role of theories), here we distil its deeper features.

- The theory outlines how new methods and tools are a *necessary condition for major discoveries, and commonly the central driver—they are the common thread driving our biggest breakthroughs across science and history.* While supporting factors—like education, institutions and psychological features—foster breakthroughs, they vary by discovery. Without a new method or tool to detect or measure, no amount of funding, collaboration, training or luck will enable a new breakthrough, and theories need them to be meaningfully developed or tested. We spark significant shares of discoveries without these other factors—factors that are more context dependent and not always necessary (Chapters 4 and 6).
- It is an *explanatory and predictive theory of scientific progress* that holds across time and fields. It explains the causal mechanism of how we accelerate science—the pace of generating new methods is key to forecasting future breakthroughs they make possible.
- It explains how new methods are powerful in disrupting research because we can *apply methods widely to topics beyond their initial scope*. Unlike theories or findings often bounded by context, methods are more transferable and stable when we import them into unexpected domains—such as transformational statistical, computational or experimental methods. They better defy disciplinary silos, empowering us to break free from a field's inherent assumptions and provide an entirely new lens.

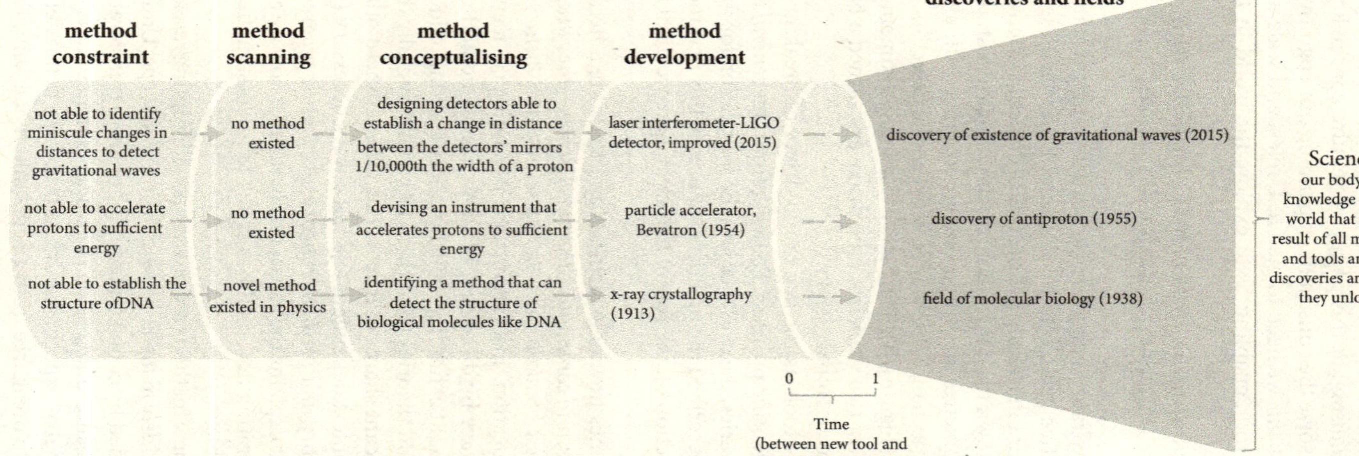

Figure 6.4 *The five method-driven steps we take to trigger new discoveries and fields*

- It may arguably be, through the new method-to-discovery logic, the closest we can get to a *unified theory of science* or scientific progress. While disciplines differ widely in content and culture, they are consistently powered by the same engine: new methods. If a universal theory or principle exists that cuts across physics, biology, chemistry, psychology and beyond, it does not reside in what we study—but in how we study the world in new ways (Chapters 1 and 9).

Surprisingly, examining science's major discoveries we find that we cannot spark them by just observing, experimenting or theorising using an existing method, but method innovation is essential (Chapter 1). For the increased power of a tool generates new conditions for what we can observe, experiment and theorise—with greater complexity, precision, quantity, temperature, energy and pressure. New major tools commonly produce large positive externalities far beyond the original discoverer or research team: they spill over to other scientists in the same and other fields. At times, new tools are repurposed to generate a wave of discoveries across disciplines that their inventors never imagined (Chapters 1–5). Take the simple PCR method: it accelerated the sequencing of the human genome, transformed forensic science and enables rapid medical diagnostics like HIV, tracing ancient DNA in palaeontology and tracking ecological evidence of biodiversity through soil and water. Strikingly, none of these applications were foreseen by its creator, Kary Mullis.[128] And yet paradoxically, researchers do not often capitalise on these spillover effects, because research is focused on experimental results and theoretical insights—not tool inventions. But our methods are what connect our experiments to discoveries, our data to theory. *This tools-first view flips the traditional narrative of discovery: it shifts the core source of scientific change from scientific advances to method advances that enable them.* It is inventing new tools that has been the missing link in explaining new breakthroughs.

This new tools-driven discovery principle represents a general rule for triggering breakthroughs. It is a general logic of discovery in which scientific progress is broadly driven by method innovations—not narrow methodological rules that, applied correctly, would spark breakthroughs (Chapter 2). This tool-first principle of discovery moves us beyond leading explanations in understanding discovery: historians like Kuhn emphasise paradigm shifts in theoretical frameworks as the central parameter driving science,[15,16] scientometricians and network scientists point to team dynamics, careers, research productivity and networks,[3,4,9,14,38,39] economists stress funding and incentives[6–8] and sociologists like Merton focus on social dynamics.[262] Each plays a role in some but not other advances. No explanation has yet established the common mechanism sparking science's major discoveries and highlighted the direct, key role of new methods and tools (Chapter 1).

So why do historians and philosophers emphasise the importance of theories as the central metric of scientific progress? Why do scientometricians highlight scientists' citations, economists stress public investment, psychologists turn to creativity and serendipity, and so on? The answer lies in the nature of their own methods and training, with each field's research gravitating towards a particular metric that shapes the questions they ask and the kinds of answers they can see.[82] Researchers are trained to see with certain lenses—but not others. That is why stepping back and adopting a more comprehensive perspective is so important. Here, the integrated approach we adopt enables us to overcome shortcomings of relying on one method or factor, to uncover the broader insights of how breakthroughs actually emerge (Chapters 1–6). Ultimately, a theory that aims to explain scientific progress must place new tools at the centre of focus.

Scientific table of methods: mapping and fusing methods and tools of discovery

How does the new method-to-discovery model make us think or rethink about how science is structured? Mendeleev transformed chemistry not by discovering a new element, but by developing a new structure: the periodic table. He arranged the elements—by weight and chemical resemblance—so his students could better understand their properties. This remarkable new structure led to predicting and uncovering unknown elements. Similarly, science is organised to make it more accessible. Science textbooks categorise fields into neatly bounded disciplines—physics, chemistry, biology, psychology, history—and present the scientific method as 'the collection of data through observation and experiment, and the formulation and testing of hypotheses' (Chapter 3).[(196)] This traditional architecture is widely adopted across science. But is it a natural categorisation of science? It is simply one among other ways to structure science. In fact, it can constrain us by overlooking two key features of how science actually works and progresses: first, the sophisticated methods and tools that do heavy lifting behind science's advances; and second, science's often interdisciplinary nature. We need to restructure the scaffolding of science. So what if we restructure science not just around its subject matter or its abstract ideals, but also around the practical methodological structure that drives science and unlocks discovery? Here we tackle this question.

To begin, what are the most cognitively demanding methods and tools in science? At one end of the spectrum, we find complex *modelling and simulation methods*—techniques that demand elaborate formalisation, iterative refinement and greater abstraction. Whether modelling climate systems, biological processes or economic behaviour, these methods force us to make critical decisions that shape our findings: what should we simplify, exclude and mathematically represent? *Mathematical methods* like calculus and algebra demand much theoretical reasoning, often far removed from everyday intuition—think of Newton's laws expressed using calculus.[(263)] *Statistical inference methods*, too, are cognitively intense and involve navigating multiple layers of complexity. Moving from raw data to robust conclusions and predictions demands using methods like frequentist hypothesis testing, Bayesian inference and regression analysis—all requiring reasoning about uncertainty, probability and margin of error in the results. These methods continuously evolve, with ongoing debates over p-values, effect sizes and, more broadly, the replication crisis driving their refinement.[(187,189)] In turn, *quasi-experimental methods* are used when controlled experiments and randomisation are not feasible. In fields like epidemiology and economics, researchers can at times rely on natural experiments or instrumental variable methods to approximate causal effects from real-world data.[264] By contrast, *controlled experimental methods*—like randomised controlled trials—are built on clearer foundations but remain cognitively heavy, requiring careful technical judgment in isolating causal effects and managing bias.[234]

Sampling methods are central in data-driven science and occupy a unique methodological middle ground. Take well-designed longitudinal studies that follow the same individuals over time—these demand meticulous planning to minimise problems of representativeness, attrition and weighting the sample. Dealing with small populations or big data makes study designs more complex and increases cognitive load. We then move to *experimental tools*—such as centrifuges, PCR and chromatography devices—that uncover data not directly accessible to our senses. Their output often later feeds into statistical analysis, even if operating the tools can at times become routine with practice.

At the other end of the spectrum lie *indirect observational or measurement tools*, like x-ray crystallography and spectrometers. These instruments enable us to infer structures not directly observable like atoms and molecules—such as DNA's double helix. In lab settings, their use can become streamlined and routine through standardised protocols. At the simplest conceptual level are *direct observational or measurement tools*—light microscopes, telescopes, thermometers—that extend our sensory reach in intuitive ways resembling human senses. They do not demand deep abstraction but amplify our senses, making them generally cognitively simple. This continuum highlights a rich hierarchy of cognitive demands: from many methods—like complex experimental methods driven by elaborate statistics—demanding several years to master and complex decisions in each study, to some tools that require much less effort and minimal training to use—like light microscopes and thermometers. Many education systems reflect this gradient, introducing observational tools early and progressing towards more complex methods.

So why do serendipitous discoveries tend to arise from new tools, not from new methods? This classification here sheds light on this fascinating pattern. Physical instruments like microscopes and x-ray methods can suddenly reveal what no one knew existed, often differing fundamentally from mathematical and statistical methods, often leveraged to understand, quantify and predict phenomena after we uncover that they exist. Unlike a sudden serendipitous observation, breakthroughs grounded in statistics and mathematics often emerge from slower, more deliberate cognitive effort (as we uncovered in Chapter 2).

To better guide how we catalyse innovation across science, we develop here a taxonomy of science's methods and tools—a kind of *periodic table of methods*. Surprisingly, no framework yet exists that systematically maps methods and tools across science—from telescopes to t-tests. By mapping the method landscape, the taxonomy reveals gaps and possibilities—helping guide where underused or unused techniques can be applied to catalyse discovery in unexpected domains. This science-wide classification can also offer researchers, educators and innovators a structured way to think and rethink about the range of scientific methods and how we can link, adapt and recombine them in innovative ways to generate new kinds of knowledge. We can classify—and compare—each method and tool by:

- *Type and purpose*: we differentiate between physical instruments (like microscopes) and analytical methods (like randomised controlled trials). We also categorise them by their function (observational, experimental, statistical) and their sub-categories (like direct or indirect observational tools).
- *Interdisciplinary scope*: we categorise how transferable methods are across domains—from field-specific impact (like radio telescopes) to interdisciplinary impact (like microscopes and PCR) and even (nearly) universal impact (like statistical and AI methods).
- *Cognitive demand*: we distinguish across the range of methods, with some techniques requiring increasing analytical effort, constant refinement and much judgment (like complex statistical modelling) while some tools are increasingly standardised and intuitive across researchers (like light microscopes) (Table 6.2).

Key insights emerge from this classification of science's toolbox. One, the taxonomy can help accelerate discovery by guiding us to scan for powerful tools from other fields—tools that have surprisingly not yet been adopted in neighbouring domains. Understanding how specific tools have unlocked breakthroughs in one field helps us pinpoint what kinds of tools are underused or entirely missing in other fields—revealing unexplored territories where our current methods fall short. The solution to a problem in one discipline often already exists in another—as we highlighted in Chapter 1. Two, the taxonomy reveals which methods are most versatile and high-impact across fields, helping institutions and funders allocate resources and set research priorities more strategically towards the most adaptable, scalable and effective methods. Three, it can help overcome an overreliance on one or a few familiar methods—biasing research directions when other powerful tools are often available. Four, the taxonomy can open a new meta-research agenda—one that tracks method diffusion in and across fields (or failure to diffuse) and identifies bottlenecks in innovation and what factors enable or stifle the spread of transformative tools. It opens new avenues for interdisciplinary work and training, where researchers are methodologically multilingual and can move across domains fluidly.[82] While not all methods are transferable, many are far more flexible than we assume.

In short, as Mendeleev's periodic table revealed hidden relationships among chemical elements, a taxonomy of scientific methods can help reveal powerful methods: uncover underused methods outside disciplinary borders, expose gaps in the method landscape, spotlight new combinations of tools and realise opportunities to reinvent new techniques—ones that can unlock the next big advances. Science's best innovators and discoverers have stumbled across these approaches—for a particular breakthrough—but now we can intentionally pursue them head-on: within a systematic methodological framework and strategically extending our methods of discovery. Think of how medicine and public health have generally far outpaced

Table 6.2 ***Scientific table of methods: a taxonomy of science's methods and tools—by type, purpose, interdisciplinary scope and cognitive demand***

purpose:	***Observation (or measurement)***			***Experimental design***			***Statistics and mathematics***		***Hybrid method combinations***		
	a single observation (or few) is possible			*multiple observations (data driven)*							
type:	***direct observational tools***	***indirect observational tools***	***experimental tools (in-/direct)***	***experimental sampling methods***	***controlled experimental methods***	***quasi-experimental methods***	***statistical inference and analysis methods***	***modelling and simulation / mathematical methods***	***tool-tool combinations***	***method-tool combinations***	***method-method combinations***
interdisciplinary scope (used in many fields → used in one or few fields)	light microscopes thermometers telescopes	x-ray devices spectrometers electron microscopes particle detectors radio telescopes	lasers centrifuges chromatography electrophoresis PCR* CRISPR gene-editing particle accelerators	longitudinal studies structured surveys stratified sampling qualitative methods	randomised controlled trials (placebos, double-blinding)	natural experiments regression discontinuity instrumental variable methods, difference-in-differences methods	machine learning methods frequentist testing Bayesian inference systematic meta-analysis	calculus, algebra (theoretical modelling)** mathematical modelling (differential equations, game theory methods, network models) computational simulation (agent-based models, Monte Carlo simulation)	light microscopes + spectrometers electron microscopes + lasers x-ray devices + chromatography spectroscopes + telescopes	imaging (x-ray, spectrometers) + machine learning methods MRI + statistical and computational methods chromatography + AI-methods CRISPR gene-editing + simulations telescope data + quasi-experi mental methods	randomised controlled trials + meta-analysis PCR + Bayesian inference difference-in-differences methods + network modelling
	tools (physical instruments)						**methods** (analytical techniques)				
	at times less			***cognitive demand***			*at times more*				

This taxonomy is developed here based on the methods and tools used to spark science's over 750 major discoveries—featured throughout the book—and includes how broadly they are used across fields. Some tools and methods—like lasers and machine learning—can fall into different categories, with lasers for example applied as both an observational tool (laser interferometry in physics) and an experimental tool (surgical lasers in medicine). Experimental methods—like controlled trials—produce data that flow into statistical analysis, linking experimental design with statistical inference. *Some tools are often described as methods or techniques, such as PCR and CRISPR, and vice versa—though these are not methods in the abstract sense like statistical inference. **Some mathematical methods are used without relying directly on data but indirectly on previously collected data and findings. Methods can also be further classified by features like scalability, automation, reusability, precision, data type and known limitations.

neighbouring fields in pioneering rigorous, scalable methods: from randomised controlled trials (RCTs) to meta-analyses and natural experiments. The RCT method is so powerful that adopting it later to economics received a Nobel prize.(133,135,265) Yet many fields still rarely take advantage of the remarkable RCT method. Imagine its underexploited capacity in the science of science, evaluating the effects of changes in researchers' methods, universities or fields on their impact—or in environmental science, assessing pollution-reduction or forestation programmes across countries.(82) Many opportunities hide in plain sight.

Ultimately, a scientific table of discovery methods is a map of explored and unexplored method options that can help identify and predict untapped combinations of tools and where the next breakthroughs can come from. It captures several key features:

- Like elements in the periodic table, scientific methods and instruments can be thought of as foundational units.
- Just as we combine elements from the periodic table to create entirely new compounds, we can fuse a vast array of tools across our method landscape to create new compound methods that trigger breakthroughs.
- Just as there is no fixed number of chemical compounds we can form from over 100 known elements, there is no set limit to the number of powerful composite methods we can invent from science's developed methods and tools.
- We can bind, mix and match two elements or methods, or three, four or more.

Techniques like randomisation, blinding and/or placebos for example can be fused together to form stronger controlled experiments; their outputs then feed into statistical analysis; these in turn can be aggregated through meta-analysis—each methodological element linking together, like Lego pieces, towards greater scientific rigour. The key insight can be powerful: *each fusion is a potential catalyst for discovery.* Many combinations are remarkably straightforward: most methods and tools generate data (imaging tools, PCR, surveys) that can be linked to those that process and analyse data (statistical inference, machine learning) and to those that model behaviour and systems (simulations, mathematical models), and so on.

The combinations are nearly endless—AI methods plus spectroscopy; MRI plus statistical and computational methods; lasers plus machine learning methods; the CRISPR method plus Bayesian inference or simulations (across populations or species); the PCR method plus network modelling (of how viruses spread), and so on. Exploring science's major discoveries, we can imagine countless fusions that hold high potential: combine chromatography with machine learning to predict new compound structures or optimise separations through learned molecular patterns. Link radio telescope data to quasi-experimental and causal inference methods to separate astronomical signals from noise and distortions. Connect centrifuges to AI imaging to train models to detect abnormalities in samples, and so on (Table 6.2).

What are other ways a scientific table of methods can help bridge methodologically siloed fields and spur innovation? For researchers with open minds, the possibilities are immense. On one end, *data-driven fields* that rely heavily on statistical modelling and large datasets—from economics to the science of science—could benefit enormously from adopting a range of physical tools. In medicine, for example, wearable devices like smartwatches transformed how real-time data are collected on patients to track heart rate, sleep, blood sugar and other health risks.(266) Why not apply this same remarkable technology across the economic and social sciences to run experiments that monitor consumer decisions under stress in economics and emotional states in psychology? Why not increasingly employ neuroimaging tools like MRI to observe how the brain can react to incentives, risks and cooperation or use advanced facial recognition to help interpret emotional responses to economic decisions?(267) These tools could push towards an era of embodied social science, where we do not just study human behaviour, but track its physiological and neural foundations in real time. By expanding our toolbox, highly data-reliant fields can evolve beyond statistical inference and abstract models to collect other rich, real-world experimental data.

Another novel example is systematic meta-analysis—a method that synthesises findings across independent studies to reveal overarching trends. In medicine, this method has become indispensable to determine the overall effectiveness of treatments. But why not leverage this transformative method across many fields? It would have a vast impact if fully exploited across fields, from scientometrics and economics, to the science of science and even philosophy of science. For it helps fields move from isolated results scattered across individual papers—or even entire domains plagued by conflicting or inconclusive findings—to establishing greater consensus by synthesising a field's broader findings.(265) Surprisingly, this often reflects low-hanging fruit for generating more reliable, cumulative knowledge—the method is already available and just needs to be applied.

On the other end, *tool-driven fields* that rely heavily on physical instruments to make observations and measurements could advance more rapidly by better harnessing more complex statistical, computational and AI-based methods.(268) Take fields like biology, chemistry and geology that generate images taken with tools like microscopes, x-ray devices, spectrometers and telescopes. Advanced image analysis tools powered by deep machine learning can rapidly analyse images to detect and interpret previously hidden patterns in vast datasets across fields—patterns invisible to the human eye.(269) Take simulation techniques that have been transformative in fields like climate science, computer science and economics.(90,268) Why not increasingly use cutting-edge AI technology to simulate, model and predict complex systems and processes across fields? Imagine unexplored frontiers across every domain: in the science of science, why not aim to predict how new methods and discoveries spread or how these changes impact new innovation? In public administration, why not simulate the dynamics of institutions and government crises? In philosophy of science, why not model scientific knowledge, practices and communities to see how they gain and lose prominence over time? And so on.

If we intentionally cross-pollinate tools and export methods across disciplines, the very structure of scientific knowledge could change. In the future, we may conceive and define disciplines not by what they study but by how they study: by common method and tool clusters. From this method lens, we can already see this in our current scientific landscape. Think of physics, a field that conducts vast research across domains spanning the sciences—from biological systems, ecological networks and weather prediction models, to quantum biology, neurophysics, computational chemistry, machine learning and the science of science. These are not studied because they are inherently 'physical' topics, but because physicists often receive more rigorous training in quantitative methods including mathematical and computational modelling, differential equations and network analysis, and in instruments including lasers, particle detectors and high-resolution imaging tools. These equip them to model and simulate complex systems across domains. The reach of physics is defined not by topics, but by the power and portability of their methods that unite these applications.

Take also economics, a field that conducts studies on nearly all domains across the social sciences—from happiness, crime, racial bias and political polarisation, to education, transportation systems, energy and climate change. Again, it is not because these topics are intrinsically 'economic', but because economists often have more extensive training in statistical and causal inference methods—including regression discontinuity designs, instrumental variables, and randomised experiments. These empower them to quantitatively analyse complex social phenomena. Often far from economic or physical theory and far from formal training in the topics they study, it is statistical and mathematical methods that provide the common thread across these diverse applications. Physics often appears to dominate the natural sciences, much like economics dominates the social sciences. Yet within this method-driven framework here, we can see what actually drives this remarkable dominance in science: exportable methods, not subject matter. Powerful tools and their combinations are often the key elements driving progress, not just theory or subject matter. The frontier of science is commonly the unexplored and empty spaces in this evolving periodic table of methods.

A deeper insight emerges here about *path dependency* in science: the discoveries we have achieved so far were made possible by the tools we have come up with so far. The very trajectory of science is shaped by what we have had at our disposal to measure, see and test. If we had created different—or more—methods up to now, science would likely have developed in different—or more—directions. Some of science's leading methods, from statistical methods to randomised controlled experiments, could have turned out differently from how they did. Placebos, double-blinding, p-values or meta-analyses are not inevitable and the only methodological solutions to the problems they help address.[(234)] Had we invented other techniques first, our understanding of evidence and causation may look different today. Our methods of science are created under particular constraints to solve particular problems. And just as our methods are invented, they have evolved, been merged and vastly expanded—and will turn out differently from how they currently are. Recognising this fact is critical: the future of science hinges not just on new data or new theories, but on our capacity to invent bold new methods. (Yet methods are certainly not arbitrary as they commonly produce consistent, rigorous findings and technologies that improve our lives.) This opens our minds to the undetermined power of the methods we create and the possibility of generating a range of previously unimaginable methods to tackle new real-world problems and challenges we face.

Methods labs and *methods hubs*: why we need to establish global incubators of innovation

Small, unexpected connections between method-curious researchers can prove important. Ernst Ruska, who created the groundbreaking electron microscope (Box 1.1), had interactions with Max von Laue, the inventor of x-ray diffraction (Box 1.2). Kary Mullis designed the extraordinary PCR method while working at the biotech company Cetus whose co-founder was Donald Glaser, the pioneer behind the bubble chamber (particle detector). Glaser was intrigued by the potential of molecular biology that was making exciting new findings, so he made the rare shift from physics to establish this biotech company (Box 1.3). Remarkably, Arne Tiselius developed electrophoresis using the ultracentrifuge that his doctoral supervisor Theodor Svedberg recently designed, who allowed Tiselius to pursue his own independent research. And later, Richard Synge—the co-inventor of partition chromatography—spent a year researching with Tiselius (Box 6.1). These are not footnotes in the history of science—they are catalytic tools that each enabled dozens of discoveries.

These few exceptional inventors did not directly work together, but what unites them is that they took place within the same environment that could inspire focusing on new techniques. These few researchers, with a methodological eye, deviated from conventional research paths—and in doing so, they reshaped the course of science through their new tools. Imagine what would happen if such extraordinary, unexpected interactions were not the rare exception, but systematically cultivated and the norm. There is an enormous untapped potential of triggering innovations in tools: we need to begin intentionally creating dedicated spaces within and beyond universities around the world where method-focused minds from different fields, backgrounds and toolkits can converge to invent, merge, adapt and share new methods—what we call here *methods labs* and *methods hubs*.

Just being embedded in a methods hub—surrounded by researchers designing the next generation of instruments—can inspire synergies and breakthrough innovations, even among researchers working independently. By examining science's major methods and discoveries, we reveal that Cambridge's Cavendish Lab stands out as the best and closest the world has come to a true methods lab. In just a few decades in the early 20th century, it produced a cascade of world-changing inventions. At the Cavendish Lab, Charles Wilson invented the iconic particle detector (cloud chamber) in 1911. Francis Aston constructed the mass spectrograph in 1919. Patrick Blackett developed the Wilson cloud chamber in 1932. John Cockcroft and Ernest Walton built an improved particle accelerator that same year. Yet it does not stop there. Pyotr Kapitsa then devised a method to produce liquid helium at scale in 1934. And Martin Ryle designed an enormous radio telescope system in 1954. These innovations did not emerge by chance, but were engineered in an exceptional method environment. J.J. Thomson, the Nobel-winning physicist and director of the Cavendish Lab, mentored Aston, whose mass spectrograph was built on Thomson's earlier 1913 prototype.[(95)]

Across the Atlantic, Bell Labs later served as another hotbed of innovation. In the span of about four decades, its researchers transformed key parts of science and technology: Shockley, Bardeen and Brattain invented the groundbreaking transistor in 1947. Arthur Schawlow engineered Doppler-free spectroscopy in 1958. Jack Kilby pioneered the microchip in 1959. Willard Boyle and George Smith conceived the CCD sensor in 1970. Steven Chu developed Doppler cooling in 1985. And Arthur Ashkin devised optical tweezers in 1987.

Other smaller examples include Scripps Research in California, where Benjamin List invented an environmentally friendly tool for constructing molecules—organocatalysis—in 2000, and where Barry Sharpless developed click chemistry in 2001. And at Berkeley, Ernest Lawrence built the first particle accelerator in 1929, and Luis Alvarez created the hydrogen bubble chamber in 1959. All of these innovative scientists won a Nobel prize for these breakthrough inventions.[95]

The Cavendish Lab—peaking from the 1910s to the 1950s—and Bell Labs—from the 1940s to the 1980s—were not just two productive labs, they were the two most prolific methods incubators in the history of science. Remarkably, none of these institutions was founded with the goal of advancing scientific methods. Yet they became, for a period in scientific history, the world's leading methods powerhouses—without a systematic, science-wide understanding of how transformative tools speed up progress. These innovations made in the Cavendish Lab—a part of Cambridge's department of physics—and in the Bell Labs were all made by physicists—except for Jack Kilby, an electrical engineer. But the vision of a network of methods labs worldwide is far broader: it is not confined to physics or limited to any specific field, but spans across science and connects cross-disciplinary toolmaking and infrastructure. Today, no such methods powerhouses exist. The Cavendish Lab, Bell Labs and also Uppsala Lab—described in Box 6.1—are the nearest we have gotten but are field-specific exceptions of the past. Yet they made clear that the invention of better tools can be just as transformative, if not more, than the theories they later uncover.

Nearly all labs today are problem-driven or data-driven—but tool-driven labs offer a very different kind of power: they multiply and extend what scientists can do in tackling problems and analysing data. Just as we build research centres around scientific topics—like climate change or cancer—it is time we equally build methods centres: incubators of tool innovation designed to accelerate discovery. The goal of methods labs would be to speed up the creation of cutting-edge tools by fostering a flexible environment of innovation—by merging, adapting and testing methods and tools in new ways to solve today's challenges and anticipate tomorrow's. Think of methods labs as *method accelerators* where researchers can freely explore hybrid approaches, prototypes and experiment with new tools across disciplines more strategically and rapidly. These labs would deviate from traditional, disciplinary structures, where scientists are often constrained by existing methods and norms within their fields. These toolmaking labs would be where breakthroughs often begin—before the experiments and findings can even start. Here we envision how such labs could be organised—places designed not just to answer scientific questions but to invent the tools that open entirely new questions and answers in the first place (see Figure 6.5).

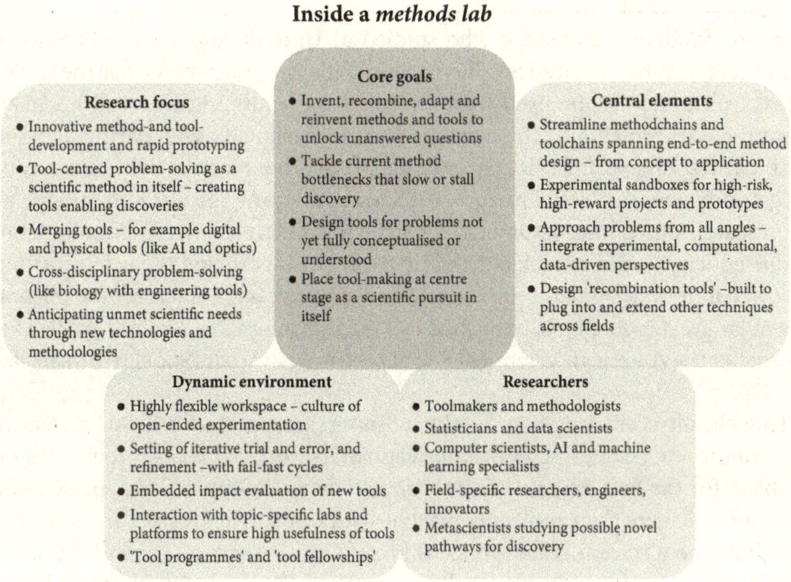

Inside a *methods lab*

Figure 6.5 *Creating methods labs as powerhouses of innovation and discovery*

The concept of methods labs is developed here by synthesising—and expanding on—patterns observed in the methods and tools that have been created to trigger science's over 750 major discoveries.

Ultimately, it is difficult to even imagine trying to do research or make a discovery in genetics without methods like CRISPR or PCR, in archaeology without methods like carbon-14 dating, in neuroscience without tools like MRI, in particle physics without particle accelerators and detectors, in epidemiology, public health and applied economics without statistical modelling and controlled experimental methods. This logic applies across all fields of science. This makes the necessity for methods labs that expand our tools of discovery clear—and long overdue. In short, establishing global networks of methods labs would not just help speed up the pace of discovery—it would redefine how discovery can happen strategically by turning today's method limitations into tomorrow's possibilities of discovery.

Box 6.1 ***A small lab in Sweden enabling many discoveries by developing the ultracentrifuge and electrophoresis—one of the world's first methods labs***

It is no coincidence that two of science's ten great method advances were invented in the same lab—a modest lab at Uppsala University in Sweden. We now zoom in on this small lab that made many groundbreaking discoveries possible—before zooming back out on the broad patterns across science. The story begins with

Theodor Svedberg, a chemist who studied at Uppsala and stayed there for his entire career. What inspired Svedberg to build his new instrument was the recently created ultramicroscope used to study colloids—leading him to call his instrument the ultracentrifuge. The ultramicroscope, designed by Richard Zsigmondy, enabled studying particle sizes in colloids beyond the reach of light microscopes. Zsigmondy won the Nobel prize in 1925 for developing the ultramicroscope and his research on colloids applying it. Yet the ultramicroscope too had limitations and Svedberg set out to tackle them. In 1924, he constructed the ultracentrifuge, a powerful tool designed to investigate and measure such microscopic phenomena in far greater detail. He published this breakthrough in a seminal article in the Journal of the American Chemical Society, earning him the Nobel two years later, in 1926.[143,270]

How the ultracentrifuge works is by spinning samples at over 40,000 revolutions per minute. By generating enormous centrifugal force, this ingenious invention enabled, for the first time, to separate and study for example blood, proteins and cells and calculate their molecular weight with unprecedented accuracy. To visualise how the ultracentrifuge works, think of a milk separator: as milk spins, the heavier components (skimmed milk) are pushed outward and separated by centrifugal force while the lighter components (cream) remain in the centre. This same principle is also how plasma is separated from blood. The potential of this new extraordinary tool to advance biochemistry was quickly recognised. Svedberg then obtained one of few grants available at the time to help further expand his research, receiving 25,000 Swedish crowns in 1924 (about 123,700 US$ in 2025 prices).[270,271]

In Svedberg's lab, researchers did not devote their work to one branch of chemistry or biology, but rather developed methods and tools to separate molecules—a true pioneering lab for method advances far ahead of its time. The lab was extremely progressive, building the tools that would unlock the secrets of life's most complex molecules. It uniquely treated methodology itself as a central research goal.

Just one year after Svedberg crafted his groundbreaking ultracentrifuge, Arne Tiselius—at only 23—began working in Svedberg's lab in 1925 (Picture 6.1). The new ultracentrifuge inspired Tiselius, his doctoral student, who devised a transformative new method to separate and study proteins using electricity: electrophoresis.[272] The principle is deceptively simple: place a charged molecule in a fluid and apply an electric field; the molecule moves. He developed this ingenious and simple device at a low cost and consisted of a U-tube, a thermostat and a microphotometer (Picture 6.2).[273] He published this invention in his 1930 PhD thesis, *The moving boundary method of studying the electrophoresis of proteins.*[142]

These extraordinary tools sparked discoveries from lysosomes using the centrifuge, to DNA sequencing using electrophoresis. This small Swedish lab has had an enormous impact on scientific progress. This method-driven research environment attracted the attention of the Nobelist Richard Synge—the inventor of partition chromatography, another one of science's top ten tools. And Synge spent a research year with Tiselius at Uppsala in 1946 (Box 5.2).[236] The early pioneering methods lab—where tools, not theories took centre stage—has unfortunately remained an exceptional rarity in science. Yet in 1958, another nobel-prize-winning instrument, electron spectroscopy for chemical analysis, was also born at Uppsala by Kai Siegbahn. Though they did not collaborate, he was possibly inspired by the legacies of Svedberg and Tiselius in developing cutting-edge instruments at this institution.[95]

Picture 6.1 *Tiselius in the lab.*

Reproduced from Uppsala University Library via Alvin.

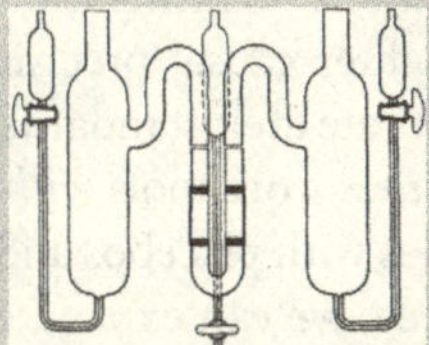

Picture 6.2 *The device Tiselius created for his electrophoretic research.*

Reproduced from Hjertén 1988.[273]

Eight pathways to invent and reinvent our methods and tools—the making of discovery-triggering tools

We now turn to what is the most overlooked force behind scientific progress: method-making and tool-making. With this new *method-to-discovery* principle, we can begin to strategically plan and target ways to accelerate science: rather than waiting for opportunities to arise, we can actively develop new tools and adopt them from other fields. Innovation in methods follows two primary approaches. The most common strategy is to extend our tools, recombine them in novel ways or invent entirely new tools—with capacities never before achieved. The other strategy is to scan other fields for tools to tap into to solve problems in one's own field. Yet because science still does not have the infrastructure set up to systematically foster tool-building (methods labs and hubs around the world) and rapidly spread knowledge about new tools across fields when they emerge, most breakthroughs continue to be made in an ad-hoc way. There is more chance and less design in discoveries than necessary. (Those readers more interested in discoveries than how we design methods can skip this section.)

So what general steps increase our chances of breaking new ground? To answer this, we scan science's over 750 major discovery-making studies to identify patterns in how method innovations behind them came into being. We uncover eight key recurring pathways of how we create new methods and tools. To bring these to life, we illustrate these general pathways using what is often seen as the leading method across the medical, behavioural and economic sciences: randomised controlled trials (RCTs). This powerful experimental method randomly divides individuals into a treatment or control group to isolate and test the causal impact of for example a treatment, drug or policy intervention. To date, more than two million RCTs have been conducted worldwide—on topics from vaccine safety to poverty alleviation policies.[234] So what are the eight pathways we can take to design future breakthroughs?

One, *we can search other scientific domains for tools* that we can leverage and repurpose in our own field. Strikingly, the RCT method, first conceived in medicine in 1948, was only adopted decades later in neighbouring disciplines like economics and psychology.[133,265] When the method was finally imported in the early 2000s, it sparked a methodological revolution—shifting these fields away from looser observational correlations towards tighter causal precision.[135] This method transplant forced entire disciplines to recalibrate their standards of evidence. Two, we can extend our tools by *combining new features* from tools *within the same scientific domain.* In RCTs, we link blinding techniques with placebos and other controls—design features that drastically reduce bias. Three, we can expand our tools by *combining new features* from tools *in other scientific domains.* In RCTs, we can randomise the entire sample of participants before the intervention even begins—a common technique in economics but not yet in medicine although often feasible. Pre-randomising participants ensures more reliable results by balancing background influencers between the trial groups.[234,265]

Four, we can extend tools by *refining a feature.* In RCTs, we can adopt not only common double-blinding (where participants and clinicians are unaware of who the treatment is assigned to), but also employ triple blinding or even *full* blinding. This ensures that all individuals involved in a study are blinded, including the data analysts and administrators. Nobody then knows which group the participants are assigned to as each individual can unconsciously influence and bias results at each level.[234] Five, we can expand tools by *developing an entirely new feature.* Rather than testing just one intervention at a time, RCTs can compare multiple treatments side by side. Imagine a trial comparing increased exercise, improved diet, not smoking and a medical treatment simultaneously. This allows researchers to evaluate not just whether interventions work, but also reveals which combinations are most effective in improving our lives. Six, we can *invent a completely new tool* through major innovations (at times in pathways two to five). Remarkably, the RCT method itself is such an invention: an extraordinary convergence of design features into a powerful, aggregate methodology.[133] Imagine the AI-driven RCTs we can conduct: patients could be randomly assigned for example to either receive a diagnosis from a traditional physician using conventional tools, or from a physician using an AI-assisted diagnostic system—trained to detect likely disease patterns using data from thousands of similar cases. The trial would not just test the treatment, it would test the tools of diagnosis themselves.

Seven, we can critically identify *methodological constraints, assumptions and biases* of our tools—and then design strategies to reduce them. These strategies in RCTs include testing a new treatment against both a placebo and a conventional treatment to improve the relevance of the findings. They involve fully reporting both the treatment's positive and negative effects and clearly assessing how generalisable the findings are—how the results apply beyond the trial context. Eight, we can detect *cognitive, sensory and social constraints* and biases we face—and then devise new techniques to minimise them. In RCTs, we apply experimental controls, randomisation, intention-to-treat analysis, and blinded peer review—each helps tackle deeper human error and bias.[234,274] Ultimately, through continual methodological innovations, the RCT method has become one of science's most remarkable methods for protecting our lives and improving our health—by assessing medications, cancer therapies and countless other interventions.

These strategies reveal a powerful insight: method innovation is not a rare cognitive moment, it is a systematic process. When we treat method-making as a central research goal, we can actively design discovery itself. There are countless opportunities to stretch and upgrade nearly all existing tools through vast combinations of features across disciplines. The sheer number of unexplored method configurations is enormous—especially when we think of the vast range of techniques across the experimental, statistical and computational sciences. Entire domains are unlocked by rethinking how we configure the tools we already possess—and new tools we design. Yet any path we take requires shifting our focus to leveraging tools in new ways that open new avenues of exploration (Figure 6.6).

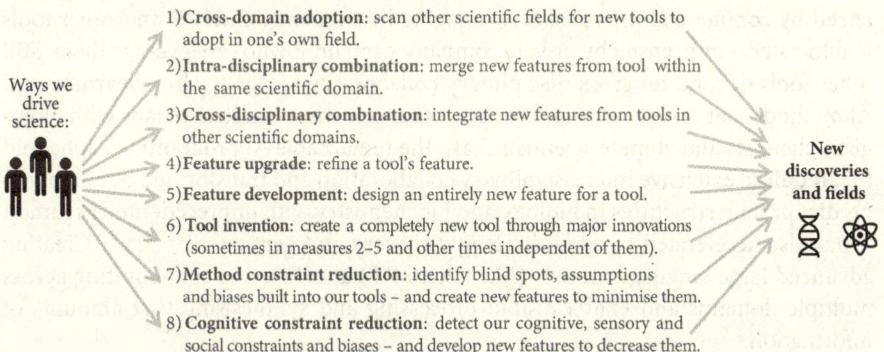

Figure 6.6 *Practical guidelines for how we extend our methods and tools: Eight pathways*
The eight pathways were identified by screening the over 750 major discoveries—and they can overlap in practice. For pathway six—tool creation—we commonly invent a new tool with an objective in mind. But in exceptional cases, tools are built with no direct aim or necessity at first, created as ends in themselves. The maser is such an example but this 'purpose-free' tool soon found wide application in science.

These eight practical strategies are not just a map of how past discoveries have been sparked—they serve as a roadmap for future breakthroughs. Any young student or established researcher can use these strategies to drive innovation in their work. Gaps in our tools do not just stifle progress, we find that they are its core inspiration. Because the limits of our tools largely determine what we can achieve experimentally and theoretically, it is crucial to tackle our tool constraints—and do so more efficiently. There is vast potential here: if researchers begin adopting this mindset, tool-making may emerge as the most powerful meta-method in science and discovery. It would shift how we view scientific innovation: from sudden to systematic inspiration—from intuitive to intentional design.

Nearly all fields can benefit from integrating general-purpose methods—for example new machine learning techniques, advanced statistical methods or big data analysis. These methods can generally be accessed freely by researchers anywhere in the world with a computer and training. Some breakthroughs hinge on coupling cutting-edge computational and machine learning methods with established tools, from microscopes and spectrometers to x-ray scanners. These recombinations hold vast potential to generate new perspectives. Yet this requires computer scientists, data scientists, experimentalists, microscopists and the like at times to collaborate not to analyse findings, but to expand our *scientific toolbox* that generates new findings. Scientists need to think of themselves not just as problem solvers, but as method architects—designers of the very tools that shape their ability to know.

Yet some new tools can face some resistance. Some encounter ethical constraints, others involve much technical complexity and still others are entangled in economic or political hurdles. A striking example is CRISPR, the extraordinary gene-editing method that some scientists are hesitant to adopt for certain purposes—raising ethical fears over possible 'designer babies' and irreversible ecological impacts.[114] Other powerful tools demand greater training that can affect the pace of discoveries. Advanced computational methods and super-resolution microscopy promise greater efficiency, but their complexity can slow their adoption. Tools are transformative when more than a handful of scientists can use them. Some tools can be further influenced by commercial interests. New imaging instruments, CRISPR and other tools tied to patents or licensed by private companies can limit who can leverage them. Still other tools depend on cross-disciplinary collaborations, like machine learning and AI methods that bridge the worlds of computer scientists and statisticians with biologists, chemists and climate scientists. Take the remarkable AI programme AlphaFold that required extensive interdisciplinary collaboration and transformed our ability to predict protein structures in biology and biochemistry with unprecedented accuracy. It has also accelerated research in drug discovery and genomics.[90,95,275] Creating advanced large language models (like ChatGPT) also involves collaborating across multiple domains and enables rapid processing and synthesising vast amounts of information.

And of course, while scientific discoveries and tools have vastly benefitted our lives, some have unintended risks—like contributing to environmental degradation and climate change. Others also raise fears about some unforeseeable consequences of using AI tools in science. These may contribute to automation and job loss—not just in manufacturing, but in research and education—and to some feeling more socially isolated as machines can replace human interaction and even human judgement. To navigate these challenges, regulation and ethical oversight are essential. Governments and scientists must ensure the tools we build serve the public good.[90]

We later unpack the actual constraints holding back the top ten tools of science—and how tackling them can vastly expand the research frontier (Chapter 11). It goes without saying that designing methods—like doing science itself—is often a process of trial and error. Many initial methods, when first developed, may not be useful enough or too limited in scope to survive and spur advances. Just like experimental results and theories, we only see methods that make it into publications.

So, when are new methods most useful and have the biggest impact? It is when they solve a critical methodological bottleneck—visualising or measuring something we could not before. When they significantly improve the efficiency or precision of our current methods, unlocking completely new perspectives, questions and even entire research areas. And when we can apply and scale them across disciplines, catalysing insights at the intersection of fields. New machine learning methods, electron microscopes, advanced statistical methods, etc meet all of these conditions.

Developing the new field of *Methodology of Science*—and the science policies needed to accelerate scientific progress

What if we treated the design of scientific tools across science as a field of research in itself that speeds up discovery? Most scientists do not commonly focus on studying and extending the very methods that enable them to do research. But if researchers and science institutions began devoting much greater attention and resources to tool innovation, we could unlock many important questions at the very core of scientific progress: How can we best incentivise *methods projects*, not just *science projects*? What research programmes and networks can we restructure or create to catalyse tool innovations—especially among early- and mid-career researchers? How can we reorganise scientific research to tackle the hidden bottlenecks within our current tools that hinder progress? How can we best support interdisciplinary research that combines method developers, tool innovators and scientists collaborating together? How can we reallocate part of existing funds to pioneer new tools that push the frontier? Mobilising more funding, training more scientists or building larger teams—without addressing such questions—will not solve our challenges. Without identifying such strategic methodological areas for advancing science, breakthroughs risk being more driven by chance than by design—ultimately slowing the pace of progress.

The central paradox of science and its incentive system emerges here: the very researchers building the tools that define the frontier of research are often overlooked by the institutions meant to advance it. Targeted research on tool-building would vastly accelerate science, so it is surprising that it is not yet valued greatly across science and society—at least not nearly as much as scientific discoveries and theories. It is astonishing that, despite centuries of scientific expansion, no field has emerged that systematically studies tool development from all angles and perspectives across science—building the very methods and tools that make science possible. Students in many fields are required to take a course in research methods or statistics, but these courses typically focus on using existing tools, not on how to innovate them and tackle their limitations.

So why has a field devoted to innovations in tools across science not yet emerged? One explanation lies in a lack of general awareness about how profoundly methods and tools actually shape scientific progress. Another is a lack of incentives for researchers. Science's reward system—journals, universities, major awards and grant agencies—prioritise scientific discoveries. Science textbooks and public media also spotlight these final outputs. In short, science and society celebrate experimental findings and theories over foundational inputs—the method innovations that enable them. Like some areas of basic research, research on tool innovation can be less appealing for scientists. Since the scientific ecosystem rewards those who publish striking results quickly, researchers often have little motivation to invest time in method-making.

Yet because our available tools largely determine what we are able to know and discover, what could be more important than a research field devoted to expanding our scientific tools—our very means of knowing? This makes a field like the *Methodology of Science* essential—one that does not just study knowledge, but invents the ways we can produce new knowledge.

With no field yet existing that aims to advance tools of discovery across science, we propose such a field here and outline what it can look like:

> The *Methodology of Science* is a field dedicated to understanding and designing the scientific methods and tools that enable better ways to discover, measure and explain the world. It studies the foundations, limitations and advancement of our tools: from observational and experimental to statistical and computational tools. The field does not view methods and tools as technical supports, but places them at the centre of scientific progress—because how we investigate determines what we can discover. It identifies the constraints, assumptions and biases facing tools and develops ways to tackle them. The field continually scans across disciplines for new method combinations and maps how new tools shape knowledge production and redefine disciplines. It builds systems for interdisciplinary method-building, offers a platform for rapidly disseminating promising tools across fields and trains researchers how to adapt and invent new tools of discovery. Ultimately, the field offers a framework for how to make method innovations—not just to conduct science, but to build science and discoveries themselves.

The field would represent a new domain of basic research. It would serve as a cross-disciplinary bridge, bringing together toolmakers, methodologists, experimentalists, computational scientists, engineers and field-specific researchers in a culture of innovation. Existing research to improve a method or tool—such as statistical techniques or microscopy—has been domain-specific and fragmented.[137,238] But this field would rise above such disciplinary silos as a kind of meta-method discipline, integrating and recombining entirely new kinds of tools and methods from across the sciences to accelerate progress. The vision is to reframe tools not as background instruments, but as evolving systems of perception, reasoning, exploration, imagination and innovation. And the hope is that developing more sophisticated telescopes, computational methods and x-ray techniques would no longer be seen as less important than discovering new planets or decoding DNA or protein structures—the breakthroughs these very tools make possible. The new methods-driven discovery theory can provide a theoretical foundation for the *Methodology of Science.* For the theory explains how the foundations, limits and advancement of science are largely determined by the foundations, limits and advancement of the tools we have developed so far—that produce the data enabling our theories.

What is the best test to assess a new field? We can ask, how useful is it to address unsolved problems? To open pathways and solutions to problems not yet formulated? To reveal systemic blind spots to progress? And to advance new discoveries and entire lines of research? On these measures, a new science of methodology has great potential: because the greatest accelerations in scientific progress throughout history have followed major method breakthroughs (Chapters 1–5) and because of the range of constraints and untapped opportunities of today's leading tools (Chapters 6 and 11).

So what would it take to establish the field of Methodology of Science and accelerate the pace of discovery? We propose seven key reforms: developing a broad method community, embracing tool experimentation, rethinking our conception of science, introducing new incentives for tool creation, building dedicated method institutions, recalibrating citation practices, and training the next generation of tool developers.

1. *We need a methodological shift in the scientific community towards a much larger share of researchers dedicated to tackling method bottlenecks and making tool innovations.*

About nine million scientists work across science worldwide.[(167)] Yet if we ask scientists to name researchers dedicated full-time to developing or improving methods and tools, most would not be able to—or, in some fields, not even know they exist. Such researchers are rare and largely not visible. While tools power every experiment, model and dataset, few researchers are explicitly devoted to building them. Our ability to investigate the world is generally not constrained by the lack of funding or collaboration—for most discoveries; but by the scarcity of tool-builders. To accelerate progress, we urgently need a methodological shift: a structural change that moves many more researchers towards solving barriers facing our tools. What obstacles prevent us from observing the world more precisely? What measurement systems and instruments do we need to redesign to expand our scope to the world? How can we strategically foster the diverse methodological pathways—described in the previous section—to spur new research areas? Answering these questions systematically across science requires more than ad-hoc methodological curiosity—it requires a growing community of method innovators and tool architects who treat method-building as a core scientific endeavour.

Imagine if just 1–3% of the global scientific workforce, a modest fraction, was devoted only to tool development, as proposed here—dedicated to the *Methodology of Science*. This shift alone would give rise to 90,000 to 270,000 full-time researchers building the tools of discovery. This is a small share but the impact could be enormous, since every breakthrough fundamentally relies on the strength of our tools. Such a shift would catalyse a generation of tool architects who reimagine not just what we study, but how we study. Some can for example develop new AI systems that integrate with any scientific data, helping researchers generate experimental designs and make better predictions. Others can invent hybrid tools that blend simulation, visualisation and real-time analytics, enabling advances in fields from neuroscience to climate science. By embedding toolmaking more deeply into the scientific workforce, we can build a faster, more self-correcting and more adaptive research system. Hidden bottlenecks would be flagged and outdated methods would be upgraded, allowing us for the first time to more strategically redirect slower or stagnating fields.

2. *We need to make testing and expanding our best methods a regular part of what we all do—often as an end in itself.*

Making our most powerful methods in statistics, spectrography and chromatography even more powerful, making our best telescopes and particle accelerators even better has consistently paved the way for new discoveries. Many big advances in science are not guided by a specific hypothesis that led to the specific discovery, but are sparked by experimentally developing a new tool applied to do exploratory research and then seeing something entirely new (Chapter 2). Just as we run experiments in science, we need to run instrumental and method experiments—we need to design research not to generate findings or test a theory, but to test a tool. What if labs began running *method trials*—like clinical trials, but for scientific tools—where a control group uses a conventional tool, and experimental groups test new tools to compare which are the most effective in achieving advances? It is essential that we treat tool developments as a primary scientific goal even before we know what discoveries they could yield—while of course always taking our planet and human wellbeing into account. Just like with experimental research, tool innovation requires us to try new things and be open to inevitable failures, as learning opportunities that refine approaches.

Once the scientific community becomes fully aware of the fact that new tools trigger new breakthroughs, we can begin to target research programmes and exploratory experimentation to catalyse tool invention across science. This shift would reframe how we understand discovery itself: from its traditional focus on just experimental and theoretical triumphs to equally focus on the tool discoveries that make them possible. In sum: method innovation needs to be a core pillar of science.

3. *We need to reconceive science and science policy around methods-driven research that is exploratory—not just question-driven research.*

In science, articles and grant proposals are structured around a research question: 'what causes x?' or 'how does y work?' This structure reinforces the idea that all science should start with a hypothesis or puzzle that needs solving. But many of science's transformative discoveries have begun not with a question but with new tools through exploratory research (Chapter 2). We need to revise the traditional view that science must always be question- and hypothesis-driven, it must also be methods-driven—open to the unknown and sometimes questionless. We can for example opportunistically exploit new methods created in one's own and other fields. What happens when we apply new imaging methods to biological or ecological systems to observe what they show? Or when we test a new statistical technique or new machine learning method on climate or psychological data with no hypothesis in mind? Tool-driven research has enabled spotting patterns, anomalies and structures—and even formulating many of our biggest questions in science—that we did not even know to explore. It comes with uncertainty and less clear outcomes, but the payoff for method opportunists is often discovering what no one else has thought to look for.

Science agencies need to equally support exploratory research projects—not just traditional projects that already have consensus among established research communities. A methods-first mindset, where researchers invent, adapt and apply methods to new domains and shifting challenges, can be just as powerful as traditional hypothesis testing.

4. *We need to fix science's reward system so journals, prizes and funding bodies equally incentivise method experimentation and innovations.*

When asking scientists what brings recognition in their field, the answer is generally straightforward: the main incentive in science is to chase immediate scientific discoveries. This focus is deeply embedded across science's institutions: journals and science prizes reward groundbreaking findings, universities hire and promote based on breakthroughs, and funding bodies support the greatest prospect of discovery.[3,4] But what if we have been overly incentivising the final part of the discovery process? Here we provide an alternative logic: if we reform science's institutions to value not only results, but equally developing new tools and methods that make those results possible, we can accelerate discoveries. Rather than largely only prioritising the finish line, we must prioritise tool-building: the roads to get to the finish line and often the key part of the finish line itself. Foundational research in methods, at first glance, may seem to even distract our attention—deter from publishing experimental and theoretical studies: the central measure in science.[3,4] But it does not delay gratification, it lays the groundwork for getting there faster and is often a multiplier: opening new paths. It can often be thought of as a difference between harvesting a tree and planting an orchard—one focuses more on today, the other more on the broader future.

Today, the greatest progress in science commonly comes not from chasing prestige, but from bold new ways to explore and measure—like highly advanced microscopes, machine learning methods and the CRISPR method. These enormous breakthroughs are made possible by method-makers: researchers who often step back from the race for headlines to redesign the track. Our current research system overwhelmingly invests in *science projects*—experimental and theoretical research—as we see looking at projects supported by science's largest funding agencies like the European Commission and the National Science Foundation. The current system does not yet systematically invest in *method projects.* Yet we need to begin reallocating a large share of current funding to *method projects, tool programmes and tool fellowships* and also give priority to science projects that include method innovation. Funding currently disproportionately bets on established researchers and elite institutions. But we need to disproportionally redistribute resources to bold original tool innovations to unlock discoveries faster, independent of prestige and institution. For our current incentive structure is at odds with advancing science more rapidly. It is time we recognise the inventor of a breakthrough telescope—the *tool of discovery*—not only the researcher who discovers the planet using it. And the taxonomy of scientific methods—spotlighted in Table 6.2—highlights the need to prioritise the most scalable, transferable tools.

Just as in basic research, researchers dedicated to the foundational research of designing tools deserve equal recognition within the scientific community as those generating scientific studies and theories made by leveraging them. To speed up progress, we must shift science's priorities: experimenting with new tools and method combinations needs to be seen on par with the experimental findings they produce. Method-makers need to equally be able to publish method advances in top-tier journals, win prestigious grants and prizes and secure tenure-track positions. We need professorships in methodology, in tool innovation, in integrated machine learning. The hope here is that method-inclined researchers receive the same opportunities, resources and positions as those carrying out experimental and theoretical research using their powerful innovations. These measures are crucial to incentivise young researchers to enter this basic research field and explore high-risk, high-reward tool-building projects—projects that do not fit into conventional disciplinary silos but can reshape their disciplines.

The way we publish research does not help: many leading journals like *Nature* and *Science* follow an 'introduction–main results–discussion' format, while relegating the methods used to the appendix or a separate 'supplementary material' file. This sends a signal: they view methods as subordinate or supplementary. Journals need to correct this bias. They must also vastly increase featuring high-impact tool advances. They should begin requiring a methods section with much greater detail in all studies, outlining all techniques, protocols, study context and method limitations throughout the research process. This would empower other researchers to improve and build on those methods, understand and reduce method constraints and biases and push the field forward faster. It would enhance method awareness.

So this can help us rethink questions about science: should we prioritise funding discoveries, or toolmaking that sparks them? Should we chase new findings, or expand the foundations that enable them? And is it better to fund research or development—to foster new knowledge or expand existing knowledge? For each question, the answer—of course—is both, but we are currently biased much too far towards the former. Supporting method-makers directly gives science its infrastructure of discovery. It empowers researchers to pursue opportunities as they emerge, to shift and re-shift their focus as new problems arise and to take risks that conventional structures do not yet reward.

5. *We need to develop new Methods Labs, Methods Hubs and Departments of Methodology of Science—as incubators of innovation.*

Just as scientists work in science labs, we need to establish *methods labs and hubs*—spaces designed to invent new methods and tools of discovery. These labs would provide, for the first time, a systematic infrastructure for tool innovation across fields—and house the growing share of method-makers among the scientific community. These labs can connect tool and method innovators, statisticians, experimentalists, AI researchers and domain-specific researchers to work side by side on breakthrough instruments. These labs would streamline *methodchains* and *toolchains* that span the full method design—from the early concept design to real-world applications (Figure 6.5). No such systematic methods powerhouses exist today. But the Cavendish Lab and Bell Labs—the two most productive methods incubators in history—make clear the enormous effect of such labs on driving progress. And unlike those physics-centred labs of the past, the future of a global network of methods labs lies not in one field but spans across science and bridges diverse toolmakers across disciplines. It unites synthetic biologists with data scientists, neuroscientists with roboticists, ecologists with machine learning experts and statisticians.

We could measure the success of the Methodology of Science by establishing methods labs, hubs, institutes, centres and departments that train early-career scientists to become toolmakers. It would include creating a society, journals and conferences across fields focused on improving how (not what) we discover. Communities of method-makers could then pool resources through open-ended experimentation with innovative and adaptive tools that tackle our methodological barriers to progress. In a landscape where science grows increasingly fragmented into narrower subfields, methods labs could reconnect disciplines by giving them a shared language of innovation. This is important because current trends of dividing science into ever more narrow units of knowledge can slow different paths of discovery (Chapter 4). But methods labs worldwide would not only build bridges for using new general-purpose and field-specific methods across disconnected fields. They would also export groundbreaking method innovations—through shared method and tool repositories that open access worldwide. The greatest accelerator of science may not be any one discovery, but a strategic shift: where institutions and scientists begin to treat toolmaking itself as a key priority. This is not just a reform—it is a redesign of how science builds its future more efficiently.

What could Departments of Methodology of Science look like? Imagine a department where the logic, mechanics and innovation of science itself are studied, taught and advanced. It is an academic space of scientific self-reflection and invention. Researchers here can explore questions like: what features make a research method more likely to generalise across fields? How can we design early-warning systems to detect flawed or outdated methods in a field? What constraints are introduced by statistical, computational and experimental methods? What are the best ways to measure a method's impact—on clinical trials, climate models or public policy? How can AI and automation be best coupled with research methods? Can we map the genealogy of science's major research methods—in a periodic table of methods—to better identify methodological blind spots and missing combinations quicker? Such departments can also collaborate with all fields to detect and design new method and tool combinations. They can also work specifically with medical schools to co-develop trial designs, with engineering departments to prototype sensors or robotic tools, with social science institutes to rethink survey and causal inference techniques and with AI labs to explore automated tools.

Courses in Departments of Methodology of Science could also range widely. A course on *Tool Innovation: Designing, Building and Testing* can cover how to devise, prototype and evaluate method innovations—from experimental designs and new types of sensors, to AI systems, measurement instruments and software platforms. A course on *The Architecture of Scientific Discovery* can introduce how methods emerge, spread and evolve, and what the hidden barriers are behind measurement, modelling and experimentation. A course on *Method Transfer and Translation Across Fields* can explore how methods from one discipline (like advanced statistical and machine learning methods) can be adapted for another (like ecology or epidemiology). A course on *Bias, Inference and Error: A Statistical Toolkit* can cover statistical reasoning, alternative inference frameworks (like Bayesian methods and causal inference) and teach how to detect and correct methodological errors. A course on *Meta-Science and the Practice of Science* can explore how the scientific system works and cover topics like incentives, research, publication systems, funding systems, replication crisis and how to reform scientific institutions. These courses would prepare a new generation of tool builders—those fluent in the scaffolding of tools and discovery themselves. Creating such departments would make the scientific enterprise more self-correcting and more innovative.

6. *We need to shift our main measure of scientific success from just outputs (especially article citations) to focus equally on inputs—method innovations.*

In today's scientific landscape, getting hired, promoted or funded is overwhelmingly tied to how many citations and publications scientists accumulate.[14] This metric is easy to measure but overlooks something crucial: scientists do not yet cite the full range of methods, tools, techniques and protocols that underpin their work. Instead, they primarily just cite other research papers—not the methodological scaffolding. This creates a distorted picture of what matters in science and discourages developing new instruments and techniques. A rethinking is needed away from using common ex-post indicators like citation counts [3,14,38,40] towards ex-ante indicators—the process of science—that recognise tool innovations.

Today, many view articles as high-impact if they, depending on the field, receive one or a few hundred citations. But what if we begin to systematically reference the methods and tools applied in each study? The foundational role and impact of methods would become powerfully clear. The most important method-making articles in history, such as Ruska's article on the invention of the electron microscope,[112] Tiselius' electrophoresis method,[142] Martin and Synge's partition chromatography method[237] or Bloch and Purcell's NMR spectroscopy[240] have only received a few hundred to a few thousand citations. But each is mentioned in millions of publications on Google Scholar, making them the real giants in science. If citation practices accounted well for tools, then inferential statistics, x-ray crystallography, computational methods, randomised controlled experimentation and lasers for example would each have received millions of citations. For they are mentioned millions of times on Google Scholar but are not properly cited. This far surpasses the most-cited scientific studies of all time that rarely give (citation) credit to the discovery of the tools they use. The paradox is clear: our best scientific tools are so foundational to science that they are largely taken as given, without the need to even cite them across most—but not all—fields.

By relying on citations as our primary measure of success, we foster an output-oriented, retrospective view of knowledge. Many new breakthroughs would go unseen when just focusing on citations, and new discoveries are important precisely because they have never been seen and cited before.[4,178] Citations not only just capture past success—in a crude way—but, more importantly, cannot explain why or how breakthroughs emerge.[38,40] For that, we need to shift our attention to methodological and scientific innovations themselves. When our incentive system recognises the indispensable power of methods in driving discovery, researchers will also recognise the enormous value of making innovations in tools.

7. *We need to train the next generation of tool-innovators—and integrate deep methodological training into university degrees across science.*

In fields from biology and medicine to earth sciences, graduate students are typically required to take at most only one course in research methods or statistics during their degree. And yet, these are the very tools they later rely on to do and publish their research. It is not difficult to see how this gap can constrain research. With the purpose of graduate degrees to train students how to do science, we need to place much greater focus not only on training how to apply but also improve methods. Limited method training—both in breadth (the range of tools) and depth (their limitations, assumptions and biases)—narrows a scientist's ability to recognise the flaws facing the tools used to develop our evidence and theories and to fix them. Yet this ability is critical in sparking discoveries (Chapter 1).

Without first learning how to detect our methodological blind spots, we cannot tackle them—with the pathways to do so laid out earlier. Most education systems instead focus on the *outputs* of science—facts, theories and discoveries—rather than the *process* in which we generate them. We need to reform this imbalance: shifting much greater weight on studying the complexities of how we implement and fine-tune our tools to gain that knowledge. Scientific training must be recentred around the *how* of discovery, not just the *what*. This means immersing students in the complexities of measurement, experimental design, inference, calibration and innovation—not as side topics, but as the core of scientific literacy. A critical barrier here is hyper-specialisation in education, a barrier that leads to long time lags in developing and adopting tools, especially across fields (Table 1.1). This leads to path dependency, a kind of method inertia: a field-wide anchoring effect on the specific methods that researchers already use, while powerful new techniques from other fields often go unnoticed (Chapter 1). Path dependency, disciplinary silos and a lack of cross-pollination stifle innovation. To break this inertia, education systems need to systematically expose scientists to the broader universe of tools and techniques. Just as modern biology has been transformed by techniques from computer science and physics, many of tomorrow's breakthroughs will come from such cross-disciplinary fertilisation—from researchers fluent in multiple toolkits.

To make this shift sustainable, we must go one step further: we need to institutionalise the training of method-makers. To establish the field of *Methodology of Science*, tool and method innovators would need to be formally trained, just as chemists and psychologists are. We cannot assume that some researchers, who happen to develop an inclination for method and tool questions, will find the time to tackle them. Without structured support and training, many potential innovators will never get the chance to build the tools science needs. Historically, only a few exceptional scientists have strayed away from their original academic training as the top ten method outsiders did—science's greatest architects of discovery (Table 6.1). But our most transformative innovations in science should no longer depend on chance. They must be systematically built into the system. Establishing dedicated training in this field will empower early-career researchers with the necessary methodological skills to push the boundaries of what we can investigate in the world. And method workshops and training will help equip established researchers across fields with these skills. Only when tool innovation is valued and taught can science truly accelerate its capacity to discover the unknown. Imagine a generation of scientists trained not just to apply existing tools, but to question and reinvent them. This would be nothing short of a shift in how knowledge is created.

Conclusion

If we were to adopt these seven reforms, we could trigger a *method revolution* in science—a deep transformation in how science progresses and at what pace. Implementing these reforms would break from the status quo: the era *before the method revolution*, where we until now commonly made advances in methods and tools in a surprisingly ad-hoc and unplanned way with large time lags between new methods and discoveries. It would spark a shift towards *the method revolution*, where we would design and develop methods and tools in a planned, strategic and targeted way—with minimal time lags. Historically, the development of new tools and the development of new scientific discoveries have been riddled with long, unnecessary delays—simply because we have not yet systematically coordinated and fast-tracked tool development across science. Advances in tools would then move from the background to the foreground in understanding how we drive scientific advances. We can then begin actively engineering the conditions for making breakthroughs. *When we dig deeper, we find a key pattern among science's major discoveries: many did not begin with chasing answers, but with building better ways to ask questions that sparked those answers.*

Yet today, science still lacks a general understanding of how we create new tools and research agendas for tool development—a general theory of the main force for its own progress. Without it, researchers until now have had to experiment and work out on their own—researcher by researcher—how to trigger a breakthrough and which new tools we need. Much time is lost, researchers continue in stagnant fields and breakthroughs go unrealised, just because the current system has not yet had a roadmap for building better tools. *Relying on familiar methods simply because they are familiar is the reason why much research does not advance science.*

Future scientists may view the advent of this revolution as the moment at which the entire scientific community became self-aware about how it progresses. It is the point at which we began to deliberately refine, combine, restructure and invent methods and instruments that accelerate breakthroughs. *The beginning of the revolution would be the moment when we collectively realise the fact that methods and tools themselves are perhaps the most powerful frontier we have yet fully explored.* No more would our tools be studied by a small group of methodologically interested researchers who diverged from their initial training in established fields. The tool revolution in science would be driven by a new understanding across scientists: tools are not neutral, and advancing our tools changes the very kinds of questions we can ask and answers we can imagine. As the periodic table transformed chemistry by mapping its fundamental building blocks, science now needs an evolving periodic table of methods: a living map of the methods, tools and their vast recombinations that power discoveries across domains.

The proposed field of *Methodology of Science* has the potential to transform science by speeding the pace at which we can produce new breakthrough research. But realising this transformation depends on a bold reorientation that challenges the status quo—the current scientific system that deeply prioritises scientific outputs, not inputs. These large reforms may face resistance especially from those established institutions benefitting most from the traditional, output-oriented system. Academic inertia can be a challenge to large-scale reform. Once the first institutions, journals and agencies begin to prioritise innovations in tools, others will more easily follow suit. Once methods labs and hubs begin to take root and spread, a new scientific culture can emerge.

Ultimately, the pace of scientific progress is dictated by the pace of method progress. Science has traditionally taken an indirect route to discovery: researchers have explored scientific questions that we often did not quite yet have the methods and technologies to answer. But imagine what would happen when we flip the research direction: when researchers pursue method questions first, designing and selecting new tools not just to answer today's questions, but to unlock tomorrow's. This is frontier science and would be a direct route to many discoveries. Our tools should no longer be seen as just means to do research, but as active sources of innovation and inspiration that continually open new questions. We sketched the pathways we can take to drive innovations in tools. But this opens new questions: how can we begin to better assess the impact of new methods and discoveries on society, policy and people's lives? How can we better evaluate their benefits on new medical treatments, technologies and environmental solutions?

Until we place our powerful toolbox at the centre of science, we will continue to think of scientific discoveries as the most important feature of science—and not also the incredible tools that make them possible and largely determine their scope. We will keep viewing scientific discoveries as just being made by brilliant discoverers—and not also the brilliant tool inventors who enable triggering them. Our great toolmakers are often the unsung discoverers and can no longer be pushed into the footnotes of history. Establishing the field of *Methodology of Science* would mark a vast leap forward for science. It would signal that we finally recognise methods, tools and their combinations across fields as a frontier in their own right—deserving the same priority and intellectual rigour as any other scientific domain. After all, with what we know about the world so far largely driven by the tools we have developed so far, the *Methodology of Science* is—by expanding our very tools of discovery—as important as any other scientific field and long overdue.

PART II
THE ORIGINS AND FOUNDATIONS OF SCIENCE AND DISCOVERY

In Part I of the book, we asked the big questions about how we drive new discoveries, scientific fields and science in general: the answer lies in developing powerful new tools. Our ever-expanding *scientific toolbox* is what triggers our transformative breakthroughs, can explain much serendipity in science, makes scientific progress highly cumulative and fundamentally defines how we do science. In a logical circle, we then asked the key question: how do we upgrade our powerful toolbox? In answering that, we uncovered the dynamics of toolmaking and how it powers science.

By tackling these foundational questions of scientific progress, new fascinating questions arise: where did science itself and our ability to develop tools come from in the first place? What are their early roots? And looking forward, what are our current boundaries of science? Does extending our toolbox play as powerful a role in explaining science's deeper past and its future as it does the present? Understanding how we became able to create knowledge and methods and start science helps us understand how we arrived at the present and can vastly advance science in the future. Tracing the broad history of science and discovery helps us understand their evolution, providing key insights into how we can build a stronger scientific system for tomorrow. In Part II of the book, we now turn to the deeper origins of science—where it began and why it emerged. In Part III, we then explore the current limits of science and how we can push these limits.

7

The origins of our toolbox

How our mind's method-making abilities have driven science—and civilisation

Summary

Is the engine of science today the same as it was in our past? What are the origins of science and our powerful methods? We find that the key turning point in scientific history came with the unexpected invention of tools that enhance human perception: the first microscope and telescope. They opened unimaginable new worlds and sparked most major discoveries of the 17th century. These two optical breakthroughs put science on a no-return path of rapid discovery, triggered wider curiosity and kick-started modern observational and experimental science. *Before* their invention, science was limited by ancient tools and human senses—but their arrival broke those barriers, giving us direct access to microscopic and cosmic realms we never imagined. These extraordinary instruments did not emerge from scientific institutions, a large scientific community or formal scientific method. Instead, they inspired them—*after* observing the extraordinary discoveries these tools made possible. But to understand how we got there, we go further back—to the origins of scientific thinking. This leads us to big and challenging questions: how did we develop science—and even civilisation—in the first place? We cannot study what triggered our advances throughout human history as systematically as we can study contemporary discoveries using statistics. But we can trace what has driven dozens of the most transformational early breakthroughs. While we focus on science throughout the book, we will uncover how its origins are inseparable from the origins of technology and civilisation. To grasp this, we broaden our lens here. We explain how our mind's evolved methodological abilities—to observe, solve problems and experiment—and develop ever more complex methods and tools, from mathematics to systematic experimentation, have directly enabled us to create vast knowledge. They have allowed us to make vast innovations, from plant-based medicine to pyramids. Throughout history, we have actively driven the major advances—the agricultural revolution and later scientific, industrial and digital revolutions—by leveraging these abilities in new ways and inventing better methods and tools: through *method revolutions*. Better understanding the origins of science, we can better understand their limits and how to push them: our evolved methodological capacities and the tools we develop are both the foundations and bottlenecks of progress.

The Engine of Scientific Discovery. Alexander Krauss, Oxford University Press. © Oxford University Press (2026).
DOI: 10.1093/oso/9780197829790.003.0008

Overview

How did we develop science, technology and civilisation? These are some of the most foundational and oldest questions of science. Scientists attempt to address these questions mainly by taking one of three ways. One way is studying the evolution of human intelligence. Evolutionary biologist Kevin Laland and anthropologist Joseph Henrich highlight how our cognitive capacity for culture has been the central driver of our progress.[276,277] Developmental psychologist Michael Tomasello also stresses the power of social cooperation.[278] A second way scientists approach this puzzle is by exploring geography. Physiologist and geographer Jared Diamond points to favourable ecological conditions—climate, agriculture and domesticable animals—that gave some societies a head start in making these achievements.[279,280] A third way is examining a particular point in history. Common explanations for the birth of science often revolve around 17th-century Europe and include the expansion of capitalism and economic wellbeing,[281] Christian values[282,283] and the printing press[284]—catalysing the work of scholars like Galileo and Newton.[285]

These three main approaches focus on *what broad factor* can explain science and civilisation—two developments that are deeply interconnected. While these broad factors played an important role, they are often beyond our *direct* control. And they do not explain how individual breakthroughs emerge. What is missing is explaining the overall enabling conditions needed for science, technology and civilisation and especially the set of human capabilities that allowed us to actively shape and advance them. So in this chapter, we take a deeper step back in the history of science. To understand the present—and how we arrived here—we must understand the past. This principle applies across most of science, but especially when examining science itself: its cognitive foundations, methodological evolution and our mind's persistent constraints that shape discovery today.

Yet this foundational question of the origins of science and civilisation represents an enormously complex and interdisciplinary challenge. Only with an integrated, big-picture approach can we see how the different pieces fit together—and identify what underpins them. Only with such an approach can we tackle the deeper puzzle: how are we even capable of reasoning about the world and developing the methods and knowledge needed to create these tremendous advancements? How did our mind's method-making abilities evolve over time? What eventually made the work of scholars like Aristotle, Newton, Darwin and Einstein possible in the first place? And did our early ancestors—stretching across our evolutionary history—rely on the same reasoning capacities to gain knowledge as we do today? To tackle these questions, we draw on evidence from evolutionary biology, cognitive science, methodology and anthropology to reconstruct how our ancestors began to reason about the world.

In the first six chapters of the book, we examined *scientific methods and tools* systematically applied in research, such as computational methods, x-ray devices and electron microscopes. In this and the next chapter, we broaden the scope to study *more basic, general methods and tools*—those used far beyond the narrow scientific context covered thus far. This wider view helps us see how our early ancestors observed, experimented and solved problems to create practical techniques and early basic tools: developing techniques for plant-based medicines to reduce illness, experimenting with agricultural techniques to grow more food, and inventing systems of writing and numeracy to extend our limited memory. We accumulated experimental knowledge about nutrition, botany, natural pharmaceutics and ecology. These advances were not the result of chance—they reflected strategic cognitive adaptations that allowed our ancestors to iterate, refine and systematise their observations. Over time, we learned how to engineer boats, bridges and later pyramids—advances that require methodically measuring, experimenting and developing geometry and principles of engineering. And we eventually created complex systems of mathematics and astronomy.

These early methods helped us better understand and control our environment and biology. They were later refined and scaled up by contemporary scientists using these same evolved method-making abilities, forming the foundation of scientific thinking over history. Science today is defined by formal methods, advanced instruments and controlled experimentation—tools that enable us to observe, measure and test in much more systematic and far-reaching ways.

Taking this big-picture, interdisciplinary perspective, we begin by laying out the set of core requirements needed to create science and civilisation. At the centre of this explanation lies a novel insight: our mind's evolved methodological abilities (to observe, solve problems and experiment) that we have developed into ever more complex methods (like systematic experimentation and geometry) have been the key driving force that directly enables us to create scientific and technological knowledge—and eventually science and civilisation. While many think of science and civilisation as separate developments, they actually share six enabling conditions in common. Civilisation lays the infrastructure—stability, systems of mathematics, division of labour—that enables developing science. The two cannot be separated: science has only ever emerged within civilisation—within a highly complex and organised culture.[(286)] We cannot understand the origins of science without understanding the origins of civilisation.

Here we lay out a new explanation—a methods-powered view—of how we created science and civilisation: by leveraging our evolved methodological abilities of the mind (our *universal* toolbox) to collectively develop more sophisticated methods and tools (our *adaptive* toolbox) that have enabled us to make sense of the world in increasingly systematic and powerful ways.

In this chapter, we dig deeper into our toolbox, exploring these universal and adaptive elements. So how well do existing explanations of the origins of science and civilisation actually fare? Can this holistic and methods-focused view here provide a clearer picture? And how have these capacities evolved and been used in ever more complex ways—from smart crows using sticks as tools, to early humans experimenting with fire, to Mesopotamian engineers designing irrigation systems, to van Leeuwenhoek discovering bacteria with the powerful new microscope that, with the telescope, drastically extended our perception for the first time in history? We sketch out our scientific origins, tracing the growing complexity of our methods—from our evolutionary niche, to the emergence of modern science with the optical revolution and then contemporary science driven by cutting-edge instruments. In short, we map out our method evolution over time. A striking insight emerges: our minds evolved not to detect distant galaxies or edit genomes or ever imagine them without tools—but to survive. Our tools are precisely what enable detecting and editing the world by stretching our senses and expanding our mind in unimaginable ways. By better understanding how we developed our toolbox over time, we are better able to systematically understand the complexities of where it breaks down, and how to upgrade it and add onto it today.

Rethinking the origins of science and civilisation, beyond culture and geography: the missing role of methods

We developed a world of hand axes, lunar calendars and agriculture—and eventually microscopes, randomised controlled experiments and quantum computers. How did we get from hand axes to quantum computers? Can we explain these remarkable developments through favourable geography[279,280] or broad cultural forces like norms and institutions?[276–278] Leading scientists have proposed these forces—geographic luck and cultural evolution—as the central driver of our progress (Box 7.1).

Box 7.1 Leading explanations on the origins of science and civilisation

Why did science and civilisation take root in some places and not others? Jared Diamond, professor of geography at the University of California, answers the question by pointing to favourable geographic conditions. He summarises his core thesis as: 'History followed different courses for different peoples because of differences among peoples' environments.' He argues that factors related to 'the rise and spread of food production is [...] what I believe to be the most important constellation of ultimate causes.'[279] In short: for Diamond, fertile land, domesticable animals and the right climate set the stage for everything from early cities to science. Kevin Laland, professor of behavioural and evolutionary biology at the University of St Andrews, argues instead that: 'our species' extraordinary accomplishments can be attributed to our uniquely potent capability for culture.'[276] For Laland, what makes our species extraordinary and explains our great advances are high levels of social learning: copying each other's behaviour.

Michael Tomasello, professor of psychology and neuroscience at Duke University, also highlights the role of culture, focusing on cooperation. His principal thesis is strikingly simple: 'virtually all of humans' most remarkable achievements—from steam engines to higher mathematics—are based on the unique ways in which individuals are able to coordinate with one another cooperatively.' For Tomasello, our species' 'most distinctive characteristic is its high degree (and new forms) of cooperation' that reflects our shared intentionality and enables our complex language.[278] Joseph Henrich, professor of human evolutionary biology at Harvard, goes further, making his central argument clear in the title of his book,

The secret of our success: How culture is driving human evolution, domesticating our species, and making us smarter.[277] He argues that 'The key to understanding how humans evolved and why we are so different from other animals is to recognize that we are a *cultural species*'. Ultimately, these scientists argue that if the right ecological conditions emerge—like high-productivity farming and food production [279,280]—or if humans develop cultural traits like cooperation, shared intentionality, language and abilities for learning,[276–278] then science and civilisation can arise.

What is clear is that culture and geography are important pieces of the puzzle. They act as basic building blocks, setting the broad stage for human progress. But they do not explain how humans—once fertile soil and social cooperation are in place—actually started building tools and methods to understand and transform the world. They do not tell us *how* we advance—which methodological steps we have taken to develop our scientific and technological knowledge. Existing researchers, because they focus on the broad backdrop, leave open the key question about the mechanism: what cognitive and methodological forces turned those broad ingredients into discovery? How did we actually begin designing the first boats, pyramids and microscopes—long before we even understood buoyancy, gravity or optics? Could it be the result of a kind of reasoning and testing that shares many similarities with science today?

Michael Tomasello aims to identify the differences 'that ultimately lead to humans' unique forms of cultural coordination and transmission (and so to telescopes and parliaments)'.[276–278] But there are vast worlds between cultural coordination or geography on one hand, and telescopes on the other. The former is a social scaffold, the latter is a technical innovation. The two operate at fundamentally different levels. So what bridges the divide? Here we provide a new perspective: shifting to our mind's methodological abilities to observe, solve problems and experiment and create better methods like mathematics. These abilities—both ancient and modern—*directly* connect these extraordinary innovations between our early ancestors who experimented with more productive farming techniques, and today's scientists and engineers who experiment with genetically modified crops. This is the missing methodological link: these evolved capacities turn more systematic observation and trial and error into refined methods and tools that create life-saving knowledge and innovations. Our species has always relied on these same methodological abilities to be able to unravel mysteries about the world and invent technologies.

This integrated toolbox perspective offers a unified explanation that connects our early evolutionary origins with the methods we use to reason, do science and understand the complexities of the world—an explanation overlooked until now. [276–279,282,287] So to explain the rise of science and civilisation we should not just focus on factors seen as the key to human success and uniqueness—like language, collective intentionality[278] or agriculture.[279,280] We must focus on this evolving set of methodological abilities and techniques we have developed over history. These are more directly observable and measurable in the tools, artefacts and knowledge we created than those broad factors.

Our early ancestors sparked major early breakthroughs through *method revolutions*

Our early ancestors did not just adapt to diverse environments, but transformed them. We actively drove the palaeolithic technological revolution and—once favourable geographic conditions emerged—the agricultural revolution. We then—once favourable demographic conditions were in place—actively brought about the civilisation revolution. How were we *directly* able to spur each of these monumental shifts and then the scientific, industrial and digital revolutions? Through our ancestors' evolving methodological abilities: through *method revolutions*—developing and refining new methods to think, experiment and control the world. To connect geography and culture to the rise of better technologies, agriculture, civilisation or science, we must focus on our hands-on methodological capacities and innovative methods and tools used to make these advances possible. To drive the palaeolithic technological revolution over 50,000 years ago, early hunter-gatherers did not stumble upon tools like bows and arrows.[279,280] They had to use those abilities in new ways: from reasoning causally about tension and trajectory to refining their designs by experimenting. That is methodical innovation.

To trigger the agricultural revolution at least 12,000 years ago, early hunter-gatherers had to experiment with seeds and develop planting techniques.[277,278] They understood the causal interplay between seeds, water, soil fertility and sunlight needed to produce crops. To bring about the civilisation revolution and the first large societies at least 6000 years ago, early village dwellers had to develop methods like systems of geometry and arithmetic that pushed the boundaries of our mind. With these mathematical innovations, they planned and constructed buildings, cities and monetary systems on a scale never seen before.

To spark the so-called scientific revolution in the 17th century, we had to invent new extraordinary tools and methods—especially the first microscope and telescope—that transformed entirely how we observe, measure and experiment in more systematic, controlled and far-reaching ways. These pioneering tools opened up entirely new realms and ways to see and understand the world never imagined before. Galileo suddenly made unexpected astronomical discoveries employing the new telescope[193] and Newton then laid a new foundation for understanding the physical world by also applying new methods like calculus.[263] To spur the industrial revolution in the 18th and 19th centuries, engineers and entrepreneurs had to invent entirely new methods to transform raw materials into mass-produced goods.[279] This leap involved trial-and-error innovation—from building steam engines to producing iron and textile spinning machines—and demanded understanding the complexities of heat, motion and mechanical pressure. To give rise to the digital revolution and digital technologies in the 20th century, scientists and specialised professionals had to work together to produce new and efficient transistors, computers and the internet—technologies that define modern society. This leap required developing new mathematical and computational methods, coding systems and continually experimenting.

Across all these major revolutions one thread runs consistently: our mind's evolved methodological capacities and inventing new tools have been the main force we *actively and directly* influence. These were not just major historical shifts, but well-defined methodological shifts. It is this *method-making engine*—our ever-expanding toolbox of cognitive strategies and technological tools—that has spurred human progress. In short: method innovation drives the continuous thread of discovery across history, connecting us to our ancestors and propelling future advances. Future revolutions, such as an AI revolution, would also be catalysed by developing new data-processing architectures, more powerful machine learning algorithms and massive datasets.[(90,249,268)] Complex AI systems rely on computer scientists, data analysts and systems engineers to optimise ever more complex deep learning programmes and computer technology. But even these cutting-edge tools are built on ancient foundations: logical deduction, systematic observation and iteration—strategies honed over millennia by our ancestors to track animals and craft effective tools. So how did it all begin? How have we even been able to start science in the first place? We begin with this question—as we progress through the history of science in the chapter.

Science of smart animals

How did our species come to develop such a powerful toolbox that not only helps us survive but also enables us to do science—to study particles with accelerators and search for life on other planets with space telescopes? To survive and meet basic needs, all animals—including us—have to learn which types of food are edible and poisonous, how to avoid predators and where to find shelter. This means observing, solving problems, categorising and recognising regularities in nature.[(288,289)] Many animals possess these cognitive and social abilities needed to use rudimentary tools and methods—through trial and error, testing hypotheses, and learning and sharing them with others.[(290)]

Take chimpanzees: they use tools for extractive foraging. They leverage stone and wood hammers to crack nuts and long sticks to extend their reach and extract termites and honey. They can use leaf sponges to collect water and levers for different tasks.[(278,290–292)] Harnessing different tools to solve different problems, chimpanzees have a toolkit that they acquire through social learning and experimenting. For chimpanzees to use these tools requires that they have an objective of the tool in mind, predict how the tool can enable them to achieve that objective and understand how to apply the tool. This is not instinct—but relies on understanding the interactions needed between the tool, their hands and the desired outcome. From crows to sea otters and octopuses, many animals possess these cognitive capacities to innovate tools and build shelters by manipulating objects for their purposes and making inferences.[(290,293)]

If the way we and other smart animals have always gained knowledge was not based on observing, trial and error, and building theories about the world, then we would constantly fall into holes, get burned by fire and even walk off cliffs (something one generally only gets one shot at). Natural selection favours brains that observe, test and remember. Remarkably, these common evolved abilities have been at the centre of how we and other smart animals think, solve problems and leverage tools over millions of years and they remain pivotal to science today. This raises crucial questions: how can we best design tools to tackle the diverse bottlenecks of our basic cognitive architecture? How far can AI systems be trained to observe, test, learn and think like clever animals do?

Science of early humans

At least about 1.5 to 2 million years ago, early human species like Homo erectus and Neanderthals created complex tools, including the iconic hand axe.[294,295] While early stone toolmakers used their hands to chip stone, it was their minds that carved the real tools using a broad set of abilities: they needed to carefully observe, experiment (beyond basic trial and error), test hypotheses, infer and predict (Picture 7.1).[296] Making such tools also requires being able to imagine and plan. A remarkable feat was achieved when early humans, about 600,000 to 1 million years ago, learned to control fire—a milestone that made cooking, protection and warmth possible. About 400,000 years ago, they then developed sophisticated fire-hardened spears.[287,288] These innovations, then and today, demand multi-step reasoning, making interconnected inferences and examining hypotheses.[277,297] And we rely on these extraordinary methodological skills when we do science today.

Picture 7.1 *Creating stone hand axes requires causal reasoning, testing a hypothesis, abstraction, planning and refining them, circa 1.5 million years ago.*
Reproduced from Joy of Museums via Wikimedia Commons.

Because we know early human species created these complex tools[288] and because recreating them today requires carefully refining the tools, we know they also used these abilities. There is also no plausible alternative to explain how early humans reasoned and gained knowledge: these cognitive strategies were the only way to produce such technologies.

Homo sapiens science (until about 11,000 years ago)

We are the result of millions of years of evolution, gradually emerging as a unique species in Africa around 250,000 to 300,000 years ago.[278] Our evolution has shaped the way we think and understand the world. It has given rise to our core cognitive abilities to solve problems, experiment, reason about causes and effects, test ideas and hypotheses and develop methods—abilities used to create stone tools in the Palaeolithic and run high-speed simulations with particle accelerators today.[290,298]

With growing social cooperation, our early ancestors developed more sophisticated language systems. These cognitive bridges enabled us to describe what we observe, share how tools work and better pass along tools and knowledge across generations.[276,278,287,299] Imagination and abstraction also expanded in our early ancestors, not only in toolmaking but also in symbolic cave paintings.[288,300] Today, across science we also use representational models to simplify complexity—from statistical simulations and mathematical equations, to models of scientific discovery (Chapter 6). Darwin himself sketched for example a model of a branching tree to illustrate evolution, emphasising that all species are related and humans are on just one branch of the same tree.[57]

Then came a massive ecological shift: the end of the last ice age around 11,000 years ago. This global climate change led to long-term settlements and catalysed greater cumulative knowledge.[279,280] With our method-making minds, we then developed remarkable techniques for domesticating animals. This relied on experimentation and understanding biological reproduction, selective breeding to foster specific traits, and animals' nutritional needs. We then did not stumble on agriculture, but created and refined innovative farming techniques over time. To cultivate crops requires us to intricately understand how seeds, rain, soil quality and erosion causally interact.[277,278] Farming emerged by careful, patient and iterative experimentation using natural controls: testing different seeds, soils, timings and planting techniques, and tracking seasonal cycles and comparing the outcomes to optimise yields. Developing agriculture goes far beyond trial and error by controlling for these variables—reflecting more systematic experimentation that resembles agricultural trials today (Picture 7.2). The payoff of this method-making strategy was enormous: it enabled stable food production, allowed more people to survive, generated a surplus of labour and freed up time for innovation—whether in toolmaking or even astronomy and mathematics that soon followed.[279]

Picture 7.2 *Early farmers required reasoning about the causal interactions between seeds and rain, and comparatively experimenting with seeds—circa 9000* BCE.

Reproduced from archaeology newsroom.

The advent of agriculture and animal domestication marked a massive leap in our ability to actively shape and manipulate the environment using ingenious techniques. They were a milestone that enabled population density, more specialised labour and methodological diversity to grow hand in hand—each influencing and amplifying the others. What explains the difference between the pace at which we accumulated knowledge over the last 10,000 years and the ever-increasing pace today is not a vast difference in our brainpower. Biologically, our brains have not changed much. Rather, the difference lies in the surge in the variety and sophistication of the tools and techniques we have invented, refined and passed along over generations.

Our evolved abilities for method-making and tool-making developed through continual dynamic feedbacks between our physical environment, our cognitive capacities and our social structures (Chapter 8).[276,287] These have enabled us to build ever more complex knowledge systems, invent tools, organise societies and ultimately lay the groundwork for science long before the 17th century. Our tools are products of our method-making capacities harnessed to solve problems more effectively. These abilities and methods evolve through a cumulative, bootstrapping spiral: each method we create enhances the next, creating a reinforcing cycle of innovation.

So a natural question arises: why have our early ancestors largely been forgotten in the history of science? Because we do not have written records describing the methods they developed and the theories they held. They did not leave us behind direct documentation of their systems of language and numeracy—as the Sumerians, Egyptians, Greeks and others did. So we return to the puzzle: if the deep foundations of science—basically all the necessary methodological abilities to be able to reason and do science—existed at least 11,000 years ago, why did we not develop modern science then? Quite simply, these abilities alone are not enough.

Early civilisation science

As hunter-gatherer clans shifted into agricultural life, growing into villages and eventually cities,[279,280] we developed tools and knowledge at an unprecedented rate. Our improved methods and tools changed most aspects of our lives: from the crops we cultivated, to the more complex structures we built and lived in, to the early medical and technological advances we made. With growing populations in the first civilisations, we took large leaps towards science: around 6000 years ago, we developed the earliest known systems of written language and mathematics.[279] These were enormous cognitive upgrades. Babylonian mathematics developed fractions, algebra, Pythagorean theorem and quadratic and cubic equations (Picture 7.3). These remarkable systems marked a transformational shift in our thinking from oral to written knowledge. These systems make recording and organising what we observe far more efficient. They reduce our cognitive constraints in memorising everything, processing information

Picture 7.3 *Babylonian mathematical tablet, circa 2000* BCE.

Reproduced from public domain.

and making mathematical calculations—and even more accurate comparisons across generations. Once ideas could be written down, tested and shared without distortion, we could more efficiently build on our past methods and tools. It marked a leap in how knowledge evolved: human thought and tools became more testable. These advances can then be more easily shared (without forgetting) from generation to generation. Creating systems of writing reflects a pivotal point in cumulatively developing science, but also in understanding how science developed.

Around 5000 years ago, developing geometry can be traced back to Mesopotamia and Egypt. It involved principles of areas, lengths, angles and volumes—used to survey land, measure building materials, manage irrigation systems and study the stars.[301] Geometry helped us plan, measure and map the world and the sky. Creating astronomy provided a strategic advance: it enabled us to use stars and the moon as a clock for time keeping, a compass for navigating and a calendar for planning crop cycles, weather and temperature. At least 4500 years ago, ancient Egyptian civilisation and Norte Chico civilisation in modern Peru constructed massive pyramids. To construct a pyramid, we require—both then and today—developing principles in engineering, architecture and geometry that are grounded in systematically measuring, planning and experimenting. It reflects the kind of multi-domain thinking that underpins science today. Early systematic thinking in medicine also emerged. A surviving Egyptian medical textbook from around 1600 BCE outlines surgical procedures and instructions for treating fractures, tumours and wounds. It follows a clear experimental structure: diagnosis, intervention and prognosis.[302] It is not modern medicine—but it is methodical, corrective and cumulative. These early systems—writing, geometry, astronomy and medicine—paved the way for science by increasingly freeing our minds from the limits of oral memory and isolated trial-and-error.

Surprisingly, our ability to design controlled experiments stretches back at least to the Old Testament.[303] The book of Daniel (1: 12–13) describes an experimental trial with control groups that tests the influence of a vegetarian diet: 'Test your servants for ten days. Give us nothing but vegetables to eat and water to drink. Then compare our appearance with that of the young men who eat the royal food [and drink wine], and treat your servants in accordance with what you see'. This is a classic controlled experiment: two groups, two diets, one observable outcome. The core logic of experimentation is clear: isolate a variable, control the conditions and observe the effect—asking 'what happens if we change just one factor?' To conceive and design such a controlled experiment—whether in ancient Babylon or in a modern hospital—we have to integrate multiple cognitive skills. It demands systematic reasoning to test whether a potential cause (a vegetarian, water-based diet compared to a non-vegetarian, alcohol-based diet) has an observable effect on people's physical appearance. It requires carrying out a trial and recording and comparing the outcomes of the physical appearance between the two groups after 10 days, and then drawing inferences from the trial outcomes to inform people's diets in the future (Picture 7.4).

Today, we rely on these same method-making skills when designing clinical trials, but we amplify them with advanced tools like randomised controlled trials and statistical modelling. In the 19th century, we then combined controlled experiments

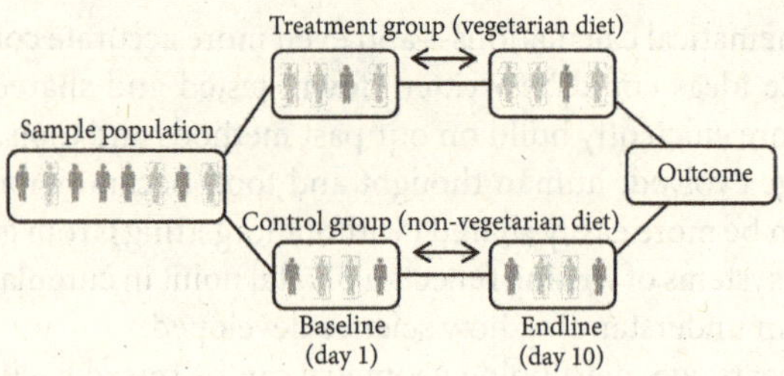

Picture 7.4 *Controlled experimentation using a treatment and control group described in the Old Testament, 2nd century* BCE

with newly developed methodological features like blinding and randomisation to reduce human bias.[304] So how did we become able to do this kind of extraordinary experimental reasoning in the first place? It is the remarkable flexibility—plasticity—of our cognitive architecture that enables us, often through trial and error, to develop techniques to test interventions and isolate causes from correlations. From testing a hunting strategy to testing a crop rotation, our ancestors had to isolate causes, adjust interventions and watch the results—over and over again.

What is astonishing is that civilisations across different regions of the world—the ancient Babylonians, Chinese, Mayans and others—developed agricultural, mathematical and technological systems largely independently, by using these method-making abilities and tools in increasingly experimental ways. Because they developed them largely independently illustrates that these enabling conditions are robust features of the human mind. These early systems were lenses through which we began to see the world in more structured, analytical ways—seeing patterns and possibilities, not just events and outcomes.

Ancient Chinese and Ancient Greek science

With a remarkably hands-on, experimental mindset, the ancient Chinese pioneered—as the first or independently—an impressive array of experimental advances that outpaced earlier civilisations and even their Mediterranean counterparts, the ancient Greeks.[195] From immunisation techniques, magnetic compasses, negative numbers and the 'Pascal' triangle, to sophisticated astronomical records of supernovae, seismographs, advanced irrigation systems and quantitative cartography, Chinese innovation was driven by a pragmatic commitment to method and measurement.[195,305,306] Ancient Chinese also devised papermaking and printing that preserved, spread and accumulated that growing knowledge—while ancient Greeks used slaves to hand-copy texts. They created a more complex system of astronomical records than any previous culture—including detailed star catalogues and observations of eclipses and supernovae. Because of these meticulous observations, our records in contemporary science are able to go back millennia and link to observations we make today.[195,307]

These advances allowed the ancient Chinese to predict and control aspects of the natural world with striking accuracy through new methods.[308] Take their smallpox immunisation technique: the groundbreaking method involved making a powder-based vaccine out of smallpox scabs from infected individuals that was administered through nasal inhalation or small incisions into the skin of healthy individuals (Picture 7.5). This produced a milder, controlled infection that offered immunity—drastically reducing mortality from smallpox by 20–30%.[195] This incredible method mirrors today's immunology and vaccine logic: introducing a weakened form of a pathogen to train the immune system. It reflects the complex outcome of experimental strategies and a causal understanding of disease transmission. This striking depth in methodological thinking—applying methods to tackle chance—was not confined to medicine. The Chinese state actively supported experimentation across domains—from engineering and geology to technology and alchemy/chemistry.[307]

Picture 7.5 *Ancient Chinese developed a smallpox immunisation technique that required understanding the causes and effects of the disease and how to control it.*
Reproduced from Wellcome Library, London.

Take the invention of the mechanical clock engineered in the 8th century China—and later in the 13th century Europe.[285,309] What does it take to create such a sophisticated technology—not just then, but now? We require a deeply methodical, cumulative process: observing mechanisms at play, experimenting to test different designs, reasoning about causes and effects to understand how gears and springs interact, and testing hypotheses to ensure each piece fits and functions precisely. Inventing a mechanical clock also demands abstract thinking to plan and imagine it, predictive reasoning to foresee its use for timekeeping, mathematical reasoning to divide time into repeatable units like 24 hours, 60 minutes and 60 seconds, and deep understanding of the complex interactions of the different parts as a whole. It is about building a remarkable system that precisely measures and regulates the world around us. It even suggested that nature could be thought of as operating like a mechanical machine.[220] The same kinds of reasoning and methodical thinking are at the heart of scientific reasoning today. And the future of AI largely depends on designing systems that can flexibly do and combine such diverse reasoning. In fact, developing mechanical clocks can require as much complex reasoning, experimenting and knowledge as needed for example for Galileo to discover the moons of Jupiter using a telescope or establish that objects of different weights reach the ground at the same time by dropping and testing them. A powerful insight emerges: what matters most is not the object, but the method.

Take also the ancient Greeks. Erasistratus for example tested weight loss by repeatedly weighing a captive bird while controlling its food intake—another one of history's many early controlled experiments.[310] Archimedes uncovered the principles of levers and buoyancy—fundamental principles in the fields of physics and mechanical engineering.[102] The Pythagorean theorem, developed around 540 BCE, was applied as a cornerstone for developing Euclidean geometry around 300 BCE. Euclidean geometry then enabled Eratosthenes to make the remarkable calculation of the Earth's circumference around 240 BCE at 252,000 stadia—astonishingly close to the actual measurement of 40,075 km.[311]

In the Islamic Golden Age, the 11th-century polymath Ibn Al-Haytham made a vast methodological leap: collecting rich experimental evidence to pioneer elaborate theories of vision, light and colour. His groundbreaking work not only covered optics, but also inertia and mechanics—centuries before Newton. One of his most influential works, the seven-volume *Book of Optics* (Kitab al-Manathir), was completed in 1021,[312] becoming one of the most important scientific texts of the medieval period. After its translation into Latin in the late 12th century, it circulated widely across Europe and influenced the work of scholars including Newton himself. This legacy earned Al-Haytham the title of the father of optics.[313] He also described and applied the classic scientific method, namely doing controlled experiments to prove a hypothesis applying verifiable procedures and mathematical logic.[312] He is often seen as the greatest physicist of the medieval era. Yet today, despite his monumental influence in past centuries, most have likely never heard of Al-Haytham.[313]

There is a striking pattern here. The key condition for innovation across these diverse contexts was not just geography or culture,[276–279] but the patient, deliberate investment of attention and intellectual effort: method has emerged where we dedicate the time to observe and experiment systematically.

That is how, over 1.5 millennia up to the 1500s, Chinese thinkers were able to design mechanical clocks and engineer the chain drive and long iron-chain suspension bridges—through systematic experimentation and ever more precise tools. This sustained institutional memory of existing technologies over time had a spiral effect. It enabled creating branches of physics—optics, magnetism and acoustics—and hydraulic engineering. It fed into pharmacology, publishing the first official book of medicinal drugs that outlined techniques for how to use them and their effects. It triggered developing one of the most ambitious intellectual projects in history: the world's largest physical encyclopaedia, the *Yongle Encyclopedia*—a state-sponsored project to codify and preserve the vast corpus of accumulated knowledge.[195,314] By the 15th century, the Chinese had built the most sophisticated scientific and engineering system that the world had ever seen—fuelled not just by innovation, but by a foundation of long-term social stability.[306] Importantly, the Chinese architects of these remarkable methods and technologies did not necessarily rely on mythology, religion or the supernatural. But they had to rely on our general abilities of systematic observation, iterative problem-solving and methodical experimentation.

So we return to a puzzle raised in Chapter 3: who then actually invented the classic scientific method? We cannot meaningfully talk about the founders of the classic scientific method in any general sense. Yet if we were to, then the first hunter-gatherer groups who developed methods and tools may be the best candidates as the founders—because they were the first to systematically observe, reason causally, experiment and imagine to be able to develop the methods and technologies that science relies on. As long as we have existed, we have used these methodological abilities in ever more systematic ways. We see them in the work of Archimedes and Democritus in ancient Greece, laid out in systematic detail in the 11th century by Al-Haytham,[312] and reiterated again in the 13th century by Grosseteste,[315] Roger Bacon[316] and Aquinas;[317] and later in the 17th century by Francis Bacon,[204] Galileo[318] and Newton,[263] alongside others. The term the founder of the classic scientific method has been attributed to each of these scholars.

How did we develop modern science around the 17th century?

This is one of the great unsolved mysteries in the history of science.[285,319] Common explanations point to a mix of external factors: the role of Christian worldviews, the spread of capitalism and wealth, the printing press and greater political and social liberties—conditions that supported scholars like Galileo and Newton. [281-285,306] Some argue that Christianity fostered science through a belief in laws of the universe governed by God—mirroring scientific laws.[281,282] But Christianity emerged over 1.5 millennia earlier and did not trigger more scientific and technological advances than the Chinese or Arabs.[195] Protestant values also do not seem to explain the scientific change,[283,285,320] as the earliest of these European scholars, Copernicus and Galileo, were in fact Catholics.

The rise of contemporary capitalism and growing economic prosperity helped free up time for more people to study the world.[281,285] But wealth does not explain why and how they studied nature in more methodical ways. Paper and printing certainly played a critical role in spreading new methods and knowledge.[284,320] But these technologies had been developed and widely used in China for nearly two millennia, and movable type printing was only later brought to Europe in the 13th century.[321] As for political liberties, Chinese scholars—despite lacking Europe's democratic institutions—developed more remarkable advancements before Europeans up to the 15th century.[195,285,306] So what changed in Europe? Leading researchers remain divided on the causes of this scientific shift, and historians of science highlight the persistent lack of consensus. [220,285,319,320,322] Broad external factors (greater wealth, freedom and printing) supported the spread of knowledge, but it was our tools that created the new knowledge itself. So external factors do not explain the direct spark itself. Yet did science suddenly begin in the 17th century or was its greater potential rather unlocked—with the right tools—finally being invented, within one of several stable social environments? Researchers have largely overlooked what enabled us—internally—to actively catalyse this scientific change: inventing new methods and tools that directly triggered the major breakthroughs.

We find a striking pattern: six groundbreaking methods and tools set off the chain reaction of discoveries that defined the 17th century—the first remarkable new invention was the microscope in 1590 and sparked the creation of the telescope in 1608, followed later by the barometer in 1643, the vacuum pump (air pump) in 1659, early statistics in 1663 and calculus in 1675. Each one extended our senses in unimaginable ways, sharpened our reasoning or amplified our ability to measure and predict. What changed in the 17th century was not just *what* we studied, but *how* we studied. More systematic observation, more precise measurement and more controlled experiments were made possible by these new powerful methods and tools. *These tools visualised, detected and measured in entirely new ways—as the key difference we see in the work of the pioneering scholars at the time compared to their predecessors.*

Modern science—with its emphasis on formal methods, rigorous testing and sophisticated instruments—was born by leveraging these six extraordinary innovations. Galileo, Hooke, Boyle, van Leeuwenhoek, Newton and their contemporaries did not stumble on their insights using just our mind. But only by deliberately applying one or more of these new tools were they able to observe and explore entirely new domains—expanding our understanding in astronomy, biology, physiology, pneumatics, mechanics and optics. The fascinating telescopic breakthroughs quickly spread across Europe, making clear the power of systematic observation that is replicable.(193) About a decade later, Bacon then highlighted these features again in describing scientific methodology in 1620.(204)

These remarkable new tools mark this transformative change—from early science to modern science—because they enabled an entirely new access, reach and understanding of the world. The microscope made it possible to peer into a world invisible to the naked eye. The telescope opened up the solar system and enabled us to map out new stars and planets. The barometer allowed us to measure invisible forces like atmospheric pressure. The vacuum pump let us isolate gases. Statistics gave us a way to grapple with hidden patterns and uncertainty and understand probabilities. And calculus enabled us to express dynamic systems precisely and provided a language for change and motion. Together, these innovations enabled the deeper experiments, the more reliable measurements and the more robust theories that characterise the breakthroughs of the 17th century. These new tools are the key engine of the scientific transformation at the time. In short, our adaptive toolbox expanded rapidly, vastly extending our universal toolbox with these more precise methods. So rigorous, controlled science—modern science—was made possible through:

- *New methods and tools*: we developed more sophisticated instruments—by building on existing tools—that drove the major discoveries of the 17th century.
  - o *Mathematical measurement*: we specifically used improved mathematical methods to explain nature in quantifiable terms—seen in some major discoveries in the 17th century.
  - o *Theoretical methodology*: we then deepened our methodological and theoretical understanding of science (how we do science using methods)—seen in some major discoveries in the 17th century.

The scientific revolution was, at its core, a method revolution: sparking discoveries by improving and systematising our tools. We can best understand the scientific change through this method change. Geography, culture and religion did not suddenly change in the 17th century. These broad factors cannot directly explain the sudden leap in the sophistication of our methods seen in the work of Galileo and Newton—seen in the new ways they could study the world. With science sparked by new tools, what then sparked this surge of new tools—and why in Europe? How and why exactly did scholars observe, measure and experiment with more systematic tools that enabled the chain of new discoveries?

The invention of the microscope and telescope: the key turning point that sparked multiple groundbreaking discoveries and modern observational and experimental science

Our evolved capacities of human perception and reasoning could only take us so—with progress slow over centuries. Science, bounded by our mind and senses, largely plateaued. Then came a turning point in scientific history: the invention of two transformative instruments—the microscope in 1590 and the telescope in 1608. With these tools, we opened up entirely hidden worlds far beyond human reach and imagination: from extraordinary microscopic organisms to distant planets we never knew existed. These two tools triggered a chain of surprising major breakthroughs: from discovering cells, capillaries and bacteria, to the moons of Jupiter, Saturn's rings, the motion of stars, and galaxies beyond our own. Each changed our understanding of life and the universe, like nothing that came before. These innovations arguably accelerated science and human understanding at a pace unmatched by any other development in history up to their invention. They triggered a feedback loop: better tools led to better discoveries that fuelled the demand for more precise and powerful instruments. This runaway dynamic put science on a self-reinforcing path of growth and exploration. These tools inspired wider curiosity in studying the fascinating invisible world and spurred the emerging scientific community.

We argue that these two path-breaking inventions at the turn of the 17th century marked the critical tipping point: the beginning of modern observational and experimental science. *The microscope and telescope were the key spark, when we began putting ever more powerful tools in our hands that vastly surpassed the human mind as the central driver of perception, exploration, imagination, reasoning and theorising.* Science rapidly shifted from largely mind-driven to tool-powered: tools that enabled outseeing, outmeasuring and outthinking what came before. These groundbreaking tools proved just how transformative the power of integrating observation with instruments could be.

But some might ask—if contemporary science is driven by new scientific tools, and tools are made by scientists, is that not circular? Surprisingly, the microscope and telescope were not in fact made by researchers but by two Dutch eyeglass shop owners: Zacharias Janssen and Hans Lippershey—who did not do scientific research. Simply trying to enhance eyesight beyond what ordinary glasses could offer, they stumbled upon these innovations without any research purposes in mind.(46,140) These two craftsmen just happened to live in Europe—in the small town of Middelburg in the Netherlands—and just happened to create simple magnifying extensions to glasses in their eyeglass shops.(46,140) Lippershey did apply—once its potential became more clear—for a patent for the telescope in 1608 as he thought it could perhaps be useful for military or naval purposes.(140) But others—Galileo in Italy and Hooke in England—began using these optical curiosities to study the world around us.(111,193)

Galileo turned the telescope up to the sky—marking the fundamental transformation in modern, perception-enhanced science. Hooke then pointed the microscope down into the biological world. Once they published their new groundbreaking observations, the potential of these 'seeing devices' spread rapidly. Only later were the tools formally named the microscope and the telescope. What is striking is that tools built with no scientific goal at all ended up rewriting science itself.[46,140] Looking through those first lenses, we saw an extraordinary world never perceived, predicted or imagined before. A key insight emerges: the microscope and telescope were the critical spark enabling most groundbreaking discoveries of the 17th century—more than individual cases of genius, formal funding or large teams.

These two unexpected research tools sparked a chain reaction of unexpected discoveries—laid out above—without previous knowledge or theories about the phenomena they revealed. These fascinating optical curiosities pushed science across a critical threshold—a point of no return that had not been crossed before and captured the attention of scholars, the public and even governments. Remarkably, these two Dutch eyeglass shop owners—crafting practical optical devices rather than theories—were key catalysts in unintentionally spurring and consolidating modern science. They gave us entirely new lenses to see the world, while some scholars were still relying heavily on their mind to reason about the world.

A deeper methodological insight arises here: these groundbreaking discoveries were beyond our reach and imagination before creating these instruments, but the odds of discovery jumped once they were invented. Using quasi-experimental logic helps understand the cause-and-effect relationship. These transformational tools were not built intentionally as scientific tools to make discoveries, so the tools—*the intervention*—were at first not directly related to *the outcome*: the discoveries. The causal effect is straightforward: *before* we invented the tools, these discoveries—from red blood cells to Saturn's rings—were not possible; but *after* we created the tools, we could directly observe and make these discoveries by using them (Chapter 1). The inventions came first, and the insights followed their use. These instruments were like chance interventions—introduced into a system without initial intention, yet they became the key causal driver of the major discoveries at the time. Without them, the breakthroughs could not occur; with them, they became vastly more likely. And in some cases they were almost inevitable if put into the hands of researchers. In short: when these tools were created, the odds of modern science emerging grew rapidly.

But what about the few exceptional cases like Copernicus's heliocentric model, developed before these two tools? Copernicus's model, proposed in 1543, was a bold hypothesis, but speculative and conceptual. He developed his model using a quadrant, astrolabe and mathematical calculations. But his model lacked convincing observational evidence. It was the telescope's discoveries—especially Galileo's observation of the phases of Venus and Jupiter's moons, orbiting something other than Earth—that provided the powerful evidence that celestial bodies do not have to revolve around us.[186,193] With telescopic findings, the abstract heliocentric model gained traction.

Think of the fascination many of us experience the first time we look through a telescope or microscope—the thrill of seeing an unknown world with our eyes. This same sense of wonder likely inspired many 17th-century scholars—observing star systems and miniscule living organisms not ever seen before—to explore nature and the unknown using these fascinating tools of discovery. These tools marked a break from thousands of years of understanding the world using only the naked eye—often with speculation outpacing evidence. These optical breakthroughs are key in explaining the divide between scholars before the 17th century and those who followed and got their hands on these new lenses. These tools of exploration are not just likely, but necessary, to explain how the course of scientific methodology and history drastically changed.

Interestingly, telescopes became the most widely owned of these 17th-century tools, once commercially available, generally for wealthy and educated citizens. Even some hobby astronomers made major (non-nobel-prize-winning) breakthroughs. One of them, William Herschel, originally a musician with only a secondary school education, discovered Uranus in 1781. His sister, Caroline Herschel, discovered comet 35P (Herschel-Rigollet) in 1787. Neither was affiliated with any institution when they made these discoveries.(140) They are examples of what the right tool in curious hands can do. Other existing methods—like geometry or quadrants—were confined to a few specialised scholars over millennia. They lacked a broad, intuitive appeal to become widespread tools for exploring the world.(186) This also helps explain why, even today, households in developed countries rarely own a quadrant or astrolabe, but are more likely to own a telescope or microscope.

This leads to another crucial point: the tools used to study the world in the 1500s and earlier—such as the quadrant, astrolabe, armillary sphere and geometric methods that Copernicus also used—were developed in antiquity, generally millennia earlier.(186) These tools were mainly used for measuring navigational and astronomical positions and as geometric aids that organised what we could already observe with the naked eye. *No tool capable of vastly extending our perception emerged in human history preceding the microscope and telescope—which did reveal entirely invisible phenomena and soon led to other perception-enhancing tools like the barometer and vacuum pump.* Magnifying lenses only improved visibility of already observable objects.(46,140) The contrast could not be stronger between pre-1600 tools that studied what was already visible, on one hand, and the microscope and telescope that opened entirely unimaginable realms of observation, on the other.

This marks a striking and overlooked insight: *the pre-1600 tools trace back to ancient times, while the microscope and telescope—and other perception-enhancing tools that followed them—trace forward to science today. This key transition—from inherited tools to newly invented tools that augment our perception—represents the most fundamental divide in the history of science and the first rapid acceleration of making new discoveries they enabled. Other factors do not compare in impact in making those new breakthrough observations possible.* But surprisingly, this instrumental turning point—the new tools that made the discoveries possible and their unexpected origins—has been largely ignored by researchers studying the origins of modern science.(281–285,306) This instrumental revolution sparked the scientific revolution by seeing and measuring the world in entirely unthinkable ways, like nothing before them.

These two breakthrough inventions inspired and triggered the development of a series of other successful tools that could also expand our mind and senses—the barometer, the vacuum pump, early statistics and calculus. Together, *this set of six fundamental tools—the first systematic scientific toolbox in history—created an entirely new foundation for studying the world.* This toolbox also set off a cascade of further inventions—including the thermometer. While our evolved human mind remained extraordinary, it became clear just how limited our raw senses had been. No longer did we rely largely on our unaided mind. Aristotle's era of largely mind-based knowledge began to give way to a new source of inspiration: our hands-on instruments.[323] These not only extended our perception, they stretched what was possible and our very imagination. Seeing microbes or Saturn's rings was not just extraordinary observational knowledge—it reconfigured what we thought the world was and vastly expanded our imaginative world. These tools transformed what we believed was worth exploring.

Uncovering these new phenomena made it impossible to rely just on Aristotle. Science ultimately took off when we stopped relying solely on human intuition and started building instruments that stretch what we are capable of knowing. And science would accelerate rapidly when the scientific community, as a whole, recognises the enormous power of new tools and methods and devotes greater effort and time to develop, test and refine them. For breakthroughs do not come from just inspiration and new questions. But history shows that new instruments are commonly the key spark of inspiration, new questions and unexpected answers—as became clear with the six new pioneering tools of the 17th century. The main driver of science shifted from our minds to our external experimental tools. Understanding this shift—from mind-led to tool-triggering science—helps us understand why fields at times hit a plateau. Think of much of theoretical physics, theoretical economics and philosophy of science today. Progress stalls generally because they rely too heavily on the unaided mind. When methods stagnate, so do our questions and answers, which the major scientific shift before and after the 17th century revealed. To break the stagnation, we need more powerful tools of resolution, precision and thought—enabling us to probe deeper and reframe problems.

So why, after thousands of years of creating tools, did we suddenly begin inventing such powerful ones at the turn of the 17th century? It was serendipity. Dutch eyeglass makers happened to experiment with better lenses and ended up building these optical devices: devices that unexpectedly happened to vastly expand our senses and completely transform how we understand nature with a simple lens. A natural question arises: could modern science have emerged centuries earlier if lens-makers had happened to experiment with magnifying lenses sooner? It is plausible because the wonder-inspiring telescope and microscope uncovered invisible worlds and catalysed fields like modern biology, physiology, astronomy and optics. What is clear is that these primitive, but powerful initial tools sparked a snowball effect of curiosity and reshaping our understanding of nature unlike anything before.

Over time, we gradually applied new tools not just to problems with clear practical payoffs (technological knowledge), but also to questions whose significance was not always immediately clear (scientific knowledge). Developing technology and science go hand in hand, with reinforcing feedbacks. New tools opened new frontiers, and those frontiers demanded even better tools—the heart of the scientific revolution. The story of the telescope and microscope exemplifies this well: what started as optical curiosities quickly evolved into the central engine of discovery.

Greater use of mathematics to describe what new tools uncovered

Newton is often seen as the greatest of 17th-century scientists—portrayed as a lone genius relying on mathematical methods for his transformational breakthroughs: the laws of motion, universal gravitation and optics.[220,322] Yet his work was not a product of sudden insight. It emerged from the broader tool revolution in science sparked by new instruments that enabled observing and measuring the world with unprecedented precision. This transformation culminated at the end of the century in Newton's *Philosophiae Naturalis Principia Mathematica* in 1687. Newton inherited a scientific landscape already transformed by instruments and experiments. He built on powerful new tools like the telescope and was deeply influenced by Galileo's experiments on motion, Boyle's work on pressure using the vacuum pump, and Hooke's meticulous empirical studies—all of which relied on new measurement tools and repeatable experiments. He also inherited rapidly expanding mathematical methods, including analytic geometry developed by Descartes, and early forms of calculus pioneered by Fermat and Wallis.[263]

Newton built extensively on Galileo's experimental findings on acceleration and on rich observational data of planetary motions. Recognising the telescope's extraordinary power, Newton himself developed a reflecting telescope. He also conducted prism experiments demonstrating that light consists of a spectrum of colours—with theory following experiments. This toolbox helped lay the groundwork for mathematics to describe what these instruments helped reveal: motion, force and light. It was observation and measurement that demanded better mathematics to express the motion of falling bodies and celestial objects. Newton developed calculus—as Leibniz did independently in 1675—as a means to formulate his theories and make better predictions.[263]

Remove one of these foundational elements—the telescope's observational power, the experimental data from these other new instruments, or the mathematical methods for expressing dynamic relationships—and Newton's achievements would not have been possible.[263] Newton's ability lay in synthesising previous research, and experiments using instruments, with these mathematical methods.

Newton could see further because he, in his own words, 'stood on the shoulders of giants'—and those giants had access to the right new tools.[263] In the centuries leading up to Newton, engineers, architects, surveyors, navigators and watchmakers were using quantitative methods to solve real-world problems.[281,320] Leveraging these successful mathematical methods, a small but growing group of scholars increasingly integrated the methods into science. We can trace these scientific roots even further back. In Newton's seminal *Philosophiae Naturalis Principia Mathematica*, we see how he applied our mind's evolved abilities to observe and experiment, he adopted the numerical system created by ancient Indian mathematicians and brought to Europe through the work of Islamic scholars, he used geometry refined by the mathematician Euclid in ancient Egypt, he wrote in Latin created by the Romans and on paper invented in China, he employed algebra developed by various cultures across millennia, and he was influenced by Ibn Al-Haytham's pioneering theory of optics (Picture 7.6).[263,313]

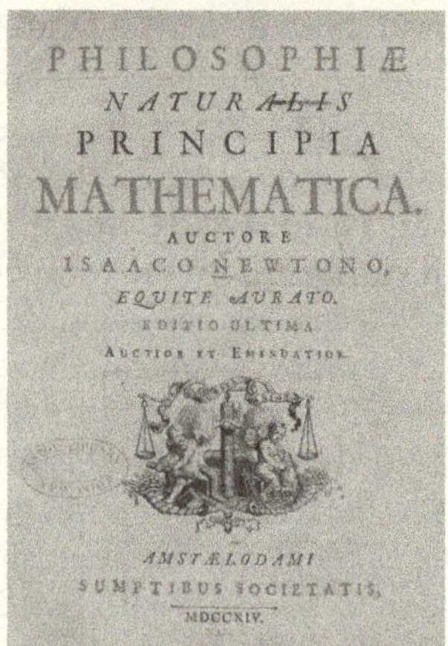
PHILOSOPHIÆ
NATURALIS
PRINCIPIA
MATHEMATICA.
AUCTORE
ISAACO NEWTONO,
EQUITE AURATO.
EDITIO ULTIMA
AUCTIOR ET EMENDATIOR.
AMSTÆLODAMI
SUMPTIBUS SOCIETATIS,
MDCCXIV.

Picture 7.6 *Newton's seminal book is the outcome of cumulative methods and knowledge developed over time, 1687.*

Reproduced from New York Public Library.

Physicists often credit mathematics—especially as embodied in Newton's work—as the central driver of the scientific revolution. But mathematics was one method. These powerful experimental tools and observations opened up entirely new realms of reality—findings that could then be described with mathematics. Many pivotal breakthroughs of the 17th century—such as biological and medical discoveries including the circulation of blood in 1628, capillaries in 1661 and cells in 1665—were derived from experiments and the new microscope.[220] Breakthroughs relying on mathematics were mainly concentrated in physics. How modern science arose cannot be reduced to mathematical advances. New microscopes and telescopes were critical in better observing and uncovering the nature of life and the universe. (These observable scientific explanations also increasingly reduced the need for religious explanations of how the world works, though Galileo and Newton remained Christian.)

Tools and methods sparked a deeper understanding of scientific methodology

As these pioneering experimental tools uncovered new layers of knowledge, scientists increasingly focused on *how* we apply a method or tool: scientific methodology. Researchers began combining different methodological approaches—integrating for example experimental methods, mathematical methods or causal inference—to generate more reliable results. This method integration allowed for experiments that were more quantifiable, testable and replicable, and helped establish clearer cause-and-effect relationships. We see this layered methodology in the works of Galileo, Boyle and Newton.[193,263] With each new tool—microscopes, barometers, calculus—methodology evolved into an expanding toolkit that more explicitly stressed experimentation and measurement. Boyle—working with the new vacuum pump—conducted controlled experiments, and became a leading advocate for controlled, repeatable experiments. Galileo and Newton merged instrumental data with mathematical reasoning. These developments spurred a broader meta-understanding of the scientific process itself: not just *what* but *how* we should apply science's tools and methods. At its core, scientific methodology is about improving how we use tools. The success of the experimental breakthroughs made clear: a more systematic, more controlled methodology of studying the world could uncover truths of nature once thought unreachable.

Centuries earlier, Al-Haytham in the 11th century[312] and Grosseteste[315] and Roger Bacon in the 13th century[316] had also stressed the need to unify observation, experiments and methods with theory. What they lacked were the powerful tools for observing and experimenting that 17th-century scholars finally had in their hands—ultimately enabling scientific methodology and testable science to reach its greater potential. Scientific methodology added a meta layer to science—making how we applied our toolbox more systematic and robust. The concept of scientific methodology, in its explicit form, spread. With better methods came the hope that we could finally solve the great puzzles of stars, planets, human biology and the causes of diseases. The spark of the scientific revolution was not a new hypothesis or new experiment—it was a new way of seeing that led to entirely new types of hypotheses and experiments and to realising their importance. After the invention of these two tools, Francis Bacon, in *Novum Organum* in 1620, then reiterated the importance of methodology. He highlighted that knowledge should not come from speculation but from inductive reasoning grounded in carefully collected evidence.[204] And this is precisely what these new tools had been generating at the time. They provided a systematic, unifying framework for cross-disciplinary inquiry, from astronomy to biology—moving beyond philosophical speculation.

The core insight here remains unchanged today: great tools and methods not only make discoveries possible but also make them reliable, repeatable and robust. Science keeps refining its methodology—for example through more rigorous experimental controls, randomisation and robust statistical inference to draw clearer conclusions. These methodological features were largely developed in the 19th and 20th centuries within medicine and public health. It is here where the stakes are at times life and death and the demand for methodological rigour is especially high. A powerful new method emerged in the mid-1800s: the natural experiment. It was pioneered by the British epidemiologist John Snow in his groundbreaking study in 1855 of a deadly cholera epidemic in London.[264] Snow studied a 'natural' separation in the water supply across different neighbourhoods, comparing disease rates between areas served by different pumps. By linking the prevalence of cholera to contaminated water drawn from a single public pump, compared to other public pumps that were not contaminated, he provided an effective method of using data to infer causality in public health. This offered compelling evidence of the cause of cholera: water contaminated by sewage. His findings convinced local authorities to take action—and the outbreak stopped.[324] It was a turning point in public health: an excellent example of how combining observational data with thoughtful methodology saves many lives.

Nearly a century later, in 1948, another major method breakthrough transformed medicine and public health: the first randomised controlled trial. Developed by epidemiologist and statistician Bradford Hill, the method was applied in a landmark study testing a treatment (streptomycin) for tuberculosis in London—one of the most deadly diseases of the time.[133] By randomly assigning patients to treatment or control groups, the RCT introduced an entirely new methodological rigor: it vastly reduced bias, clarified cause and effect and set a new methodological benchmark for future studies.[134]

Then came the invention of computers—since the mid-20th century—that sparked a deep transformation in *how* we conduct science: making it far more systematic, scalable and replicable than ever before. Where scientists once relied on slow data collection and manual calculations, they can now rapidly process vast datasets, run complex statistical analyses automatically, replicate experiments with striking precision and even run many experiments at the same time—all of which were unimaginable before.[90] From basic data analysis to running millions of simulations, digital technology has made scientific methodology much faster, more methodical and infinitely more scalable. New computational methods—powered by big data, large-scale statistical simulations, machine learning and AI—have transformed entire fields, uncovering patterns and layers of complexity we never saw. The internet has also reshaped the speed and openness of science: research can now be quickly published, reviewed, replicated and improved by scientists across the globe. These developments are spurring a digital and AI revolution—one that can redefine how efficiently and quickly we do science.[90,249,268]

National scientific societies, the growing scientific community and systematic observational methods took off *after*, not before, the microscope and telescope

Modern science was seeded in 1590 with the microscope and truly ignited in 1608 with the telescope—triggering the first large-scale surge of public and scholarly interest in science. These tools laid the critical foundation and catalysed science's instrumental, demographic, institutional and conceptual development in the decades that followed. Here we synthesise the key insights. Modern science followed—not preceded—the invention of these first groundbreaking instruments:

- These optical breakthroughs were the first tools in history to fundamentally extend human perception—*breaking with millennia of inherited instruments from antiquity*. In contrast to ancient tools—from quadrants to geometry—used to study the world in the 1500s and earlier, the microscope and telescope revealed phenomena previously unimaginable, both at very small and very distant scales.[46,140]
- They fundamentally transformed our relationship with nature and how we interacted with it—shifting us from observers using our evolved naked eye to active discoverers. These tools transferred authority from ancient Aristotelian texts to nature itself—to what we could actually see through these fascinating lenses. At its core, modern scientific consciousness lies in *transcending our natural sensory limits* through technological innovations.

- These two optical instruments turned, for the first time, isolated insights into *a sustained engine of discovery*, unexpectedly sparking most of the major discoveries of the 17th century. The tools did not just uncover unknown phenomena about life and the universe—but they forced us to redefine how we understood nature, and triggered modern observational and experimental science.
- These fascinating tools of discovery sparked widespread curiosity and attracted *a growing community of scholars*. Works like Galileo's Starry Messenger and Hooke's Micrographia showcased these new instrument-driven discoveries, quickly becoming widely read and influential—and catalysing the scientific community.[111,193] Before this optical revolution, natural philosophy was a fragmented and minor pursuit. In 1600, fewer than an estimated hundred individuals, mostly in Italy, engaged in regular inquiry; by the end of the century, a community began to emerge that spread to England, France and Germany.[325]
- The exciting telescopic and microscopic breakthroughs motivated the *establishment of the first major national scientific societies*—beginning with the Royal Society in 1660, whose early members used these tools including Hooke, Boyle and Newton. The French Academy of Sciences followed in 1666 and then the Berlin Academy of Sciences in 1700. These institutions—and the first scientific journals that soon followed—mark how science had transformed into a tool-enhanced, observation-led system and demanded debate about the surge of these new breakthroughs.[103]
- These powerful tools *made systematic observation the gold standard*, redefined what counted as valid evidence and enabled science to truly become replicable. They gave science its first universal language of evidence: scholars across countries—for the first time—could observe the same phenomena, apply consistent tools and methods and validate each other's findings. The success of these instruments catalysed *how* to use them better. Indeed, the tools themselves embody methodology: systematic observation, inductive evidence and experimentation. In 1620, about a decade later, Bacon emphasised these features again in describing how science should be conducted.[204] These better instruments enabled better observations and demanded better hypotheses.

In summary, *before the invention* of the microscope and telescope, science had not yet surpassed the limits of ancient tools or transcended our evolved sensory limits—but their arrival marked the moment we finally did. *With their invention*, we gained direct access to entirely new realms of reality at microscopic and astronomical scales that were previously unimaginable. And *after their invention*, we observe three large scientific developments: these instruments did not emerge from scientific institutions, or from a large scientific community, or from a formal scientific methodology, but they preceded and inspired them. Their transformative impact forced science to evolve and grow, to create new institutions that could share and publish these breakthroughs and to apply more systematic methods for observation and confirmation. More so than other factors, the transformative power of these new instruments better explains the rise of modern science by enabling these very breakthroughs.

And what is key is that they did not arise from science at all, but unexpectedly from two eyeglass makers—Janssen and Lippershey—experimenting with lenses; yet they ultimately made modern science possible. These two instruments did not help science, but fundamentally redefined it: forcing natural philosophy to transform into modern science. They illustrated what science was and could become. They completely changed the sophistication of science's major tools before and after this decisive point. In short, science did not take off because we asked better questions, but because these tools finally offered what natural philosophy had lacked: reliable access to new evidence of unimaginable phenomena—triggering new explanations and theories like nothing before. And this same dynamic powers science today, driven by modern instruments that build on that first, transformative leap in sensory enhancement that supercharged discovery. From scanning electron microscopes that reveal the nanoworld, to deep-space telescopes that peer across the universe, from particle accelerators that probe into matter, to quantum sensors that detect what was once undetectable, each new instrument keeps pushing the boundaries of what we can observe and think. Each forces science to rethink, rewrite and evolve. With each major leap in observation comes a major leap in understanding what is revealed.

The enabling conditions for science and civilisation: a framework

At this stage, we can pull together the bigger picture and synthesise the insights: nine main conditions enabled modern science—and six of them also formed the foundation for civilisation. Here we step back and trace how scientific thinking has evolved over time: comparing the conditions and methodological abilities that make it possible—across time, species and human cultures. From early hunter-gatherers to Newton and Marie Curie, all humans have drawn from the same evolved capacities. These are what allow us to observe, experiment, create methods, develop technologies and unlock knowledge about the world. The capacity for methodological reasoning can be seen on a spectrum: from smart chimpanzees using sticks to extract termites, to early humans experimenting with farming techniques, to Egyptian engineers designing pyramids with remarkable geometric precision, and ultimately to discover new planets with the telescope, and then new protein structures and cancer drug candidates with AI-powered tools. Science is a deeply-rooted evolutionary, historical and cross-cultural enterprise embedded in our expanding ability to think methodically—embedded in our method advances.

About 11,000 years ago, many of the ingredients for civilisation were in place: the necessary biological, cognitive and social abilities, as well as favourable ecological and demographic conditions. These made it possible to domesticate plants and animals, build permanent settlements and eventually develop complex cities. With civilisation came the necessary external systems that made it possible to accumulate vast tools and knowledge and pass them across generations: written language, mathematics and social stability embodied in institutions. By the time the first civilisations emerged in Mesopotamia and Egypt, we had already built the basic scaffolding for long-term scientific development. Yet researchers studying the origins of science focus on a particular broad factor (like culture or geography) or a few particular scholars (like

Galileo and Newton)—not the extraordinary new tools these scholars inherited. And extending our toolbox in innovative ways enables us to continue unlocking new, unknown frontiers (Table 7.1).

So how can we explain the big shifts in how we have come to understand the world so deeply? The answer lies in our capacity to extend our universal toolbox. What ultimately distinguished our species from other animals is expanding our universal toolbox to better combine our cognitive abilities—from trial and error to pattern recognition—and increasingly use logical reasoning and develop diverse tools. What separated ancient Greeks from our earlier ancestors is broadening our toolbox to include theoretical thinking and formalising abstract reasoning. What set 17th-century discoverers apart from their predecessors is an upgraded toolbox with perception-enhancing tools that made science vastly more powerful, systematic and reliable.

The prominent scholars at the time did not invent our mind's evolved methodological abilities or create these needed external conditions for science. *But these scholars—from Galileo to Hooke—did have access to the new and unexpected powerful tools that made the key difference by accelerating scientific progress in ways unimaginable before.* Early pioneering scientists such as Al-Haytham and their methods were equally on the right path as European scholars at the turn of the 17th century, but without perception-enhancing tools in their hands, modern science could not take off—the counterfactual.

In fact, the conditions in Al-Haytham's time in the 11th century were arguably as favourable—if not more—than in Copernicus's time in the mid-1500s and up to 1600. Al-Haytham worked within a larger, more established community of scholars supported by the institutional infrastructure of the Islamic Golden Age, while the Prussian-Polish Copernicus worked in greater isolation among fewer, more fragmented scholars in Europe. The Islamic world at the time arguably had more developed academic institutions, more extensive libraries, translation centres and scholarly networks than 16th-century Europe.[(326)] But neither Al-Haytham nor Copernicus had access to the instrumental power of the microscope or telescope. And introducing these two unanticipated inventions is what ultimately changed the game entirely.

Our methodological abilities are made possible by our deeply evolved biological, basic cognitive and social abilities. These are the three enabling conditions that allow us to do science. But these traits alone are not enough to create modern science. Let us describe these conditions: our core methodological capacities including logical reasoning (conditions 1–4) evolved—throughout much of our species' history—mainly beyond our control. Favourable environmental conditions also emerged out of our control (5). Yet over time, we have gained increasing (if still limited) influence over these abilities. And we do have greater control in shaping demographic and social conditions. Although we may be able—in principle—to do science without complex division of labour (6), complex systems of written language and mathematics (7) and long-term social stability (8), science has not ever been and would not be sustainable in the long run without them. That is why individuals throughout history—who had the needed cognitive capacities and at times created sophisticated methods and knowledge—did not develop powerful precision tools, vast discoveries and bodies of knowledge that characterise modern science. To go from intelligent individuals to

Table 7.1 The enabling conditions for science and civilisation

History of science		Methodological abilities used across species to gain knowledge (basic cognitive abilities)	Main enabling conditions for science and civilisation	Bodies of knowledge	Time period
Science of smart animals (Chimpanzees, crows, dolphins)	1–3 (2–3 only basic)	Observing, problem solving, experimenting (including trial and error), tool use, reasoning causally, testing hypotheses and abstracting	1. ***Biological abilities**: vision, other senses* 2. ***Basic cognitive abilities**: observing, problem solving, trial and error* 3. ***Social abilities**: language, learning* Universal toolbox **3 conditions for doing science**	Basic zoology and botany (knowledge of edible and poisonous foods), ecology (recognising regularities in nature)	Millions of years ago—present
Science of early humans (Neanderthals, Homo erectus)	1–3			Basic botany, zoology, ecology, nutrition, natural pharmaceutics	Millions of years ago—9000 BCE
Homo Sapiens science (until about 9000 BCE)	1–6 (4 used practically)		4. ***Complex cognitive abilities**: logical, rule-based reasoning* 5. ***Ecological conditions**: favourable for agriculture, greater labour surplus* 6. ***Demographic and socioeconomic factors**: population density, division of labour* **6 conditions for civilisation**	Botany, zoology, ecology, nutrition, natural pharmaceutics, agriculture, architecture	300,000 years ago—9000 BCE
Early civilisation science (Mesopotamia, Egypt, Maya)	1–7 (4 used practically)		7. ***Written language and mathematics systems**: external memory, notation*	Early astronomy, medicine, engineering and geometry, agriculture, architecture	4000 BCE—500 BCE
Ancient Greek science	1–8 (4 used increasingly theoretically)		8. ***Social and political stability**: for accruing vast bodies of knowledge over centuries*	Physics, geometry, astronomy, formal logic, natural history, philosophy, botany	600 BCE—200 BCE
Ancient Chinese science	1–8 (4 used practically)			Medicine, geology, chemistry (alchemy), geography, technology and engineering	1050 BCE—50 BCE
Imperial Chinese science	1–8 (4 used practically)			Developed also branches of physics (optics, magnetism and acoustics), hydraulic engineering	100s—1500s
17th century science (largely Europe)	1–9		9. ***Develop new methods and tools that enhance our perception**: by using methodological abilities (1–4) in systematic ways* Adaptive toolbox **9 conditions for modern science**	Physics, mathematical astronomy, calculus, experimental medicine	1600s
18/19th century science (largely Europe, USA)	1–9			Chemistry, genetics, palaeontology, bayesian statistics, economics	1700s—1800s
20/21st century sciences (global)	1–9			Natural, behavioural and social sciences (quantum physics, molecular biology, AI, cognitive-, climate-, computer-science)	1900s—present

The categories are not exclusive; other features can be included under these nine conditions, and other sciences—from Islamic to Indian—can also be included. The sources and evidence for each condition are provided in each section of the chapter.

civilisation, we required the first six conditions—and to launch modern science we needed all nine conditions.

These building blocks are so foundational for science that most people take them for granted. And they are not tied to one culture or continent. Just as civilisations emerged independently across the world, so did parts of scientific reasoning—at different times, in different forms. What we call modern science could have developed elsewhere—and in some places, it nearly did. Islamic civilisation came impressively close with the remarkable work and methods of pioneering scholars like Al-Haytham.(312) So what if contemporary science were to be lost in the future—lost to environmental devastation or war (which led to the decline of past civilisations)? To rebuild science from scratch in the future, we would not have to wait for a Christian worldview to emerge,(282) a particular capitalistic system(281) or a renaissance of Aristotelian thought.(327) What we would need is to recreate these nine general conditions once again.

When we view our methods as the main trigger enabling us to *directly* develop scientific and technological knowledge—and so science and civilisation—we unlock a powerful new lens for understanding science and the history of science. And the conditions offer a roadmap—not just to explain the past, but to enable us to predict the development of science in the future. If another intelligent species evolved in the future (on Earth or beyond) and met these nine criteria, science—systematic, cumulative and transformative—would be expected to follow. Understanding these conditions can help us think about how to strengthen science today and how to future-proof it for the challenges to come—for example when specific governments put social stability and science in danger.

So we can trace a continuity in human reasoning and problem-solving—from our evolutionary past to today's leading scientists—revealing how our methodological capacities have scaled with our tools. Our early ancestors observed patterns in nature—tracking animal migrations, adapting to seasonal cycles and navigating using the stars—to solve survival challenges; today's scientists also observe patterns in nature but using powerful instruments like microscopes and mass spectrometers. Where our early ancestors relied on trial and error, controlled fire, invented basic notation, made early tools and explored hypotheses, today's scientists build on that logic by designing controlled experiments, inventing advanced calculus and testing systematic hypotheses. Similarly, our early ancestors used causal reasoning—'if x, then y'—which we have developed into today's formalised causal modelling and advanced statistical inference, revealing hidden relationships in everything from climate systems to brain networks. *Our early tools—spears, fire, notation systems—extended our physical tools and basic cognitive abilities, while our contemporary tools—microscopes, supercomputers, CRISPR, particle accelerators, AI—vastly amplify not only our cognitive power but also our sensory reach.* Our cognitive architecture—pattern recognition, abstraction, inductive reasoning—remains fundamentally the same. What has changed is the range, scale and precision of these capacities through systematic scientific tools—tools with which we can now explore quantum, genomic, cosmic realms once unimaginable.

This illustrates the cumulative nature of our method evolution: ever more complex abilities, methods and tools over time have sparked ever more complex discoveries. History is a chain of extraordinary *method revolutions* over time (Figure 7.1).

Our universal toolbox
(our mind's evolved methodological abilities...)

Our adaptive toolbox
(...used to develop more sophisticated methods and tools)

Our toolbox

observing → observing systematically → microscopes and telescopes →

problem solving → methodical problem solving → computational and statistical problem solving →

trial and error → more controlled experiments → blinded, randomised experiments → pre-registering studies, replication using multiple methods →

sense of quantity → arithmetic, geometry and algebra → statistics → quantum statistics →

causal reasoning → concept of cause and effect → multivariate causal inference →

testing hypotheses → elaborate theories → mathematical theories →

abstraction → logic/abstract modelling → Large-scale simulations →

Technologies and discoveries made (using our methods)

Smart animals
Chimpanzees use stone and wood hammers to crack nuts, sticks to extract termites, leaf sponges to collect water

Early humans
Controlled fire, hand axes, fire-hardened spears, inferential animal tracking

Homo Sapiens
Plant-based medicines, notation systems, lunar calendars,boats and eventually domesticating animals and plants, early architecture

Early civilisation
Written language and mathematics systems, medical knowledge of injuries, fractures and basic surgeries, immunisation, pyramids, early experimentson diets

Ancient Greeks
Mathematical system, early experiments on weight loss, principles for levers and buoyancy, water mills, complex architecture

Ancient Chinese
Magnetic compass, negative numbers, the 'Pascal' triangle, algebra, seismographs, papermaking, immunisation, steel technology, cartography, records of supernovae

Imperial Chinese
Mechanical clocks, long iron-chain suspension bridges, hydraulic engineering, pesticides, the Yongle Encyclopedia, pharmaceutical experiments on animals

17th century
Kepler's laws of planetary motions, Copernican revolution, Newtonian mechanics and gravity, calculus, barometer, microscope

18/19th century
Pasteurisation, x-rays, light bulb, combustion engines, trains, chemical manufacturing, encyclopaedias

20/21st century
Computers, antibiotics, DNA, vitamins, insulin, cancer therapies, airplanes, mobile phones, internet, AI, quantum computing, CRISPR

Time: Millions of years ago – present | Millions of years ago – 9000 BCE | 300,000 years ago – 9000 BCE | 4000 BCE – 500 BCE | 600 BCE – 200 BCE | 1050 BCE – 50 BCE | 100s – 1500s | 1600s | 1700s – 1800s | 1900s –2000s | Future

Figure 7.1 *Tracing the evolution of our methodological abilities and tools*

Can the answer to the question of how we understood the world for most of history—and how we do today—be as straightforward as pointing to the incredible new tools we develop that stretch our senses and sharpen our minds? Can inventing the microscope and telescope, and then the barometer, vacuum pump, statistics and calculus, explain the first major surge of discoveries in the 1600s? Can developing cutting-edge tools like electron microscopes, statistical methods, supercomputers and particle accelerators largely explain the accelerated pace of discoveries that followed them? The answer is yes—based on examining over 750 major discoveries that define scientific history (Chapter 1).

Conclusion

The key turning point in the history of science came with the invention of two remarkable tools: the microscope and telescope. Through their lenses, we saw extraordinary phenomena, for the first time, in the 1600s—from Saturn's rings and craters on the Moon to red blood cells and sperm cells—and science changed forever. It was like discovering new planets inside our own bodies. Once we saw these hidden, unimaginable and completely surprising dimensions of reality, we could not go back—and the tools were what guided us forward. These new unexpected instruments revealed unexpected new worlds that we did not even know could exist and without them, they would not have been possible. They sparked dozens of major discoveries—more than a single scholar or research team has ever achieved. These two optical breakthroughs mark when we vastly surpassed the human mind as the central driver of perception and exploration—ultimately triggering modern observational and experimental science. These two discovery machines catalysed into the development of a toolbox of six groundbreaking tools and methods that accelerated and consolidated science, enabling future researchers to confirm the discoveries using them. This represented the birth of the first systematic scientific toolbox—enabling a cascade of discoveries that defined 17th-century science.

What explains the explosion of sustained scientific discovery is not just culture, institutions, a scientific community, genius or European geography, but especially the invention of these two new tools that drastically expanded our capacity to perceive and measure phenomena we never could before—and demanded new explanations. Technological acceleration—in a cycle of new tools that enable new discoveries and motivate improved tools—then became self-sustaining once these instruments arrived and remains the driving force of scientific progress today. The greatest discovery we have ever made may well be the discovery of the power of our tools themselves—the realisation that they are the greatest force of progress. So the legacy of Galileo, Hooke, Boyle, Newton and their contemporaries was not any individual discovery. It was ultimately the newly developed tools they used that moved us from isolated insights to scalable discoveries. It was expanding the very way we spark discoveries. Because our tools shape what we can perceive and know, the history of science is also a history of our sensory evolution. Science is no longer bound by our biology and cognition, but has finally learned how to outgrow them with new tools. With each generation of instruments, we grow new eyes and new intuitions—in a cycle enabling us to detect new phenomena that were not imaginable before.

But to understand how we got to that crucial point in history, we went further back and traced the deep historical origins of scientific thinking itself. This led us to a challenging and cross-disciplinary question: what are the origins of science, technology and civilisation? We tackled the question by outlining the conditions that enabled these achievements. We found how observing, problem solving, experimenting and developing tools are deeply embedded in how we have always—and will always—get by and uncover insights about nature. And we explained how our universal toolbox (these evolved abilities of the mind) together with our adaptive toolbox (the external tools we have invented) is the driving force that has enabled us to actively create vast bodies of scientific and technological knowledge—and eventually give rise to science and civilisation.

So it is time we rethink what science actually is—because it is not just a body of knowledge. We need to redescribe science as an evolving process powered by methods. Throughout history, we have actively driven the major advances—the palaeolithic technological revolution, agricultural revolution and later scientific, industrial and digital revolutions—by leveraging our methodological abilities in more sophisticated ways and creating better methods and tools: through *method revolutions*. Our new methods and tools better explain how we have developed science as the key driving force we can actively influence, and they are more observable and traceable in the works left behind throughout history—ancient artifacts, the landmark books of 17th-century scholars and the breakthrough papers today.

We can also take important practical steps to advance science by better understanding our evolved constraints and building the very tools to overcome them. Understanding science's past progress is crucial to improving what drives and hinders its progress today. Our expanding toolbox is the engine of science, historically and presently. It has been and will continue to be—through collective mind-method synergies—our primary catalyst to tackle our big scientific, environmental and social challenges. Establishing ways to improve science helps us shape the science of the future; and the answer commonly lies in the power of our toolbox: not in doing more of the same in science, but in upgrading how we think, observe and discover. The answer commonly lies in more effectively tapping our toolbox in innovative ways.

This sparks important questions: could science have advanced much faster over the past centuries if we had focused much more on innovating better thinking tools—rather than just studying nature with the ones we had? Why are we today not yet systematically designing better tools and methods to mimic and extend all human sensory and reasoning abilities—the full spectrum from advanced tools to expand our imagination and causal reasoning, to amplify simulations and pattern recognition? How many discoveries have been delayed or missed because we use familiar methods, instead of inventing new ones? How much of scientific progress has been slowed over decades and centuries by science's fixation on end results, rather than stepping back to ask if we even have the right tools and are asking the right questions? How can we best develop and test new tools of perception and thought—new tools that often yield more uncertain but more surprising breakthroughs? We explore in depth how we can expand our scientific toolbox in Chapter 11. But first, in the next chapter, we dig deeper into our evolved method-making foundation of science.

8

Homo methodologicus

How our unique evolved method-making species became problem solvers and knowledge creators

Summary

Throughout history, we have continually refined our abilities, to invent navigation techniques using stars, apply medicinal plants to heal wounds, devise lunar calendars to track and predict seasons—and eventually develop farming techniques and early mathematical systems. Each method gave us a survival advantage. Eventually we developed the microscope and telescope that enabled vastly surpassing the limits of our evolved mind—triggering the discovery of entirely new stars, moons and medicines. The evolution of this capacity for method-making and tool-making marks one of the most fundamental changes in human history—and it enables science today. But what made this possible? How did we shift from our unaided mind and instinct to developing increasingly cumulative methods and tools? While evolution gradually honed and refined our inherent methodological abilities to observe, think and imagine throughout human history, we eventually began honing, refining and amplifying them into methods and tools. The bigger the innovation in method, the bigger the leap in progress. So can we best explain our species' success by our unique ability for cumulative method-making? Our success is indeed a remarkable story of realising the power of leveraging this capacity to conceive better tools that enable us to think, plan, test and predict better. We explain our uniqueness as cumulative method-makers and how evolution favoured those better at using this exceptional capacity—giving rise to a new kind of species called here *homo methodologicus*. Viewing humans as complex method-makers who extend our own minds better explains how we have been directly able to meet our needs, solve problems, make vast knowledge and eventually extraordinary discoveries. Here we dig deeper into our scientific origins that are traced from our evolutionary roots to contemporary scientists who aim to tackle our mind's evolved bottlenecks—including with AI tools.

The Engine of Scientific Discovery. Alexander Krauss, Oxford University Press. © Oxford University Press (2026).
DOI: 10.1093/oso/9780197829790.003.0009

Overview

What makes humans extraordinary is cumulative method-making that extends our minds. Our remarkable capacity to make methods and tools is the product of long and complex evolutionary feedbacks between nature and our biology, mind and the culture we built. In turn, these very methods and tools became our best way to understand those forces that shaped us—nature, our biology, mind, culture and even evolution itself. This evolutionary path we humans have taken has enabled our mind to gain increasingly complex knowledge—through early and simple methods we created that we gradually developed into ever more complex methods, until we were doing things no other species had done. Only our species has taken this evolutionary path to creating sophisticated tools. Indeed, many other smart animals have abilities for a sense of quantity, testing hypotheses, causal reasoning, communication and using rudimentary tools—from crows, to chimpanzees, to dolphins.(278,290,299) But only our species has evolved to develop statistics, simulate complex ecosystems and devise controlled experiments by applying these extraordinary abilities in more complex ways. That is the leap: we no longer just adapted to nature—we learned to analyse, redesign and shape it with methods. Our method innovations let us see, measure and think in ways evolution never prepared us for.

The remarkable turning point came when we began moving beyond intuition, instinct and trial and error. We began building more deliberate methods and external tools to solve our challenges—to better navigate, more efficiently hunt, better treat injuries, build stronger shelters. The more precisely our early ancestors could observe, test and predict, the better their chances of surviving and thriving—and passing down those new successful methods over time. Our evolved, expanding toolbox deeply links the first hunter-gatherers tracking animal footprints and inferring animal migrations to today's scientists running computer simulations of global migration and climate patterns. This extraordinary method-making capacity did not just become useful—but foundational to our species. Without it, we would not have invented early fire-hardened spears, then the first farming techniques, and eventually more systematic randomised experiments (Chapter 7). Before this evolutionary leap in thinking, science and discovery were out of our reach. So what gave us a survival edge in nature in our early history is at the same time what pushes the boundaries of modern science and enables us to understand and predict nature today. (Impatient readers less curious about the evolutionary origins of science can jump to Chapter 9.)

Other animals have their own evolutionary strategies—and many are much faster, stronger and have a sharper range of senses than we do. Some birds detect and use earth's magnetic fields to navigate; bees can perceive ultraviolet light to detect nectar; some snakes can perceive heat signatures (infrared) in complete darkness.(82,278,290) We did not evolve any of these and many other extraordinary abilities—nor for very cold tundras or very hot deserts. But only we can survive in all of them—not by adapting our genes (besides a few traits like skin colour), but by adapting our methods and tools.(328) This is what sets our species apart. We gradually learned to become cumulative method-makers and toolmakers by more systematically observing, experimenting and reasoning causally over time. This gave us the key edge over nature, biology and other species.

Our ability to develop a set of methods for solving problems and amplifying our mind distinguishes us, more than other factors, from other species: our species is uniquely a complex method-making species, indeed a hyper-method-making species—*homo methodologicus. This deep capacity to design and create better methods and tools that extend our mind, to test them and to refine them over time is not just what drives science—it is what has always made us human.* It is being cumulative method-makers and tool-makers that makes us the species we are—giving us the power to vastly outsee and outthink nature, like no other animal can.

In evolutionary terms, this matters. While Homo habilis was an early human species whose name means 'able or handy man', many other animals are also able to use a stone tool. We have evolved far beyond that. While Homo sapiens means 'wise man', this general description also falls short because it does not capture what we are: systematic method-builders who reshape how we think and can control nature and our biology through methods and tools themselves.[103] We go beyond those classifications here. Instead, we are a cumulative method-making and tool-making species: homo methodologicus—using our mind's evolved methodological abilities of observing, solving problems, experimenting and developing and refining complex methods. To amplify our mind, early humans used these same abilities to create plant-based medicines and early agricultural techniques—and today's scientists apply and extend them to develop more complex vaccination methods and advanced agricultural trials. We invent new methods as we face new challenges—and to reduce our mind's constraints.

What had driven our survival advantages over predators and environments is at the same time what has driven our great discoveries in modern science: the capacity to tackle our human limits using continually better methods and tools that expand our mind. Understanding these evolutionary origins of our toolbox—from our early basic tools to cutting-edge tools—and how it continues to shape our mind and methods today, we are in a better position to overcome our evolved limits. And today's extraordinary instruments enable detecting phenomena invisible to our senses throughout nearly our entire evolutionary history: from tracing sub-atomic particles to gravitational waves and exoplanets. But by designing new instruments, we have detected them since the 20th century (Chapter 1).

We are all—whatever our genetic make-up—born with a *universal toolbox.* It enables us to perceive, apply trial and error and recognise patterns and is activated early in life as we begin interacting with the world around us. But as we grow, we also all inherit something even more powerful: an *adaptive toolbox.* This is a constantly evolving set of complex, invented methods that allow us to collectively make deeper knowledge and groundbreaking discoveries—many that improve our survival and others that now expand our understanding without an immediate benefit. So our methods have an evolutionary and an adaptive component. Together, we can think of them as our core and extended toolbox—and is the foundation of science itself (Figure 8.1).

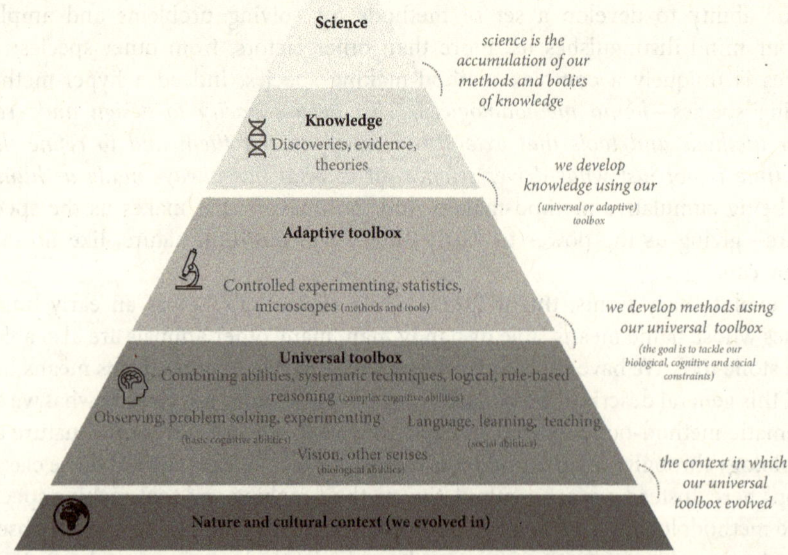

Figure 8.1 *Methods drive science pyramid: how we use our toolbox to develop knowledge and science*

Where does the mind end and the methods we create begin?

With our remarkable method-making mind, a natural question arises: where does our cognition stop and the methods and tools we invent start? We use our mind's internal abilities (observation, pattern recognition etc.) to develop external methods that stretch far beyond what our biology can do alone. We build these methods by interacting with nature, the tools we have already built, and other people (Figure 8.1). Methods, once invented, do not just exist in our heads: they become part of the outside world—external material artefacts—and can be shared and leveraged by others. Think of Babylonian mathematical tablets or today's computational and statistical programmes. Our sophisticated methods now drive every stage of the scientific process. They help researchers generate hypotheses using algorithms, collect and clean data using database systems, analyse and simulate experiments with statistical programmes and even discover potential cancer treatments with machine learning. Methods massively stretch what our mind can do. Statistical modelling lets us run experiments on climate futures. AI tools can rapidly spot patterns across millions of genetic sequences. They are like mind extensions—amplifying our brainpower with incredible speed, precision, memory and scale in ways that were entirely unimaginable just a few decades earlier.

Yet the line can be blurred between internal cognition and external tools. Our *cognition* at times incorporates the extended tools it leverages, and our *methods* integrate the internal processes they rely on. To develop new methods, we depend on our mind's methodological abilities and—in contemporary science—also existing methods and tools. Without us, methods do not have meaning, context or questions to solve.

So the important question is not just 'at what point does our cognition end and methods start?' but rather: 'how can we better understand our mind-method synergies and limits and how can we continue refining our best methods to tackle them and better understand reality?' We return to this essential question later.

It is clear that we are not biologically endowed with complex methods—in our brains. No single mind could alone come up with statistics. So how did we even develop such a powerful method in the first place? Did an individual not have to begin the process? Like most of our methods, statistics was not invented in one leap.[(329)] In early civilisations, we first created systems of mathematics. By the 1600s, European states were collecting basic population data. And mathematicians like Pascal and Fermat were laying the groundwork for probability theory. In the 1800s, Gauss and Laplace formalised statistical reasoning, and in the early 1900s, the pioneering statistician Ronald Fisher introduced game-changing techniques like randomised trials and analysis of variance. In the 2000s, scientists took these methods further by combining statistical analysis with digital computing power and machine learning—tools and methods that keep evolving today (Box 5.1).[(137,329)] What developed over many generations is a highly composite method called statistics—a multi-layered invention that drastically extends scientific cognition and much of what we can understand in the world. Our methods are typically the result of continually reworking and refining existing methods—and statistics, like the microscope, is no exception.

Since our mind and senses evolved adapting to our environmental and cultural niche, we face constraints in how we understand the world especially beyond our niche. Science pushes vastly beyond our evolutionary environment and that is where our raw abilities hit a wall—when studying phenomena from viruses to distant galaxies. That is why it is so crucial to focus more on our tools. *When we design a better method or tool, we are actually asking how we can tackle a problem by reducing a human constraint we face in understanding the world. We are asking how we can answer a question by improving our current cognitive or methodological capacities—how we can hack, or work around, the limits of our mind and existing methods.* Think of how we stretch our memory, perception or reasoning with algebra, microscopes, computers or regression analysis. Think of what it actually takes to study thousands or even millions of observations using regression analysis—for example the effects of a changing climate or new drugs for medical patients. Such analysis was not possible or even thinkable until we developed contemporary statistical methods and computers in the mid-20th century. Merging these two general-purpose methods—statistics with computational tools—transformed nearly every major scientific field: from experimental physics to medicine, psychology and economics. It marked the birth of large-scale, high-speed statistical analysis that characterises science today.

Expanding our toolbox: a strategy for survival, solving problems and gaining knowledge—and advancing science

Our species has become ever more methodologically adaptive. Unlike any other species, we have diversified and upgraded our methods as a strategy to meet our needs, tackle challenges and make sense of the world. This flexible ability—shifting between different methods in our toolbox—has enabled us to better control and discover layers of nature previously out of reach. After all, our biology, mind and senses we are born with have not changed much in the past few thousand years. The reason we understand the world much better today is not better brains we are born with—it is better tools and methods we inherit and create. Our knowledge has surged at the pace we have developed and shared more powerful methods and tools to build that very knowledge (Chapter 7).

Our early ancestors could not explain well the mysteries of shooting stars and eclipses. Today, we can—because of inventions like spectrometers and telescopes. We have gone from guessing why lightning strikes to measuring its electric charge using high-speed sensors. We no longer blame disease on the supernatural, but understand how bacteria and viruses cause disease by using microscopes and controlled experiments. But our current tools only enable us to partially understand many other phenomena—from cancer and our complex brain to the origin of life and the mechanics of the global economy. And there is so much we only poorly understand: from the size and nature of the universe to how organisms like us evolved to become self-aware and practice morality. Why do we not fully understand many phenomena? Often because we have not yet invented the tools we need to shed light on them.

Remarkably, we are both born and made into method-makers. The *evolved method-makers* in each of us has been an inevitable part of our species' cognition. Every child is born able to observe, test and count. The *trained method-makers* in each of us has become an inevitable part of our cognition in more recent history. Every child is also trained in a set of inherited, increasingly complex methods. Over the past few centuries—especially since the rise of public education in the 1800s—we have passed down increasingly sophisticated methods, from geometry and algebra to logic and basic scientific reasoning.

Each of us has gone through a massive process of method socialisation—for example to be able to read books on science like this one. This process enables us to turn a basic ability to quantify into an ability to calculate complex probabilities using statistical methods. In science and society we heavily depend on statistics and experiments, but we do not intuitively know how to do statistical analysis or controlled experiments (Chapter 7).

Our toolbox is adaptive, and grows with us, in three powerful ways. We adapt our methods over time to meet new challenges. Our methods enable us to better adapt to nature itself. And perhaps most fascinating, our methods adapt how our mind works. When we learn statistics, many begin thinking in probabilities—what are the odds of catching a virus, being affected by extreme weather or even living past 100. Our methods become thinking tools, helping us reason and imagine in new ways—in and out of science. The more we expand our toolbox, the more we stretch and reshape the boundaries of how our mind can think and imagine—and even think about thinking and imagination (Chapter 6).

Our species' most unique capacity: cumulative method-making

Current explanations of what makes our species unique and successful focus on factors like our ability for shared intentionality or cooperation[278] or developing agriculture.[279,280] Here we provide a new, overlooked explanation: what makes us human is our extraordinary capacity to invent better methods and tools that expand our mind, to test them and to improve them. Our uniqueness comes down to this: we are cumulative method-makers who amplify our own cognition. Humans are not just tool users—we are master method-makers. Our species has mastered this unique survival strategy: building, upgrading and switching between powerful methods to tackle new challenges. This remarkable capacity is not just the driver of our survival but also of scientific progress. At the heart of this capacity is our universal and adaptive toolbox: a set of evolved methodological abilities to observe, experiment and solve problems that we have collectively developed into increasingly powerful methods over time—from rigorous experimentation to statistics. Our method-making capacity is why our species has been so successful in generating knowledge and controlling our environment.

While we are not the only tool-using species that communicates and cooperates,[276–278] we are the only complex method-making species using diverse tools across domains. Only our species can leverage the shared ability for trial and error to develop controlled experiments, or go from the shared sense of quantity to create geometry and statistics, and so on. Ultimately, cumulative method-making has been the key missing link enabling us to understand the world in ways no other species ever could. It characterises our unique mind and endows us with a different relationship to the world than any other animal. With tools, we come up with better ways to represent, test and even simulate the world.

Our toolbox gradually set our species on a new method-driven evolutionary path that no other animal had yet taken. It made us increasingly unique. A new kind of species emerged: homo methodologicus, the cumulative method-making human. We diverged from other primates as our capacity to invent methods expanded, together with more complex language, imagination and cooperation. In fact, method-making enabled us to turn language and imagination into something more permanent and shareable—by creating external tools like mathematical symbols, written notation and writing systems. How our powerful methodological mind developed needs to feature at the centre of how we explain our evolutionary history. *Our ability to develop an adaptive toolbox represents arguably the most unique feature in understanding the gap between the evolution of the mind of humans and smart non-human animals—an ability that is not just likely or we can imagine, but was necessary to be able to extend our mind beyond theirs.* Without it, we could not have ever created science, technology and complex societies. Our cumulative method-making capacity is what distinguishes and defines us as the extraordinary species we have become—extending our mind beyond our DNA (Table 7.1).

Our universal toolbox: an early, unified method for gaining knowledge over human history

For over a millennium, scholars have been searching—from Aristotle to Popper—for a unified method for how we gain knowledge.[204,312,316,330–332] The nearest we can get to a single, universal method over history—for what our species has always relied on to uncover knowledge about the world—is our universal methodological abilities to observe, solve problems, experiment and develop methods. Our evolution can provide a natural foundation for this understanding. Our ability to observe is fundamental to nearly all knowledge we acquire (Chapter 6). Our mind's methodological capacities are inherited regularities that link our human nature to knowledge. We have had no choice but to use these evolved abilities. This universal toolbox view here is as close as we may get to a natural definition—a universal method—of what generating knowledge has been over our species' history. In contrast, we explored contemporary science in earlier chapters; we reframed the classic scientific method—from our mind's internal processes of observing, hypothesising and experimenting that we have used throughout human history—to the sophisticated external scientific methods and tools we more recently began developing that spark science's major discoveries (Chapter 3).

While scholars have asked for millennia how we generate knowledge and discover things,[312,323,333] the deeper question is how we even became able to develop knowledge and discoveries at all? No answer was possible as long as we did not understand the evolution of our species and mind. And many did not until the mid-19th century and Darwin's *On the Origin of Species* (Table 7.1).[57]

Science has evolved from our evolved mind. We are not just a unique species, we are methodologically unique. Our large flexible system of methods embodies our intellectual evolution: from observing, solving problems, experimenting and abstracting throughout history to observing methodically, solving problems statistically, controlled experimenting and modelling abstractly in science. *Viewing humans as homo methodologicus—using a universal and adaptive toolbox that extends our mind—better explains how we have been directly able to meet our needs, solve problems and develop technological and scientific knowledge—and give rise to complex society and science. Our powerful toolbox explains our success better than commonly focusing on geography, culture or religion as the driving force behind these great human achievements.*[276–279,282] It is how we even became able to surpass our animal nature and limits. In the final section, we now explore how our extraordinary toolbox has not just shaped our understanding but has shaped our species' remarkable cognitive evolution itself.

On the origin of science: how our mind's method-making capacity drove human evolution

What has driven our remarkable evolution to enable us to develop sophisticated methods and technological and scientific knowledge in the first place? How did we get from our early ancestors running in the savannahs and forests using trial and error, to today's scientists running complex experiments using statistics? Scientists have long debated the origins of human intelligence. They propose three main explanations. One view is the *general intelligence hypothesis*. Over time, the human brain evolved to be bigger and better at processing information than our primate relatives. Our brains are about three times larger than theirs, giving us more memory, faster thinking and stronger reasoning skills.[299,334] A second view is the *ecological intelligence hypothesis.* Our mind mainly evolved adapting to environmental challenges—just as certain birds have evolved extraordinary eyesight for high altitudes and use magnetic fields to perceive their altitude.[335,336] Darwin and Wallace supported this view: that selection favoured individuals best able to adapt to nature.[337,338] A third view is the *cultural intelligence hypothesis.* Our intelligence evolved through navigating the complex social world—through challenges we faced in complex social groups. These pushed our minds towards greater cooperation, cultural traditions and more complex language.[276,278,299,339]

Each of these competing hypotheses helps explain part of the puzzle—the broader context. But they do not explain or talk about how and why we directly became able to create methods and tools that go far beyond what nature endowed us with—and to eventually develop science. What is missing is an integrated approach that combines these views and places our evolved mind's methodological capacities at the centre.[276,278,290,299,335–339] These abilities truly set our species apart. So how did they evolve to enable our ever more complex knowledge? Here we lay out a new view—the *toolbox hypothesis*—that highlights how our capacity to observe, solve problems, experiment and develop methods evolved in increasingly complex ways. It explains how this complex method-making capacity of our *cognition* evolved adapting to our *natural and cultural environment* over time.

This *toolbox hypothesis* explains how evolution favoured those members and groups of our species who could more effectively apply our mind's method-making capacity. Throughout history, those better able to develop more effective methods for tracing animals, healing injuries, navigating using the stars, building stronger shelters and eventually growing food were more likely to survive and pass on their method-making capacities—and their genes. Over time, these early methods evolved into more complex ones. Those better at using this expanding toolbox had an edge: developing better plant-based medicines, lunar calendars to predict ocean tides and better technologies. Eventually, the best method-makers invented agricultural techniques and numerical systems to more effectively plan and produce food (Chapter 7). Each successful method added a survival advantage. Over generations, those advantages multiplied—through continual feedbacks. As our species migrated across the globe, we were forced to readapt and redesign our methods and tools over and over—facing unfamiliar climates, landscapes and threats. The result is our adaptive, method-making mind.

Yet for psychologists like Steven Pinker, one of the great mysteries of our mind and human evolution has been precisely the question: how have we—given our evolutionary history—evolved to do science?[340] How have we created calculus and surrealist art that are not directly tied to our survival? After all, evolution favours traits that help us survive and reproduce. Yet our early ancestors evolved abilities—from pattern recognition to numeracy—that helped us reason logically needed to plan hunting routes and food cycles. Similarly, we also evolved abilities for abstraction needed to carve stone tools and design shelters (Chapter 7). Today, to develop logic, calculus, abstract art or scientific discoveries, we rely on these same evolved abilities. But for purposes that often no longer immediately influence our survival. So the *toolbox hypothesis* offers an answer to this mystery: our ability to develop science and complex society grew gradually from this evolving, flexible toolbox that once directly helped us survive.

For most of our evolutionary history, the environment and biology shaped our minds—predators, droughts, diseases, as they shape all species.[276,278,287] But over time, we learned to develop an extraordinary set of tools to manipulate the environment and our biology. This is how the edge shifted in our favour: we began shaping our environment just as much as it shaped us. Over time, our method-making capacity and sharing innovative tools across generations became more important than the environment, animals, plants or disease in shaping our mind's evolution. Why? *Because the more effective the toolbox in our hands, the better we could survive and adapt to these changing external factors—through techniques to control fire and develop farming, clothing and shelters for nearly any environment.* These were buffers we built against nature. What shaped the evolution of our mind shifted from external environmental challenges over much of our past to internal feedbacks: cognition, method-making and culture. Ever more sophisticated method-making and tool-making emerged, symbiotically with greater cooperation, language and imagination. Our tools turned into new selective pressures for our brain's evolution. Evolution selected better toolmakers—through evolutionary feedback loops. In a sense, our tools began shaping and evolving us. Ultimately, we develop and use increasingly complex tools because they improve our ability to survive and adapt.

This kind of method-driven selection did not just apply to individuals, but also to groups. Tribes and communities that learned themselves or from others how to more efficiently exploit their environment were more likely to outgrow and outlive hunter-gatherers. Think of the extraordinary advantage for those who mastered experimental methods for agriculture and storing food or observational methods for predicting seasons. Some could then increasingly invent the remarkable techniques to domesticate animals, breed plants, irrigate fields and fertilise soil.[276] By changing how we thought and interacted, our expanding toolbox sparked survival rates and population growth to rise at a pace not yet experienced in history. Early cities and societies emerged and grew at the pace we expanded our toolbox—and knowledge accumulated faster than ever before.[279] These were tipping points in human history. And at their heart, better method-making was driving them.

By developing agriculture and then early civilisations came a *method revolution*. With more stable food supplies, more people could dedicate their time to testing and crafting better tools. This kicked off a method-making surge. Throughout history, human populations that rapidly increased their ability to survive and innovate generally did so by extending their methodological toolbox. These method innovations, by stretching our mind, have driven our species' advances and success. And we can observe the methods at play in each advance. (Even the satisfying thrill and eureka-like feeling when creating a new tool that works or making a new discovery may be evolution's way of rewarding us for solving complex problems, signalling they are important to us.) While social and emotional intelligence helps us bond and form communities, those with greater *methodological intelligence*—who excel at testing and refining tools—were more likely to optimise strategies for gathering food, engineer better shelters and unlock new knowledge, giving them an inevitable survival advantage.

Method-making is not just an outcome of our intelligence—it is central to what made our intelligence possible and helps define it. Our minds created better methods, and those methods in turn refined our minds—from one generation to the next. *Our method-making capacity is both an inseparable product and driving force of our evolution.* It is an essential part of explaining what made our unique, evolved species and mind possible. It shaped who we became. How our mind evolved is, more than any other species, an outcome of our own making—the result of being method-makers, designing new and better ways to solve problems. We depend, with each new method, on being taught how to use them: from arithmetic, to techniques for herbal remedies, to types of experiments.

Ultimately, this toolbox view explains why we are the only species to build science. It explains how we became able to develop complex tools and knowledge—and eventually learn to create experimental controls, solve differential equations, develop AI and make discoveries that no other species can. If our methods and tools shaped our brains in the past, then developing more powerful tools today—better statistical, computational or experimental methods—is not just scientific progress, it might also have evolutionary spinoffs.

Box 8.1 Our toolbox functions like an immune system—universal and adaptive

Think of our scientific toolbox like an immune system: both have universal parts we are born with and adaptive parts we later develop. We constantly scan our environment for problems to solve, much like our immune system constantly scans for threats to tackle. We are born into the world with a *universal* immune system—broad but unspecialised. We are initially more often sick because our basic defences have to learn. But over time, we adapt, develop antibodies and build up immunities and a stronger defence system to better tackle bacteria and viruses and better survive: our *adaptive* immune system. Our toolbox works the same way. We are also born into the world with a *universal* toolbox—also broad but unspecialised. Our mind and senses are constrained to our niche of the world we can observe, and we hit the limits of our bare mind. But over time, we adapt, create and master more complex methods and tools that enable us to tackle more complex challenges and better survive and understand the world: our *adaptive* toolbox. Our mind develops new methods, just as our body develops new antibodies—by learning from experience. Our immune system learns by facing immunological challenges (new viruses and bacteria). Our toolbox learns by facing real-world challenges (new problems in nature and society).

Our toolbox naturally divides into sub-toolkits—just like our immune system into subsystems. We use our evolved *universal* toolbox to observe, experiment and reason, and our *adaptive* toolbox to produce mathematics, microscopes and controlled experiments. Our immune cells recombine genes to create seemingly endless antibody variations to tackle immunological problems we face. We invent and recombine methods and tools to create seemingly endless method innovations to tackle scientific and technological problems we face (Table 6.2). Our toolbox is our best means to confront the challenges we encounter through our diverse methods (experimental, mathematical etc.)—just as with our immune system (cells, antibodies and proteins).

Our toolbox is our cognitive defence and survival system. It protects us against disease by enabling us to design clinical trials to test how effective our vaccines are. It helps us prevent food shortages and deal with natural disasters by allowing us to devise methods to better plan and reduce risks. (Our powerful immune system is very effective but can at times attack the body, causing autoimmune disease or failing to stop cancerous cells. Our powerful toolbox is also very effective but can at times have unintended effects, attacking science and society through environmental degradation and nuclear weapons.) Ultimately, our adaptive toolbox lets us confront vast challenges and understand the world in ways no other species can. Without it, science and technology would not be possible.

Mimicking, upgrading and outsourcing our evolved mind with better tools—including AI systems

At the heart of AI lies the bold ambition: to replicate and even enhance our core methodological capacities (observing, experimenting, solving problems) that evolution has endowed us with—our human intelligence. From complex causal reasoning and imagination to designing experiments and running statistical analyses, scientists are attempting to translate many of the cognitive capacities we have developed and honed over millions of years into computer code. It is about trying to build systems that can carry out the wide range of reasoning abilities we have evolved—systems that can think, learn, infer and solve ever more complex problems. Flexibly emulating these faculties demands an extremely deep understanding of our mind's architecture itself: how it evolved, how we learn from experience, how we solve problems we have never faced before and what it takes to try and recreate these capacities. It is about trying to reverse-engineer ourselves.[90,341]

This brings us to a central puzzle in understanding this shift—from the evolutionary roots of our cognitive abilities to designing complex artificial systems: how do we best try and reconstruct these remarkably evolved capacities? How do we best link and power our scientific tools—from statistical methods to microscopes—with current AI systems? How can we understand the logic behind such systems? Should we model our methodological abilities and tools as large language models or deep neural networks? Or probabilistic reasoning systems or dynamic network systems? Or as something not yet conceived—systems that combine all these approaches into a single adaptive architecture? Each model offers a different lens into how to capture the complexities of human thought. But none fully captures its richness. So how do we best rebuild something that evolved over millions of years and was never designed to be understood in parts? And how do we best understand the design of tools that extend—and even reinvent—our evolved mind including our extraordinary method-making capacity? It involves testing whether replicating specific capacities is best achieved by isolating cognitive functions or by integrating them into a unified architecture. These questions are not just methodological and technical, they are also deeply philosophical and ethical.[90,341]

Ultimately, this reflects a challenge at the core of science: how to understand the complexities of our mind and tools and tackle their bottlenecks with ever more sophisticated tools. We as method-makers may enter a new phase with tools like deep machine learning, where we could build methods that further build themselves: meta-tools. The evolution of method-making did not stop with stone tools, compasses or telescopes—it continues to accelerate in the digital age vastly faster than we evolved.

Conclusion

To solve the puzzle of the origins of science, we need to step back and ask deeper questions: what gave our species the power to adapt to nearly every environment on the planet, intervene in nature and domesticate plants and animals? What has driven our ability to generate scientific and technological knowledge—and eventually develop modern science? How can we overcome the evolved limits of our own mind and senses? What trait best describes humans as the unique species we are? And how have we been able to increasingly influence our mind's remarkable evolution over time? The answer that ties these diverse questions together: our mind's evolved methodological abilities to observe, experiment and solve problems—our universal toolbox—that we have developed into increasingly powerful methods and tools—our adaptive toolbox. With this toolbox, we can better explain—as the unique, cumulative method-making species we are—the origins and foundations of science and technology.

While evolution gradually sharpened these inherent abilities throughout human history, we eventually began sharpening and expanding them into ever more complex methods and physical tools. We began shaping the tools that vastly expanded our own evolved mind. We are not just a species that thinks in complex ways, we are the species that builds methods and tools to think better: *homo methodologicus*. Method-making has become its own evolutionary force—one that we can now steer. What first began as a survival advantage that only our species learned to master—as cumulative method-makers and toolmakers—became both what made us human and the very engine of discovery itself.

Fast forward to the remarkable invention of the microscope and telescope at the turn of the 1600s. For the first time, our tools drastically surpassed our evolved human abilities for perceiving and imagining like never before. This marked the crucial turning point in science: when the driving force of discovery shifted from our unaided mind to our ever more complex tools and methods themselves that enabled peering into the stars and deep into the cell. The power of these tools—as extraordinary extensions and upgrades to our mind—simply became undeniable and redefined what science today is: more tool-driven than mind-driven.

That shift has only accelerated. No one imagined or predicted how merging statistics with computing power would transform science. But today, most of science is no longer imaginable without this powerful hybrid method combination (Table 6.2). These once-unimaginable tools are now behind almost everything from vaccine trials to climate models to the search for dark matter. Yet precisely because these extraordinary tools have become so common in science, we take them for granted. Every day we use them across science, but rarely take the time to ask the crucial question: what better tools could take us further that we have not yet built? Because we are comfortable with today's tools, we direct far less attention than we should to pushing these methods even further—to inventing the next generation of tools that can unlock tomorrow's breakthroughs. Getting too comfortable with our current toolbox means we stop pushing its boundaries. *But each time we hit a wall using our tools in science—each time the predictions fall short, the electron microscope cannot go further—it is not failure. It is a critical sign: it means we are getting closer to the frontier, if we learn to upgrade those very methods and tools to break the limits of science.*

Ultimately, our species' biggest discovery is not a single tool or theory—it is realising that using our methodological abilities we can develop better methods and tools that enable us to better discover, understand and control the world. With our toolbox, we developed farming techniques and early medicines thousands of years ago. Then we eventually invented statistical surveys and controlled experiments to understand the mass population growth brought about by such agriculture and the effectiveness of our medical treatments for diseases. We invented technologies to extract many natural resources, and then eventually also built tools that produce satellite and temperature data to understand the causes of climate change that such natural resource depletion contributes to. The discovery of our toolbox and our ability to expand it has opened up the possibility for us to understand and tackle our big challenges—from feeding billions to fighting health pandemics to managing the planet's resources. But like Pandora's Box, we have also given rise to new and larger challenges. Better curing diseases leads to longer lives and bigger populations. Better controlling natural resources can speed up climate change. Again, we turn to science and its tools to tackle these very challenges.

9

The new methods-driven discovery theory

Five lenses on how tools power science

Summary

We cannot design an experiment to measure what our tools cannot detect. In science, the tools we use largely determine how and what we can study—and the very questions we can ask using them. Powerful new tools—from super-resolution microscopes to the gene-editing method CRISPR—unlock entirely new types of questions, problems and discoveries that have not ever been imagined before. Here we take a step back and look at the bigger picture, pulling together evidence from earlier chapters and expanding on the insights with new perspectives. This enables us to further develop the new methods-driven discovery theory. This general theory of science is backed by five different types of evidence from diverse domains: *Statistical analysis* of science's major discoveries reveals that behind our major leaps in science there is a powerful new method or tool that made them possible. *Evolutionary biology* explains our evolved method-making capacity. *Cognitive science* highlights how we overcome the limits of our mind and senses with powerful tools that outthink, outsee and outimagine our own minds. *Scientific practice* reveals how our available tools largely define how and what we explore and understand. *Philosophy of science* illustrates how each method has built-in assumptions, and filters how we understand and interpret the world. Together, these five lenses create a powerful new understanding of how science actually works. While the search for grand explanations of how science progresses has largely been abandoned since Thomas Kuhn's concept of paradigm shifts, we expand this new general theory of scientific progress here. Integrating the five different explanations, this unifying theory offers a roadmap for discovery—and a foundation for the fields of *Science of Science* and *Methodology of Science*. Ultimately, shifting our attention to designing new discovery tools would trigger a *tool revolution* in science: where we can vastly accelerate new breakthroughs and where labs across science begin competing to design discovery tools, not just to use them.

Overview

Discovering vaccines, anaesthesia and the structure of DNA transformed medicine and how we understand human biology. Discovering the universe is expanding reshaped our understanding of our minute place in the world. Developing quantum

The Engine of Scientific Discovery. Alexander Krauss, Oxford University Press. © Oxford University Press (2026).
DOI: 10.1093/oso/9780197829790.003.0010

mechanics and relativity theory redefined how we comprehend physical reality and explain the world on both the smallest and largest scales. Yet scientists have not yet solved the puzzle and there is still no general theory of how we trigger such extraordinary breakthroughs that is grounded across different disciplines. The current explanations of science and discovery—whether from scientometricians, historians, psychologists, computer scientists or philosophers—describe individual factors that affect some but not all discoveries. (We covered these in Chapters 1–6). Yet the best way to know if our evidence or a theory in science is reliable is commonly to examine if it is backed by independent methods from different fields. Take the theory of evolution: it is well established with rich evidence using methods from molecular biology, palaeontology, primatology, biogeography, cognitive science and archaeology—all telling the same story. The more diverse the methods that establish a finding, the stronger the finding is—and a general theory emerges.

So, how does the *new methods-driven discovery theory* developed here hold up? To further test its reliability, we explore how well the theory is grounded in multiple sources of evidence: statistical data on science's major discoveries, observational methods from biology and cognitive science, analysis of actual lab practices and research methods, and conceptual analysis. Strikingly, we find a single principle running through all these lines of evidence: our powerful tools are the driving force behind science and discovery. What emerges is an integrated theory of how we develop new ideas, breakthroughs and fields through new tools we build—tools that make new ways to think, experiment, theorise and imagine possible.

Here, we lay out these five different explanations that come from diverse domains across science. But together, they tell the same method-driven story, forming a unified theory of how discoveries emerge:

- ***Before-and-after* (statistical) explanation**: Science's major discoveries and fields consistently *follow* the creation of new methods and tools that make the discovery or field possible—and provide the entirely new perspective (Chapters 1–5).
- ***Evolutionary* explanation**: Our minds evolved with methodological abilities over our species' history to observe, solve problems, experiment and create ever more complex methods—used to gain our vast knowledge about the world. Being cumulative method-makers is what makes us human. Remarkably, it has been both the driver of our survival but also scientific progress. Our method-making species is a product of evolution, and science is its most powerful outcome (Chapters 7–8).
- ***Cognitive constraint* explanation**: Our mind and senses face many constraints, biases and blind spots—from not being able to see atoms, to overlooking patterns in data. But science progresses when we invent tools that bypass these flaws and tackle these blind spots. Tools enable us to outsmart—outsee, outthink and outimagine—our own minds (Chapters 7–8).
- ***Scientific practice* explanation**: Every experiment, study or discovery is largely defined by the tool it relies on. The method we use—whether an electron microscope or computer simulation—does not just enable but largely determines what and how we can study. It fundamentally shapes what questions and hypotheses

we can ask, what experiments we can design, run and analyse, and how we measure and collect data. It largely sets which conclusions we can draw and how others can even confirm the evidence using the method.

- ***Philosophical* explanation**: Every method comes with built-in assumptions: how we define phenomena (ontology), how we measure them (measurement system), what causes them (causation), whether we study the process, outcome, property or system (level of evaluation) and what kind of explanations and theories we count as valid (epistemology). Each method is a lens that fundamentally frames how we view and interpret reality (Figure 9.1).

A critical insight emerges here: there is no alternative explanation for how we directly develop major discoveries without leveraging the needed tools that make them possible and provide the new perspective. Here, we will tackle three questions: what exactly do these five explanations reveal about science and discoveries? How do they combine to back the new method-to-discovery view? And most importantly, how can we spark a *method revolution*—one that moves science faster and more strategically?

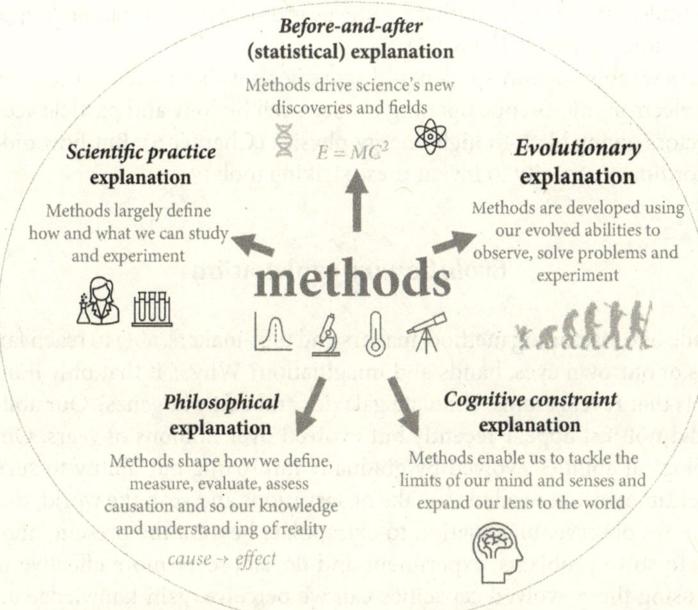

Figure 9.1 *The new methods-driven discovery theory: How new methods enable us to develop, conduct, improve and advance science*

Before-and-after (statistical) explanation

Scientific progress thrives not just when we ask better questions but when we innovate new methods that spark entirely new kinds of questions and answers—and challenge existing knowledge. The 'before' stage is marked by unsolved mysteries and existing tools that are not powerful enough to solve them. The 'after' stage marks when we devise the right tools that enable uncovering answers to those mysteries and trigger a new perspective. Take particle accelerators. Before we invented them, we did not know many subatomic particles that make up the world even existed. But after we created these extraordinary instruments, we could study their fascinating properties and behaviour. They did not just expand our understanding of physics, but they also helped fuel cutting-edge industrial technologies. Before electron microscopes emerged, we could not see the ultra-fine structural details of cells, but after we built this ultra-high-resolution tool a hidden universe opened up—from the complex structure of neurons to the inner machinery of bacteria.

Examining science's over 750 major discoveries using statistical data, we find this striking pattern across science: we made them directly after creating the powerful, discovery-triggering method or tool—and typically soon after creating them (as we found in Chapter 1). What is particularly remarkable is that many new tools do not just solve one puzzle, they spur clusters of breakthroughs unforeseeable by the original toolmakers themselves. This ripple effect often stretches across different fields beyond the ones in which the tools were designed. While scientists generally apply available tools, frontier science that produces groundbreaking research applies new tool innovations (Figure 1.1 and 1.2).

In fact, new tools not only spark new discoveries but also launch entirely new fields: from the electron microscope opening modern cell biology and particle accelerators and detectors giving birth to high-energy physics (Chapter 5). But how did we gain the extraordinary capacity to invent these striking tools?

Evolutionary explanation

What made us outstanding method-makers and tool-makers, able to reach far beyond the limits of our own eyes, hands and imagination? Why is it that only humans can build tools that reveal atoms, simulate galaxies and edit our genes? Our ability to do science did not just appear recently but evolved over millions of years. Our mind's methodological abilities evolved by gradually improving our ability to survive and adapt over time: vision needed to make observations and scan the world, memory to store what we observe, imagination to extrapolate beyond the present, and reasoning skills to solve problems, experiment and develop ever more effective methods. Only by using these evolved capacities can we perceive, gain knowledge and make discoveries. Other animals share parts of this toolbox too. But only we turned these abilities into cumulative method-making. Throughout history, our earlier ancestors leveraged our evolved methodological mind to create methods and tools: the stone tools of the Palaeolithic, the planting strategies and crop experiments with seeds and soil that sparked the agricultural revolution, and ultimately the fascinating tools that extended our perception and mind and sparked a snowball effect—triggering the scientific revolution and then industrial and digital revolutions. Each new major method in history marked a *method revolution* (as we saw in Chapter 7).

Our evolving method-making capacity transformed us: those early ancestors who could invent more and more effective hunting strategies, better track animal behaviour or reduce illness with medicinal plants had a survival advantage. From constructing pyramids by experimenting and applying geometry and engineering, to creating systems of mathematics and astronomy, it was method-makers who engineered the world around us and changed the game. Today's scientists build on these method-making skills, but with far more powerful tools (Chapters 7–8).

Let us take a step back. Our mind—like everything in nature—is the result of evolution. It developed, within our environmental and cultural niche, responding to challenges we face.[276,278,287,335,342,343] Our gradually evolving method-making capacity enabled us to develop ever more useful tools that increase our survival and vastly expand beyond what evolution alone gave us. Take our vision: it is our and other primates' most used sensory input to make sense of the world.[344,345] That is why we make so many vision-enhancing tools to tackle the natural limits of our eyesight: microscopes, telescopes, x-ray methods, spectroscopes, particle detectors, etc. (Chapter 2). These vision-extending tools do not just set how much further we can see—they enable us to imagine and understand the world much further than with our unaided mind. In science, vision is still our most relied-on form of evidence to explain the world around us (Chapter 6).

In short: our method-making mind is the deep result of evolution, and every tool we use today is an extension of our ancient mind. So we can trace the origins of science to contemporary science by tracking how we evolved to perceive, measure, theorise and invent more complex methods. But our mind faces limits imposed by nature and evolution. These help set the bounds within which we can do science—and generate new discoveries when we learn not to accept those limits but tackle our mind's constraints.

Cognitive constraint explanation

Our mind and senses—as extraordinary as they are—come with many built-in flaws, blind spots and biases to what we can know about the world. From our eyes that cannot see ultraviolet light to our ears that cannot hear ultrasound, the tools we develop using our mind enable us to overcome these limits. At its heart, this is how science moves forward. Our minds and senses evolved in a narrow window of experience and reality—the part that mattered for survival: things close to us, slow enough to track and visible to the naked eye.[276,278,290] But most of the universe does not fall into our species' small sensory window. Most phenomena in the world fall below or lie beyond what we can observe. We are increasingly constrained the further we move away from these evolutionary conditions—from the earth's surface, from our ecological niche, from our unique context. Our mind is not able, without tools, to access what is beyond our senses: minuscule atoms and photons at the quantum level, vast magnetic fields and gravitation, extremely fast phenomena like the speed of light, extremely distant planetary systems and the earth's core, intangible global economic markets and political systems, and the effects of climate change over decades. But with the right tools, we can suddenly access and pull them into view.

Today, most of what we study in science we cannot directly see with our bare eyes—from subatomic particles, to proteins and neural signals, to complex financial systems. It is the tools we create using our flexible mind that explain most of the expansion of science by uncovering what would otherwise lie far beyond our brain's natural reach.

So how we understand most of reality is determined by our tools of discovery. We can—with our naked eye—only detect a very limited range of the electromagnetic spectrum. But that narrow window widened suddenly once we constructed infrared sensors, x-ray devices and radio telescopes.

In medicine, a powerful toolbox made much of the field possible and continues transforming it: optical microscopes, centrifuges, x-ray devices, chromatography, gel electrophoresis, electron microscopes and hypodermic syringes—as we uncover when examining medicine's major discoveries. Computer-generated images inside the human body were simply unimaginable with our bare senses. With magnetic resonance imaging (MRI), electrocardiograms (ECG), computer-assisted tomography (CT scans) and cardiac catheters we can see inside a patient's body: scan our brain, track heart rhythms, map blood flow and catch disease—without a single incision. Without such powerful tools, our unaided mind could not diagnose and treat most life-threatening diseases. These tools have not just vastly improved healthcare, they have added quality and years to our lives.

Computers are vastly speeding up science and changing how we do science. Today's computing power, automation and machine learning tools enable scientists to run thousands of experiments, analyse vast data and make complex predictions at speeds and scales our brains could never imagine without them. Imagine trying to find patterns across millions of patient records by hand. Or simulate the earth's changing climate a hundred years from now. Or track a fast-mutating virus around the globe in real time. Without the massive boost of computer-assisted science, these tasks would be impossible and unthinkable. When we mapped the entire human genome, it was only possible because we combined computing systems with automated methods to process vast databases of DNA.[113] That combination also powers automated DNA sequencing, real-time weather forecasting and measuring vital signs in hospitals.[90] We think in science more with our tools than with our limited minds.

Powerful tools do not just make science more efficient, but give us entirely new eyes and open up new worlds—the microscopic world, the statistical domain, the quantum sphere and the galactic universe. To understand how tools work and shape discovery, we break scientific methods and instruments into three categories here:

- *Direct tools*: these are analogous (similar) to our mind and senses but *enhance* them—low-power microscopes, telescopes and arithmetic improve what we can already naturally do—see, measure and calculate.
- *Translational tools*: these are partly analogous to our mind and senses but stretch them further—x-ray detectors, infrared telescopes and ultraviolet spectrometers *convert* invisible wavelengths and signals further beyond our sensory range into something we can perceive and detect.
- *More abstract tools*: these are not analogous to our mind and senses but go vastly beyond them—quantum computing, network analysis, blinded randomised controlled trials and particle colliders do not resemble what we can intuitively grasp. They do not have a clear reference point with our mind or senses, but *operate far from human intuition*—on a level completely unfamiliar to our

everyday experience. Commonly, we need visual models just to interpret what the tools are doing.

How much our tools mirror human cognition is on a spectrum: from more direct, then partially direct, to more abstract tools—from sensory-enhancing tools, then signal-translating tools, to more abstract, conceptual and modelling tools. Classifying tools by cognitive distance from our evolved perception reveals an important insight. The category our tool falls into shapes how we understand—whether we rely more on our external tools or our internal mind. While all three types of tools commonly produce very reliable evidence, we often feel more confident in what we see. So consensus often forms more easily with evidence from more direct tools. To conceive and test more indirect tools often requires a radically different approach—and is often more difficult. With indirect tools, we often rely more on computational models, statistical inference and AI algorithms—and that can make consensus at times harder to reach. As we develop more indirect and abstract tools—like quantum simulators that model materials before they exist or AI systems that run thousands of experiments—we may find ourselves, in part, depending less on direct observation and more on mediated, tool-led understanding. That shift raises important questions: how do we increase confidence in tools that no one fully grasps intuitively? What does it mean for such tools—in the domains they are used—to become the principal driver of our imagination and understanding? How do we keep science grounded in our sensory world when part of science happens far outside of that world?

Think of phenomena with the least consensus like the evolution of consciousness, multiple universes and the historical origin of life.[(346)] It is because we do not yet have the needed tools to collect more direct, reliable evidence—a topic we return to on the limits of science (Chapter 10).

In short: we come into the world with a narrow range of what we can see, sense and understand—but with each new tool we push those boundaries. Computers, radio telescopes and statistical methods do not just extend but multiply the powers of our mind beyond our imagination. Most of scientific practice is distinguished by developing and leveraging tools that enable us to study parts of the world that would otherwise lie beyond our cognitive reach. In many fields today—from neuroscience, genomics and climate science, to astrophysics, economics and AI—the next major leaps will not come from thinking harder, but from designing tools that think and imagine for us. *The deeper we reach beyond our brain's borders, the more our future discoveries depend not on our mind—but on our tools. Science is, in many ways, about outsourcing experimentation, pattern recognition and imagination to advanced microscopes, x-ray machines and supercomputers.*

Scientific practice explanation—how our tools shape our very questions, problems and experiments

Science is often imagined as a logical process—one that begins with an unsolved question or problem, then applies the traditional scientific method, and ends by uncovering new experimental findings or a theory. But in reality, we are not able to run an experiment to study what our tools cannot detect—whether using advanced spectrometers or radar telescopes. In most areas of science, if we cannot observe and measure a phenomenon with our tools, it is generally invisible to our imagination too. But some domains offer some exceptions, especially theoretical physics where concepts like gravitational waves and black holes emerged—by using advanced mathematical frameworks like tensor calculus and building on existing discoveries using interferometers and telescopes (Chapter 6).

We do science using tools and methods, and those we have at our disposal are what largely define the questions and hypotheses we are even able to ask in the first place. The tools we use determine what phenomena we can explain and predict using them, and what we cannot. But how do tools do this? How do they limit and steer what we can conceive, measure and detect? Take the microscope—it transformed biology not just by discovering cells and bacteria and opening the microscopic world but by enabling us to conceive and tackle entirely new questions. For the first time, we could ask: are cells the basic unit of life? How do they actually function? Do they divide—and if so, how? The answers are essential to the foundation of biology—and these questions would never have arisen without the tool that made cells visible and experiments possible.

Inventing x-ray crystallography then revealed the mystery of the structure of DNA by enabling researchers to see its double-helix structure. This unlocked profound new questions in biology: how does DNA's structure replicate itself? How does it encode and store genetic information? These questions became the foundation of molecular genetics—and they were only possible by inventing x-ray crystallography (Chapter 1). Later, creating gene sequencing methods transformed biology again, enabling us to examine genetic variations with unprecedented precision and scale. This new tool sparked completely new types of questions unimaginable before: can we sequence the human genome? How are gene mutations linked to diseases? How can we design treatments tailored to an individual's DNA?[113]

Think of new machine learning and AI tools. These have opened new experiments, simulations and ways to detect patterns using datasets so massive that no human could make sense of them alone. From drug discovery to complex systems and climate modelling, AI tools do not just make science faster but let us do an entirely new, deeper kind of science. Innovative AI algorithms help uncover new drug candidates and antibiotics by quickly analysing millions of molecules. They spark fundamental new types of questions: how can we best find hidden patterns in medical records for millions of patients? How can we simulate billions of interactions in an ecosystem or an economy? How can we make more efficient predictions?[90,275]

Take the extraordinary invention of the spectroscope.[347] It did not just give us a new kind of vision but opened transformative new ways to explore the universe. By being able to analyse the light emitted by stars, it opened completely new fascinating questions and answers never imagined before: do the same elements that make up Earth also exist in distant stars? How did these elements form in space and come to be scattered across the universe? Where does the immense heat and energy of stars come from? What is the chemical makeup of other planets? Surprisingly, we discovered that stars and planets are built from the same fundamental elements on Earth. Different planets could then be seen as a kind of experimental control for how life could emerge. They became natural laboratories, helping us figure out what conditions make life and a liveable planet possible. The universe, at least in part, did not seem so completely distant or alien—it felt more connected, made of the same building blocks.

This logic of science—that new tools largely dictate the very scope of scientific inquiry—plays out again and again across all fields. Space telescopes open entirely new astronomical questions about the early universe billions of years ago, about the origins of galaxies and about the behaviour of gravity. Electron microscopes catalyse entirely new research areas in cell biology, materials science and nanotechnology. *New tools do not just discover what is out there—they discover what we did not know we were even missing.* Our powerful tools illuminate the deep gaps in our knowledge—they expose our ignorance. That is how many breakthroughs begin—not just by reinterpreting the old map, but by inventing a new kind of compass we had not yet imagined. So using a new tool is not just a technical upgrade, it is often actually the birth of a whole new kind of science that stretches our sensory reach.

The connection between toolmaking and the new types of questions it sparks is not just a success story for science—it is a principle for discovery. Scientific progress is driven by tool innovation and planning—not just serendipity or theories (Chapters 2 and 6). We can summarise the central point here: the particular tool we apply—whether a particle collider, statistical method or brain scanner—largely determines how we can design, carry out and analyse our study. It sets how we can define and collect data. It shapes the scope of our results, what conclusions we can draw and even how other researchers can test and confirm evidence. In other words: we tailor and fit our research around the current tools we use. We need to let go of the common belief that tools support research, they largely define it: they are our lenses that determine what parts of the world come into focus and what parts do not. The biggest breakthroughs are often less about coming up with the perfect questions and more about coming up with the right tools. In short: science is the pursuit of answers to both known and unknown questions by creating and upgrading our tools of discovery—tools that expose new realms and tackle our ignorance.

There is a deeper insight here: a tool is more than an object—it is a language of discovery. A tool like x-ray crystallography or electron microscopy does not just improve science in one lab—it transforms the kinds of questions we can ask and how we talk about and tackle them across entire fields.

While our attention is focused on science's great discoveries and theories (final outputs), our ever-expanding toolbox is what makes all steps throughout the discovery process possible. With it, we do not just uncover knowledge—we build the process of discovery itself. Here we lay out a conceptual framework of science and discovery—what we call the new method-to-discovery view (MTD). The idea is straightforward: when we extend our toolbox, we can study and measure the world in powerful new ways that enable us to develop new knowledge and discoveries—and these make up

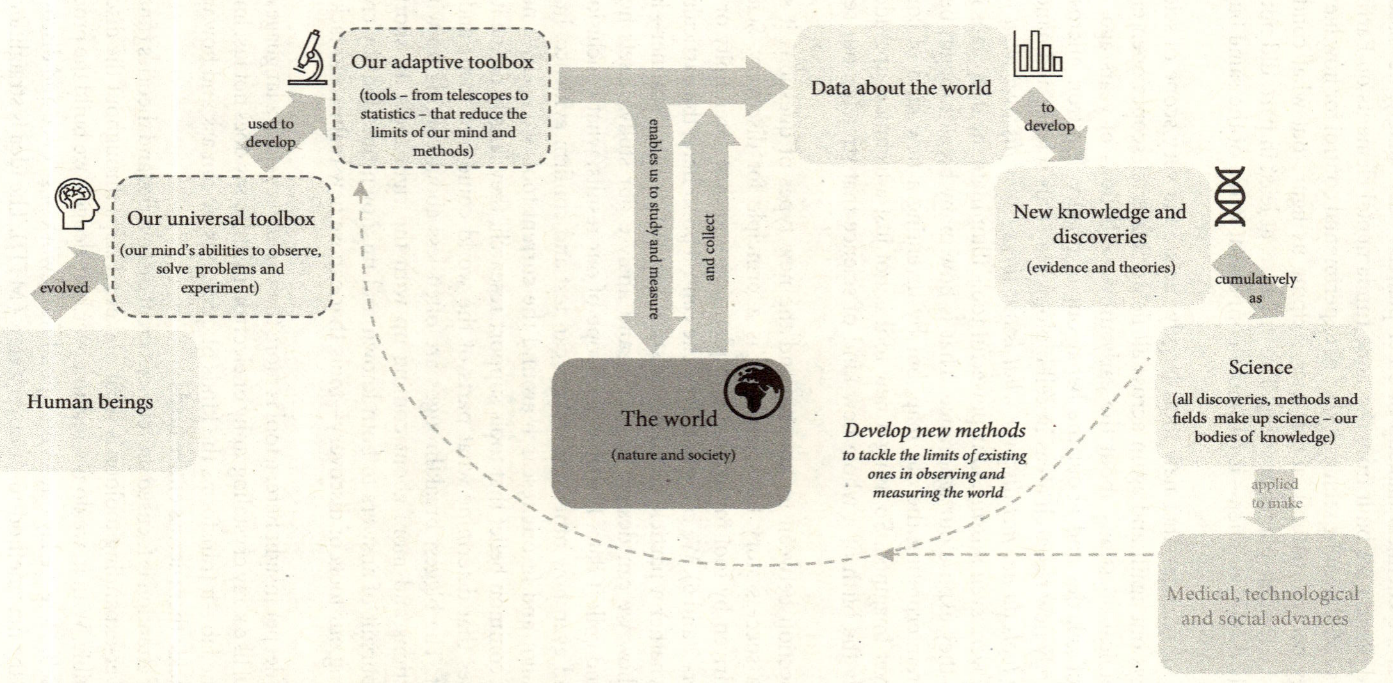

Figure 9.2 *How expanding our toolbox drives new knowledge and discovery*

science (see Figure 9.2). Here, a powerful feedback loop drives toolmaking: when existing tools fall short—whether hitting observational blind spots or measurement gaps—we build better tools, but over time those tools reveal limitations and demand better tools themselves, and the cycle begins again.

How do methods and tools connect data to results and to theory, and make both results and theory possible? How do different methods and tools produce different evidence about the same phenomenon? This MTD view offers insights: it explains how methods guide each step of the scientific and discovery process—acting as a compass steering the path to discovery (Figure 9.3). The data in a study can only be defined, collected and analysed *after* we select a method—from randomised controlled trials, to CRISPR gene editing, to gravitational wave detectors. How we assess experiments and studies also depends mainly on our methodological design. Methods define standards of evidence, replicability and precision. What we can discover in science largely begins and ends with our methods and the evidence they produce (Chapter 6).

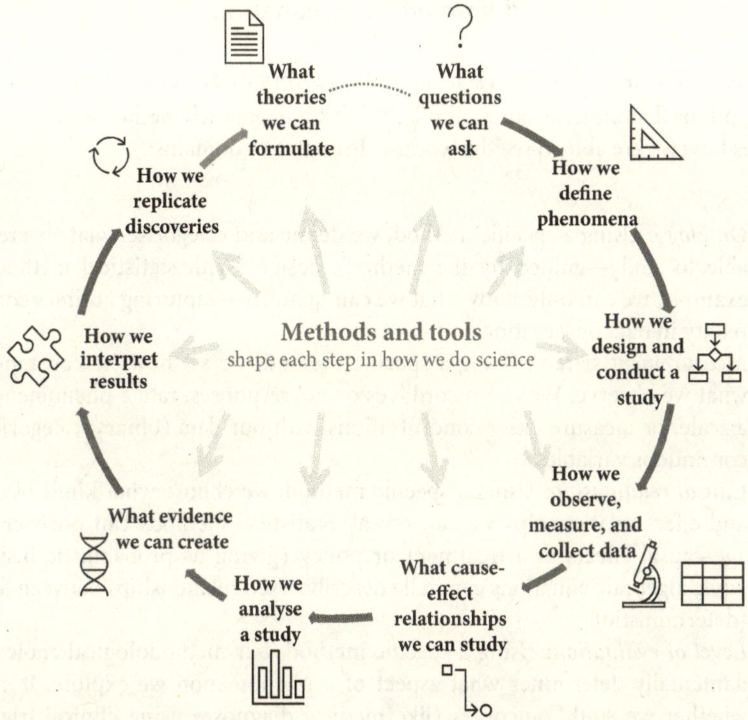

Figure 9.3 *Methods-driven science: how methods and tools shape knowledge and discovery at each step*

Other factors—like serendipity and theories—can influence part of the discovery process. But methods are the most systematic and universal framework that consistently shapes each step of the discovery process across breakthroughs and fields. Methods are the procedural backbone that makes discovery possible, testable and generalisable, while other factors generally depend more on contexts or fields (Chapter 6).

In contemporary science, we can only develop theories, directly or indirectly, by relying on the tools that uncover findings demanding an explanation; and tools are then what let us test whether our theories actually hold up (Chapter 6). Solid analysis depends on solid data. And solid data only ever comes from a solid method. That is why designing the right method is essential for running experiments. Evidence is only as strong as the methods used to generate and test it. When researchers using different methods reach the same results, our theories become more reliable. For example, climate change is not confirmed by just one method: it is shown independently by statistical climate models, satellite temperature measurements and ice core samples from Antarctica that record CO_2 levels. That convergence is commonly how our evidence and theories become established science.

Philosophical explanation

Can we also better understand science and discovery by exploring the philosophical assumptions that underlie our methods?[348,349] Indeed, each method and tool we use shapes how we are able to see the world—through five domains:

- *Ontology*: Using a specific method, we define and categorise what we are even able to study—guided by the method's design. With statistical methods for example, we can only study what we can quantify—capturing at times complex reality in data observations.
- *Measurement system*: Using a specific method, we set how we can represent what we observe. We can record 'yes or no' responses, rate a phenomenon on a scale, or measure exact concentrations with our data (binary, categorical or continuous variables).
- *Causal relationship*: Using a specific method, we choose what kinds of cause-and-effect relationships we can reveal. Statistical methods can uncover average causal effects of a treatment or policy (giving us probabilistic insights), while algebraic equations generally describe fixed relationships between factors (deterministic).
- *Level of evaluation*: Using a specific method, our methodological choice fundamentally determines what aspect of a phenomenon we explore. It shapes whether we study outcomes (like medical diagnoses using clinical trials) or processes (like tracking behaviour using observational studies). It determines whether we investigate systems (like our ecology using computational modelling) or properties (like chemical substances using spectroscopes).
- *Epistemology*: Using a specific method, we set what counts as valid evidence and explanation. The scope of our knowledge is shaped by the scope of the method we employ to create that knowledge. How rigorous our climate predictions

for example are depends on how rigorous our statistical and computational methods and satellite observations are.

Each tool has these assumptions built in. Our tools standardise and automate how we do science—making it possible for different scientists to investigate the same problem in the same way.

We can illustrate this with a clear example: randomised controlled trials. RCTs are widely seen as the best method in medicine and social science for assessing how effective a treatment, drug or policy actually is. When we apply the RCT method, we are studying one targeted cause—the treatment—and isolate its effect from the broader environment. This research design answers questions like 'Does this vaccine reduce infection?' but is less suited for exploring complex webs of causes or social systems. Using the RCT method, we measure the outcome (did it work)—not the process (how did it work). We focus on the average causal effect on everyone in the trial—not the range of effects across individuals. RCTs are best for simple, well-defined interventions but struggle with complex, interconnected interventions like urban planning or climate policy.[(234)] These built-in features of any method do not just shape our results. They shape how we see the world itself: dictating how we define, measure, evaluate, establish causation and ultimately what we count as reliable knowledge. Compared to RCTs, methods like observational studies, in-depth case studies or network analysis let us examine similar questions but from different perspectives. While the RCT method has transformed how we fight disease and design public policy, it cannot study other aspects of the same phenomena—where other methods fill in the gaps.[(234)]

We cannot emphasise this enough: the way we understand the world is inseparable from how we measure it. The cause-and-effect relationships we uncover—or miss entirely—depend on the available tools we use to detect them. Our current methods are the lenses through which we see reality—the bridge between the world and how we examine and explain it. To describe how we spark discoveries is largely to describe the methods we leverage to study, measure and reveal.

This methods-driven perspective also helps explain why scientists often argue: namely how major debates and disagreements in science arise. Big debates in biology—from the nature of life to how much nature or nurture shapes human behaviour—generally emerge from using different methods to study genes, our natural environment or social context. Big debates in physics—from the nature of matter to the universe—are often traced back to applying different methods. Think of deterministic algebraic equations or probabilistic statistics to express these phenomena. Big debates in medicine and economics—from the most effective treatments to the best government policies—typically arise from relying on different methods. Think of randomised trials, observational studies or network analysis. Different methods push certain aspects of a phenomenon into the foreground and others into the background. They frame the very way we think about and conceptualise the same phenomenon. After outlining how our powerful methods drive science from five different perspectives, we next expand on the general theory of science (which we introduced in Chapter 6).

The new methods-driven discovery theory: backed by five different perspectives

Can we reduce how we make discoveries and fields to a universal theory that we can test with evidence? Can we describe the discovery process in a way that holds up to evidence across fields? Equations like Einstein's $E = mc^2$ and Newton's $F = ma$ formalise knowledge in fields like physics—but can the method-driven process of how we spark breakthroughs also be better formalised? A good scientific explanation generally needs to combine solid evidence with a theory explaining 'how' (the causal mechanism). By examining science's major discoveries and fields, a remarkable pattern emerges: new methods and tools are *a necessary condition and commonly the key driver* for how science's major discoveries and fields emerge. Other factors like funding,[(6–8)] teamwork,[(9–11)] institutions,[(208,215)] greater education[(44,85)] or serendipity[(18,20)] support breakthroughs but are not always necessary—they vary by discovery and by context. And theories cannot be meaningfully developed or tested without building on methods (mathematical, experimental). What is at the core of progress—its common thread: the development of new tools to see and measure the world through new lenses (Chapters 1–6).

Here, we combine that evidence across science with this theoretical explanation. This new methods-driven discovery theory is straightforward: we unlock science's major *d*iscoveries and fields by using current scientific *m*ethods and *t*ools and developing a *n*ew method or tool that enables us to *o*bserve, solve *p*roblems, *e*xperiment and *th*eorise and collect new evidence of the *w*orld—in ways that are impossible without the new method or tool (Figure 9.4).[1] The discovery process can be reduced to these methodological features. The key is inventing a new method or tool that generally makes it possible to see, test and think in ways we could not before (Chapters 1–6). While disciplines vary widely in *what* they study—from chemistry to neuroscience and astronomy—they are consistently powered by the same engine of *how* they study and advance: new methods and tools. Take the nobel-prize-winning discovery of cell structure. This remarkable breakthrough was triggered using a new electron microscope to uncover cellular structures no one had seen before—while experimenting with mice and using other methods such as statistics and a centrifuge. This optical innovation transformed how we understand life at the cellular level and opened modern cell biology.

Many think scientific progress is mainly driven by new findings and theories. But digging deeper, we find that creating and applying new methods and tools are commonly what drive those new breakthrough findings and theories—by reshaping how we perceive, experiment, imagine and theorise in ways not possible before (Chapters 1–2). This is not just a theory about tools of discovery—it is a theory about science's central driver of perception, experimentation, imagination and theorising. This theory

[1] We can express the theory as: $(M \text{ and } T + nM \text{ or } nT) + (O, P, E, Th) + W(1, 2, 3\ldots) = D$

explains how we make science's major discoveries and fields—and we can express it simply as:

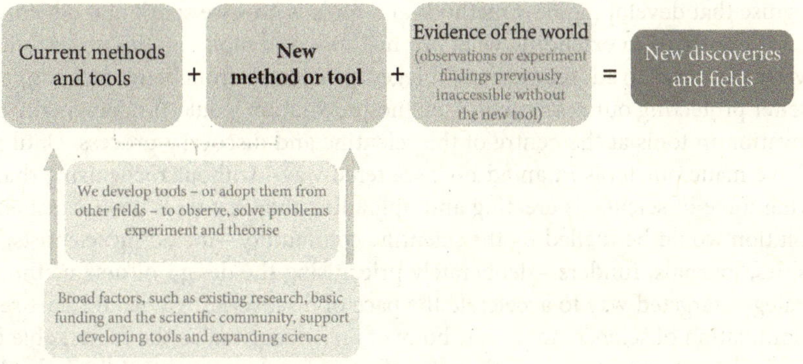

Figure 9.4 *The new methods-driven discovery theory*

This general theory links scientific progress to the current gaps in our tools. We can use the theory as a guide to spur future breakthroughs by tackling current method bottlenecks and extending our discovery tools. Our big unsolved challenges—curing neurodegenerative disease, understanding our brain's complexity, halting climate change, etc.—stall when our current tools cannot see or test deeply enough. In short: bigger new methods are more likely to produce bigger, more transformative discoveries, while relying on smaller conventional methods—ones we are familiar with—are more likely to produce smaller conventional knowledge.

A *method revolution* in science

Imagine if a method and tool revolution transformed the future of science as dramatically as six game-changing tools did for the 17th-century scientific revolution (Chapter 7). *What if scientific labs began competing to invent tools of discovery, not just to apply them? This can spark a tool revolution across scientific fields—but also across science policy, funding priorities and even the way we think about science and discovery itself.* Imagine a world where research labs invest as much time in making method innovations as they do in running experiments and collecting and analysing data. Imagine global open-source platforms where scientists prototype new research tools across disciplines—from more advanced real-time brain imaging in neuroscience, to ultra-efficient carbon capture methods in climate science, to designing materials in engineering with more advanced AI tools. Or imagine even AI tools that are built as much for designing new experimental methods as for running existing methods. *Such a transformation would fundamentally change our focus of the eureka moment: from just single discoveries to equally celebrating the moment we invent powerful new tools that often unlock many more discoveries.*

An awareness within the scientific community of this new tools-driven discovery principle would have the power to drive a method revolution in science—a fundamental change in mindset among scientists. It means researchers would recognise that developing new methods and tools is how we spark new advances. If we shift our focus to extending what our methods and tools can do, we introduce a powerful new way to address science's biggest problems—from better tackling cancer to better protecting our ecosystem. A method revolution would fundamentally place innovation in tools at the centre of the scientific and discovery process. Until now, we have made our tools in an ad-hoc, scattered way—without recognising that the driving force of science is creating and upgrading the very tools we use. A method revolution would be fuelled by the scientific community—including scientists, universities, journals, funders—deliberately prioritising the design of new methods in a strategic, targeted way to accelerate the pace of discoveries. We would start seeing the foundation of science not as our body of knowledge but as the remarkable toolbox it is that we invent and continually refine to develop new breakthroughs. If we want to better understand the world and science, we first have to understand—and improve—the ways we study the world and do science.

When we think of science, what is the first thing that comes to mind? Most think of our bodies of knowledge: laws of physics, biological mechanisms, chemical elements. But is it not just as important to understand the way we generate that knowledge—to understand our tools and how we develop them? The tools we have invented up to now largely set the scope of what we can explore and what we can know. There is an inseparable interplay between our available tools and the discoveries we can make. Before we can make a breakthrough, we first have to come up with a way to expand how we study the world. Science's enormous success is largely explained by upgrading our toolbox over and over—from massive particle accelerators discovering fundamental particles to CRISPR rewriting genes. The success of our best methods in tackling our large challenges in science, technology and society in the past highlights the power of deliberately extending the tools we have today.

A methodological shift is also needed in how we conceive science itself. We should no longer divide scientific research into two categories, experiment and theory—or experimental scientists and theoretical scientists. Methods and tools are what make both developing experiments and theories possible in the first place. They determine our experiments (how we design, run and assess them) and govern our theories (how we formulate, test, refine and validate them) (Chapter 6). That is why we introduce here a third essential dimension to science: method. Without methods and tools, science is missing a central pillar. Recognising this method-experiment-theory distinction gives us a much clearer, more realistic picture of the actual nature of science and scientific progress.

Conclusion

The world around us consists of astonishingly complex mysteries—from galaxies colliding across the cosmos to genes shaping life itself—that we can only uncover and understand using the tools we have invented. Recognising that new methods and tools are the engine of discovery offers us a new way to understand and advance science: by redirecting our attention to extending our extraordinary toolbox. What emerges is a new methods-powered explanation of the foundations of science, and a new general theory that explains how we develop breakthroughs by innovating and re-innovating new tools. This is the core of the new methods-driven discovery theory: inventing better ways to look for and find answers. It explains how our ability to access the world is constrained by the available tools we have designed so far. Simply put: we cannot do science in ways other than by using our available methods and tools. New tools shape the very direction science takes. From quantum computing that opens up an entirely new way to think about information itself, to the gene-editing method CRISPR that makes biology programmable.

Expanding the research frontier is about breaking through the bottlenecks of what our current tools and mind can handle—it is about innovative toolmaking. Science is, in many ways, about amplifying how we see, experiment, infer and imagine with cutting-edge spectrometers, telescopes and supercomputers. This methods-led principle of science is backed by multiple independent strands of evidence and methods—as we laid out here. New methods are what enable us to revise our biggest questions, answers and theories under new light—and develop entirely new ones. If we want better answers, we have to change the way we approach the questions—and that generally demands new ways to explore them.

Take the ten most powerful tools of discovery that have all been continually upgraded (Table 6.1). The great method-makers who built these groundbreaking innovations have arguably had more impact on scientific breakthroughs than other scientists. For we have implemented each of these tools to catalyse multiple discoveries (Chapter 1). Yet most people have likely not ever heard of these method discoverers—the most impactful researchers in history. But all are likely familiar with discoveries made possible by harnessing these great tools: from the discovery of DNA's double-helix structure (using x-ray crystallography) to the discovery of cell structure (using electron microscopes). These tool innovators are the forgotten heroes of science. We need to rewrite science textbooks to make them the key protagonists of science. After all, our best methods and tools may well be the greatest and most efficient things humans have ever created to understand and shape the world. The future of science depends largely on the toolmakers and method-makers who give us new ways to see, explore and imagine.

If we all became aware of the power of toolmaking on discovery-making, it could spark a *method revolution* in how we do science. If we taught young scientists to see toolmaking as an essential part of discovery—as important as finding answers. If institutions and funding agencies prioritised method innovation as much as chasing big results. And ultimately, if we take the needed steps and time to identify our own method constraints and contribute to expanding science's toolbox—to unlock new domains we cannot imagine today. But this requires us to reshift our attention from the shine of discovery to the engine that powers discoveries—so that discovery tools begin to share the spotlight.

PART III
THE PRESENT LIMITS AND FUTURE OF SCIENCE AND DISCOVERY

In Part I of the book, we asked big questions: how do we drive new discoveries, scientific fields and ultimately science? The answer: by inventing powerful new methods and tools that make new ways to see, measure and experiment possible. In Part II, we then tackled the question about the deeper origins of science: how did we start science in the first place? The answer: by using our mind's evolved methodological abilities that we have developed into ever more sophisticated methods and tools. These enabled us to develop scientific and technological knowledge—and eventually give rise to the innovative techniques that sparked the agricultural, scientific, industrial and digital revolutions. After mapping out the foundations and origins of science, other fundamental questions at the frontier of science arise: what are the current limits of science? How can we break through these limits faster? Can expanding our toolbox play as powerful a role in explaining the future of science as it does the past and present? In Part III, we turn to these questions—some of the most important questions we tackle in the book, because the answers will shape what science can do next.

Pulling the three parts of the book together, we gain a new picture of science—how our toolbox is at the foundations, origins and limits of science. Strikingly, we uncover how the foundations and boundaries of science are primarily the foundations and boundaries of the evolving methods and tools we create. We now combine these parts in the last two chapters.

10
The edge of discovery

How the boundaries of our toolbox set the current boundaries of science

Summary

Are we close to hitting the limits of what we can discover? Or can we keep pushing forward into the edges of the universe, atoms, genes and human societies? Are we nearing the boundaries in some fields like theoretical physics or macroeconomics? And what exactly shapes the current boundaries of what we can know? These are key unanswered questions, but the answers are often less mysterious than they seem. Recognising that progress in tools is what mainly drives progress in science—that spectrometers reveal the chemical makeup of distant planets, that x-ray crystallography uncovers the molecular architecture of complex compounds—gives us a new way to think about the barriers of science. Yet generally, researchers study the limits of science only through the lens of their own field; they point to *what* we are not yet able to explain: from dark matter in physics, to how life began in biology. Yet here we explain *how* our current scientific frontiers are shaped by five factors: our *scientific methods* and tools set the present limits of what we can observe, test and discover in most of science today, but our evolved *mind* shapes how we develop those methods—while the *time and place* we live in, our *human* perspective and our *social* context help shape what we study within these limits. What is key is that the incredible power of our new tools of discovery consistently break through these barriers, by reducing our human constraints and vastly stretching our scope of the world. Science's boundaries are not fixed, they are the shifting edges of a map we redraw with each major new tool. Still, there are many mysteries we cannot unlock with our current tools today: from the size of the universe and superstrings to how consciousness emerges and predicting the rise and fall of democracies. But there is a paradox: understanding what drives the current limits of science is the best way we have to expand and redefine these very limits.

Overview

We know the universe came into existence about 14 billion years ago. We know physical reality is governed by fundamental forces—gravity, electromagnetism, the strong force and the weak force. We know the chemical elements of the periodic table make

The Engine of Scientific Discovery. Alexander Krauss, Oxford University Press. © Oxford University Press (2026).
DOI: 10.1093/oso/9780197829790.003.0011

up the world. And we know our genetic material is transferred through our DNA. These discoveries form central and reliable pillars of science—so solid that we are unlikely to replace them with fundamentally different breakthroughs and theories that are as remarkable. But does that mean we may be getting closer to the limits of science?

The evidence shows that alone in the last few decades, we have made over a dozen groundbreaking discoveries that have reshaped entire fields. Think of the nobel-prize-winning discoveries of CRISPR gene editing made in 2012 that rewrites life's code—it was sparked by a new method of differential RNA sequencing developed in 2010.[(114)] Or take the existence of the Higgs particle also in 2012 that confirmed a missing piece of physics—it was detected by the large hadron collider built in 2008.[(117)] Or the detection of gravitational waves in 2015 that illuminates the mysteries of black holes—it was achieved by the LIGO laser interferometer upgraded in 2015.[(138)] Or other major non-nobel discoveries like mapping the human genome in 2003 that transformed personalised medicine today—it was driven by an improved genome mapping technique created in 2001.[(113)] These discoveries did not just add details to what we know, they redrew the borders of genetics, physics and astronomy. Each time, new tools or methods were key in breaking through the scientific frontier.

If we want to keep expanding our frontiers, we need to ask: what holds us back today? What limits our current tools to explore the world in new ways? The limits of our current tools tell us how far we have gotten and when we hit the borders of how we can presently do science. Many unsolved puzzles exist at the research frontier: what exactly makes up our universe? What are the evolutionary origins of stars and planets? How exactly did societies evolve? What exactly is the nature of memory in the brain?[(346)] Can we not just understand cancer's causes but find cures that work for everyone? What parts of the scientific process can be best carried out by AI and machine learning methods? Then there is one of the most perplexing mysteries of all: consciousness. We have advanced tools to scan brains, map neural activity in real time, detect regions involved in decision-making and language, and mimic some aspects of human thinking with AI methods. But the enormous challenge of consciousness remains unsolved: how do the physical processes of neurons create the vivid, subjective experience of being alive and aware?[(346)]

Tackling the limits of science is crucial, because even though biology and medicine have drastically reduced deadly diseases and extended our life expectancy, we still do not have cures for many diseases, cancers and viruses. Physics and chemistry have fuelled vast technological and industrial advances that built modern society—from electricity to computers—but we have not yet solved the challenges they foster like climate change, pollution and the risk of advanced weapons. Yet scientists rarely study the boundaries of science in a systematic way, at times seeing them as too big, too abstract or not subject to scientific study. A keyword search of 'limits of science' and 'boundaries of science' in publications up to 2025 in the leading journals *Nature* and *Science* results in about 250 hits. Yet we have still not explained what actually drives the current boundaries of science and how we can expand them in an integrated way.

Researchers do not focus on *how* we push our boundaries to discover entirely new phenomena; but rather generally focus on *what* phenomena are presently out of reach: puzzles in physics,[(350–352)] unsolved mathematical mysteries,[(351,353)] the bounds of our mind[354–357] and philosophical aspects underlying scientific laws.[353,358–361] The journal *Science* published a special issue *What Don't We Know?* exploring 25 of the biggest open questions in science.[346] From physicists who explore questions such as what dark matter is made of, to biologists who investigate questions such as how life began on earth. Yet it helps to think beyond disciplinary borders. In an influential book, *The End of Science*, John Horgan offers a pessimistic outlook, claiming that science may be coming to an end as science is running out of big questions.[362] But the last few decades have proven the opposite: major new breakthroughs have been achieved and many big open questions remain. Yet there is still no integrated framework that explains where our scientific boundaries lie or how we can best expand them. The best way to achieve this is by combining methods and evidence from different fields. Here we draw on methodology, cognitive science, social sciences and science of science—each contributes to capturing the larger picture.

Whenever we run into a wall with our current tools—whenever we cannot see smaller, farther, faster or deeper than our tools allow—it means we are reaching the frontier of science. If we understand what holds us back and deliberately redirect our energy to tackle it, we can break through the frontier. Here is the key insight: scientific methods and tools set the current boundaries of what we can know and what discoveries are possible for most of science today, but our evolved mind also shapes how we create those methods—while where and when we live, our human perspective and social factors help shape what we study within these boundaries. Science is something we do, using our tools and our mind, within our niche of the world. So what factors shape the limits of science and the scope of discovery? Here we describe the five main factors:

- ***Methods and tools***: Today's major discoveries that break boundaries are only possible with advanced tools—cutting-edge statistics, computers, x-ray devices, MRI scanners. These massively expand how we see, think, reduce complexity and imagine. These tools set what we can measure, model and explain in the world—and what we cannot. They come with their own blind spots and limits—from resolution limits in microscopes to biases in AI models. That is why the real challenge—and opportunity—is finding ways to push beyond those barriers with ever more powerful tools.
- ***Our mind***: While our evolved cognition and senses shape how we observe, solve problems and reason, we vastly extend them and tackle our built-in blind spots with the tools we invent. We are able to uncover those discoveries that fall into our sensory range that we are able to perceive with our methods and tools that stretch our mind in extraordinary ways—from mapping neuron synapses to detecting distant galaxies.
- ***Our place and time***: We live within a specific century and in a limited pocket of the universe—on one small planet. That means we have to deal with the data we are able to gather: here and now. This shapes part of what we can discover. The further we study the past (like the origin of our planet or the Big Bang) or into the future (like the evolution of a virus, species or galaxy), the current limits of science are shaped by human limits—the boundaries of our methods and mind. But also by the increasing difficulty of collecting reliable data as we move away from the present.

- ***Human perspective***: We humans are the ones doing the science—and developing the tools that enable doing so. We focus on questions that serve our needs and wants: we care about curing disease, understanding human origins and building better human technologies. Even the way we define a 'problem', 'solution' or 'sufficient evidence' is shaped by our human perspective—our human lens that drives incredible discoveries but also leaves blind spots in what does not matter to us as humans.
- ***Social context***: By pooling the method-making resources of many scientists, our scientific community spurs research at the frontier. Our institutions, funding and societal challenges—from pandemics to climate change—can influence some of our research priorities and tools we produce. Economies of scale, reward systems and science policy also help steer research directions, influencing some breakthroughs we chase.

These five factors together set the limits of what we know, but there is one clear, unifying thread: our methods and tools. Through innovative toolmaking, we can push past the limits of our current tools but also our senses, mind and human perspective. Our tools are generally the factor we can most deliberately improve. And only a cutting-edge telescope, computer or sensor lets us see something we could not before. Science can only ever go as far in explaining nature as the tools we have developed so far can visualise, detect and measure it.

Here, we tackle three questions: how did we, throughout history, expand the boundaries of science? How do these five factors actually shape our current limits of what we can discover? And how do the strengths and limits of today's most advanced tools largely define the edge of what is possible to uncover?

The evolution of science: expanding the boundaries of science

In the early 1500s, we viewed our physical world as geocentric and finite. But we developed extraordinary tools and methods in the 1600s like the telescope that enabled us to peer at distant planets, the barometer to measure air pressure and calculus to describe motion. These tools transformed how we thought by allowing us to explore and theorise about the world in new ways. By the end of that century, our physical world was no longer geocentric and finite but heliocentric and unimaginably vast.

In the medical world, we were not able—in the mid-1800s—to link diseases to their causes and test treatments properly. But by the early 20th century, we developed systematic controlled trials using randomisation, blinding and placebos, powered by statistical analysis. This method revolution vastly improved how we experiment, turning medicine into a rigorous, evidence-based science. We were then able to measure what actually worked—the causal effects of medical treatments—cut deadly guesswork and extend human life. With randomised controlled trials transforming how we test what drugs, vaccines and public policies work best,(234) we redefined and pushed the limits of medicine but also psychology and the social sciences. These fields became in part redefined by the limits of this new method.

The boundaries of science are not rigid but are the flexible edges of a map that we keep redrawing. When our tools stagnate, so does our progress. Just like the first telescopes forced us to revise and update what we know about our solar system, the size of the universe and our place in it, the first microscopes uncovered hidden microorganisms and eventually changed our understanding about disease and infection. Each new tool did not just add data but rewrote and redefined the very limits of physics and biology. These limits were no longer defined by the limits of our senses but became redefined by the limits of these new tools. These remarkable tools were continually improved and very successful up to the 19th century until we reached the blurred limits of their resolving power—and again the limits of science in different fields. Leading up to the 1930s, biologists and physicists were able to observe some but not other phenomena they thought could exist. A critical change came in 1933 with the invention of the electron microscope, bringing ever more minuscule phenomena into view and revealing more of the mysteries of cells, large molecules and crystals.[363] It revolutionised fields from biology to materials science, and their limits became once again reset by the limits of this new instrument.

The current edge of science is largely synonymous with the current edge of the tools we have created so far. Without progress in methods, major new discoveries are commonly not possible. Without launching improved space telescopes to search for ever better signs of new planets or even life on distant planets, we cannot ever find them (Chapter 1).

The success of science is—at any moment in history—mainly a success story of the incredible tools we have in our hands, and invented and recombined up to now. What is striking is that we have not yet given enough attention to this fundamental fact: there is no general methods-driven theory of science and the limits of science.[346,350–353,358,361,362] Instead, we have stretched the bounds of science in a surprisingly ad-hoc, piecemeal way—by individual researchers who happen to innovate a new tool enabling us to broaden our reach (Chapters 1–6). We have expanded the frontier without a science-wide understanding that our new tools are what trigger our major new advances.

We have seen how six new groundbreaking inventions unlocked most major discoveries in the 17th century (Figure 10.1). Pioneers like Galileo, Hooke, Boyle and Newton each used at least one of the newly invented tools to uncover mysteries about life, the human body, mechanical physics, light and the cosmos. Strikingly, these tools, in updated forms, remain essential today. It is not by chance but by devising new tools that make the invisible visible and the complex understandable.

Our greatest discovery may be uncovering that progress itself depends on constantly building better ways to observe, measure and test the world. Those remarkable inventions around the 17th century did not just explain nature—they planted the seeds of modern science. The roots of modern science stretch back and branch out through these few extraordinary tools (Chapter 7).

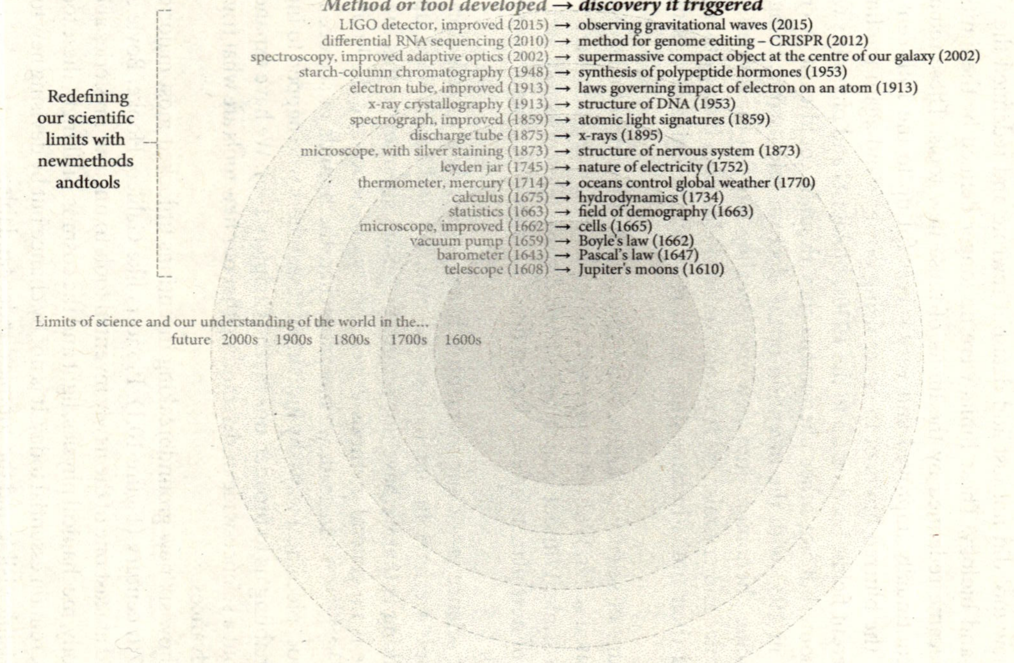

Figure 10.1 *Our scope of the world expands at the pace of our new methods and tools*

The major discoveries and the tools used to make them are based on nobel-prize and major non-nobel discoveries.[82]

The five factors shaping the current limits of science

Methods and tools

Here we unpack the five factors that both hold us back and drive us forward as we try to break through the frontier of science. In most areas of science today, a powerful statistical method, cutting-edge x-ray scanner or super-resolution microscope can generally reveal more than a million eyes or minds puzzling over the same question. Our best methods and tools do not just push the boundaries of discovery but define where we presently draw those boundaries. They determine our experiments and data. With each powerful new tool we have invented and re-invented, we have drawn and re-drawn the map of the unknown.

We have built ultrasound devices that detect soundwaves otherwise out of our reach, developed massive datasets that work as external memories, created computer and machine learning technology that examine those data to find hidden patterns. Amplifying our toolbox greatly amplifies how we can grasp reality—to observe the previously unobserved and explain the previously unexplained. Our toolbox has opened our scope to a world of microbes, synapses in the brain, elementary particles, galaxies billions of light-years away and genetic mutations that cause disease. Progress in science is largely about upgrading our instruments so we can explore parts of reality that evolution never equipped us to see.[82]

But every tool has its limits. Each MRI scanner, supercomputer or statistical method has built-in boundaries that set what questions we can study and what findings we can make (Chapter 9).[234] *Each tool defines the scale, precision and kind of detail we can uncover—leaving deeper layers of reality hidden until new instruments reveal them.* Our tools struggle to make sense of extremely complex systems where the whole is more than the sum of its parts. They fall short when trying to model emergent properties that do not easily reduce to their parts—think of weather systems with chaotic, hard-to-predict dynamics, or the intricate signalling networks inside living cells. Despite powerful tools like x-ray devices or high-energy particle colliders, there are layers of structure and interaction we cannot yet measure or image. Our best computational methods including supercomputers, quantum computers and simulators also hit real barriers to how deeply we can study, model and theorise about the world.

These limitations of our tools also constrain the very physical laws we are able to discover—and those we are not. They restrict the very theories we can conceive—because theories ultimately rely, directly or indirectly, on the unexplained findings our tools reveal (Chapter 6). We cannot properly conceive or confirm theories if we lack the methods to measure the variables or compute predictions. In neuroscience for example, we cannot fully explain consciousness because our current tools face enormous measurement constraints. Even our best brain-imaging tools cannot yet map the brain's activity with enough resolution or completeness.[346]

Our mind

Our mind is extraordinary, but it faces many built-in limits, blind spots and biases that shape how we see the world, how we reason and how we do science. As the physicist

Stephen Hawking put it, 'We are just an advanced breed of monkeys on a minor planet of a very average star. But we can understand the universe. That makes us something very special'. How we do science relies on our evolved methodological abilities to solve problems, test hypotheses and spot causes, but also abilities to remember details, use language, learn from one another, teach, abstract and imagine possibilities. Because our mind is the unpredictable product of evolution, we are tuned to a narrow window of the world: making sense of those parts of reality that mattered for our survival and our senses can directly pick up (Chapters 7–9).[276,278,287] Even our basic concepts of space and time partly reflect the limits of human perception and the way our mind simplifies a complex world to make it manageable. These built-in blind spots shape how we model the world and even the kinds of answers that seem sensible to us.

Our evolutionary history did not prepare us to see, grasp and mentally model the size and nature of the universe, the speed of light, climate systems, global financial markets, the tangled networks of neurons that produce consciousness, and huge timescales. Only in recent history have we learned to make discoveries and stretch our understanding beyond what our eyes can see or our intuition can handle: by inventing powerful new tools and methods.[82] With our evolved mind, we gather information through our vision and senses of temperature, weight, time and speed.[364] But our remarkable tools push far beyond those limits of human eyes, ears, memory and imagination. Science is mainly about identifying the blind spots in our perception and reasoning by devising new methods that enable breaking through those current limits. Each time we invent a more efficient detector, a faster algorithm or a more precise sensor, we expand the scope of what we can discover. This is why science is deeply human; it is a story of recognising our limits and refusing to accept them—with better tools.

Our place and time

How does the time and place we live in affect what we can discover? We live in a small corner of the universe, on earth, and even with our best space telescopes and spectrographs, much of the universe is out of our reach. We also live in one moment in history, in our current century, and we cannot run experiments or collect direct evidence on the distant past. Instead, we must piece together parts of a puzzle and extrapolate using the evidence we can get our hands on—from using space telescopes, to extracting samples of ice cores, to studying ancient fossils with radiocarbon dating. We cannot directly observe the Big Bang—we infer it from cosmic background radiation and galaxies moving away from us.[148] We rely on indirect evidence and models not only to study the birth of the universe but also the birth and evolution of life on earth. We do the same to predict how antibiotic resistance may spread or how our solar system may evolve in the future. Exploring the deep past or forecasting the far future always hits the limits of evidence. Even trying to explore other nearby planets or moons in our solar system faces enormous instrumental challenges that limit what we can know—let alone the hope to collect rock samples from exoplanets one day.

After all, every tool we use, experiment we run, observation we make and measurement we collect happens in a specific place and time. We test drugs in specific medical clinics, find fossils in particular geological sites, experiment on animals within local ecosystems, and observe gravitational waves with detectors from earth. Where we are in space and time constrains the evidence we can collect and the discoveries we can make about most of the universe and most of the past. Discovery is tied to the evidence we can generate—here and now—with the current tools we have.[82]

When scientists develop theories about how life first formed, what dark matter is, how the moon was created or how conscious experience evolved, they are going beyond our niche of the world accessible to us with our tools. They hit the current edges of scientific knowledge. *We generally run into our current barriers to what we can discover, the further we move away from our human domain: the farther we look back in time, forward into the future, down into subatomic particles or out to vast cosmic structures* (Figure 10.2). Why? Because our evidence gets scarcer, more indirect and less reliable as we move away from our familiar human scale.[82] When we study the deeper past—in geology, cosmology, palaeontology and archaeology—we face constraints at the borders of science. We cannot experiment on dinosaurs or observe our early solar system firsthand. Instead, we build models, extrapolate, cross-check evidence from different fields and make our best inferences. Even so, we can often build remarkably detailed explanations about the distant past.

We reach the current boundaries of science when it is no longer possible to gather new evidence with today's best tools—or reliably extrapolate from what we do know. This is why it is so important to keep designing more sensitive detectors, faster DNA sequencers, and more advanced computational models that can better spot patterns we would otherwise miss. Computer models are key in helping us test and simulate scenarios we cannot recreate in the lab or see firsthand, extending our reach beyond the here and now. Yet when we study phenomena with greater complexity, or with more depth at the micro or macro level, the present limits of science are generally shaped by human limits—our method-making and cognitive limits. Yet when we study further back or forward in time, the present borders of science are also shaped by it getting more difficult to access data. Ultimately, where and when we gather data with our methods—our present vantage point on earth—shapes the parameters and content of our evidence and the discoveries we can reveal.[82] *Yet scanning science's major discoveries we find that most breakthroughs come from what we can currently study—offering more reliable evidence: from chemical elements, matter, physical forces and the climate, to molecules, our genes, the brain, disease and immunity, to our social and economic systems.*

Human perspective

Our human perspective on the world—as the biological animals we are—also shapes what questions and goals we pursue, as we use our methods and our mind to do science. Just by being human and members of our species, do we naturally direct more attention to problems that matter to us than others? Indeed, we focus most on the parts of the world that influence our survival and interests—the niche where our species evolved and lives today. Nearly all scientists study aspects of the world that affect our needs and wants: human health, human technology, human society, human behaviour and the everyday challenges we face.

We do not just choose certain questions—we also build tools, run experiments and design theories that fit our human way of thinking and sensing. Our instruments extend and reflect the evolved limits of our senses and mind—like building x-ray devices and radar telescopes to see through soft tissue and through clouds, rather than something we did not evolve to detect at all. They are essentially enhanced 'vision'. We are constrained not just by our sensory limits but by how our cognitive architecture makes sense of information—and even the kinds of answers that seem reasonable to us (Chapter 9). Discoveries we make are shaped not only by our human needs but also by our human lens and choices. Even advanced AI systems are trained on human-labelled data—generated by humans—reflecting our choices of what is important to us and what is not. So the boundaries of science are not just out there in the universe—they are also inside us and our tools.

The world's largest science funding agencies—like the European Commission and National Science Foundation—generally require scientists to explain how their research will help society before they fund it. Most science funding worldwide is spent on studying human beings. Over half of all public research funding in the United States—at 52%—is allocated to medicine, health, life sciences and psychology. The remaining share goes to all other fields—from engineering and physical sciences to social sciences, environmental science and computer science—that often aim to improve human life in some way.[365] So why do tens of thousands of scientists focus on explaining and predicting illnesses, pandemics, mental health, population dynamics, financial markets, our behaviour and the weather but only a few scientists study dark energy of the universe, deep-sea creatures and insect minds? Because some questions have a direct impact on our lives, while others seem far removed—unless the questions can eventually help us. We humans are always our own point of reference. When we research in biology, medicine, psychology, economics or social science, it is typically to make our lives longer, healthier or easier—not to improve life for all plants and animals on earth, unless there is a benefit for us. Ultimately, we explore the world through our human lens—our anthropocentric filter that shapes what discoveries we pursue and value most. Yet this is not necessarily a drawback—our human focus has driven extraordinary advances in medicine, technology and understanding of our planet and society.

Social context

The size and diversity of our scientific community—along with the resources it has available—foster research at the frontier. A larger, more varied scientific community means more chances to develop more tools and accumulate more knowledge. No single person can invent on their own a complex method (from advanced statistics to randomised controlled trials), develop an entire field (from molecular biology to nuclear physics) or create a complex theory (from quantum theory to a theory of the origins of life).[82]

Strategic labs can have vast spillover effects on scientific progress. Think of the Cavendish Lab and Bell Labs—the two most productive methods labs and incubators in history. Inventing an array of powerful tools, these two methods labs sparked an explosion of discoveries—even though they were not founded to innovate tools (Chapter 6).

We foster cutting-edge research through economies of scale, reward systems, science policy and targeted funding.[6,7,366–368] Yet when academic careers often depend on short-term publishing metrics, scientists may avoid riskier, long-horizon research that can lead to the biggest breakthroughs. Some reward systems can help motivate scientists. Giving priority to the first person or team to publish new findings or develop a new method creates competition. This winner-takes-all system incentivises innovation.[6,95,367,368] The race to sequence the human genome was for example accelerated by competition between public and private teams.[113] Public institutions also help plan and finance frontier research, and shape part of science's direction by steering more resources to some big questions.[369] Governments set research priorities that fund at times space telescopes or fundamental physics experiments. They can support such research areas that may not pay off for decades. So for some fields and questions, our constraints can also be influenced by funding in basic research or cost-intensive research.[3,6] But it is important to stress that hundreds of major discoveries come from low-cost methods and tools: statistical and mathematical methods, light microscopes, electrophoresis, thermometers, chromatography methods, centrifuges and the PCR method (Chapter 6). Yet science is not method-neutral: researchers and institutions can be slow to adopt new methods, because of inertia with older, familiar methods (Chapter 1).

Ultimately, these five factors come together to determine our scope and boundaries of what science we currently explore. *Our methods and tools are the common thread underlying each of these factors—uniquely able to both limit and expand the frontiers of discovery more so than other factors can.* Here we see how our toolbox sets and extends—in an iterative cycle—what we are able to see, measure and discover at the edge of science (as mapped out in Figure 10.2).

All in all, what we discover—and what we even can discover—is always shaped by the power of our current tools, along with our mind and senses. But it is also influenced—varying by discovery and context—by our history, society, economy and human lens. To push past today's barriers to scientific progress, we need new tools that overcome our technical limits but also our cognitive and human constraints that can slow us down—or hold us back.

Beyond our extraordinary methods and what our mind is methodologically capable of developing, we have no other way to explore and discover the unknown. Science and its current limits move in a loop that begins and ends with our tools and methods. But we keep redrawing the edges of that loop through innovative toolmaking that stretches what is possible. To understand the borders of science we have to recognise that science is done by humans—and we face methodological and cognitive bottlenecks that draw our present boundaries.

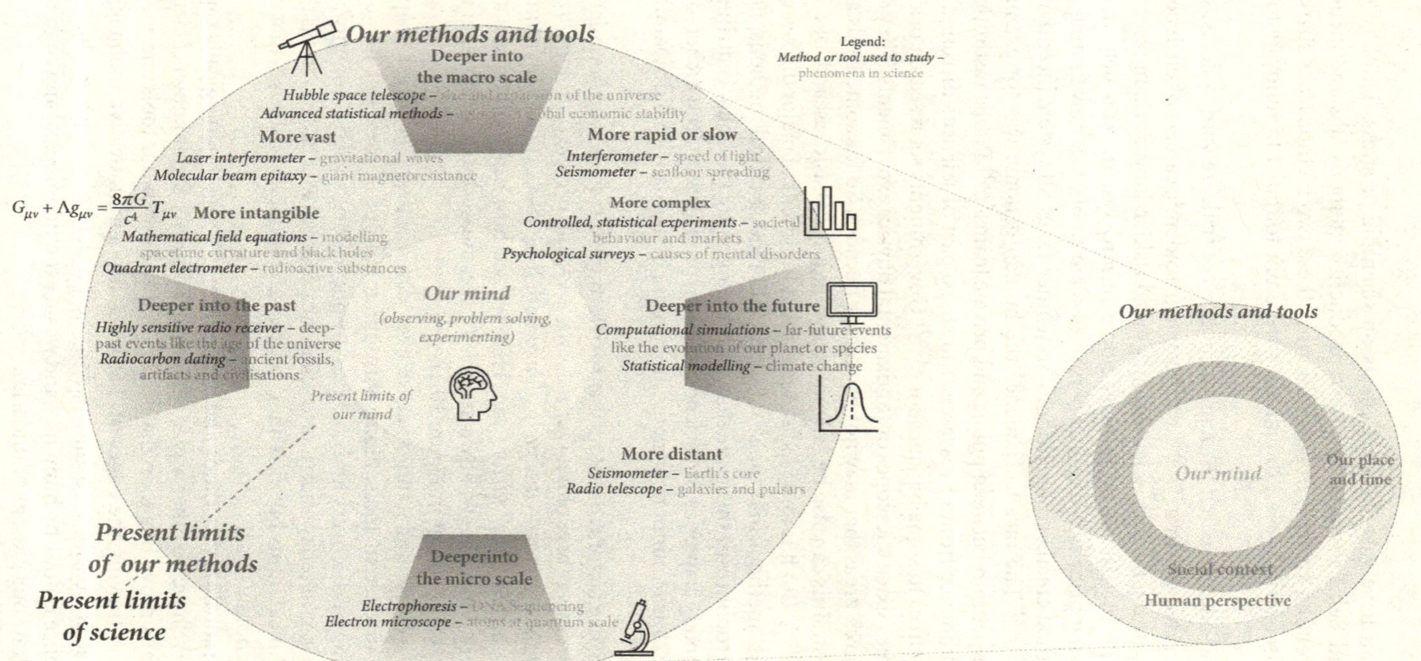

Figure 10.2 *The limits of our methods and mind shape the current limits of science*

We provide examples here for each category, but many phenomena are also studied using other tools. Phenomena can also fit into more than one category—with for example ecosystems studied on a large scale and are also very complex systems. The factors are construed broadly to flexibly capture other features of what we explore.[82]

Yet with new tools, we keep breaking through the frontier of discovery. Alone in the past few decades, nobel-prize breakthroughs in genetics, molecular biology and astronomy have rewritten what we thought was possible, but also in quantum physics and climate modelling. The pattern shows multiple breakthroughs clustered in fields where new tools enabled unprecedented observations—from the molecular scale (gene editing tools) to the cosmic scale (space telescopes) to the quantum scale (particle accelerators and quantum computers). We have made many major recent discoveries across fields—as shown in Figure 10.3 (even though it takes 21 years on average from discovery to Nobel prize recognition).

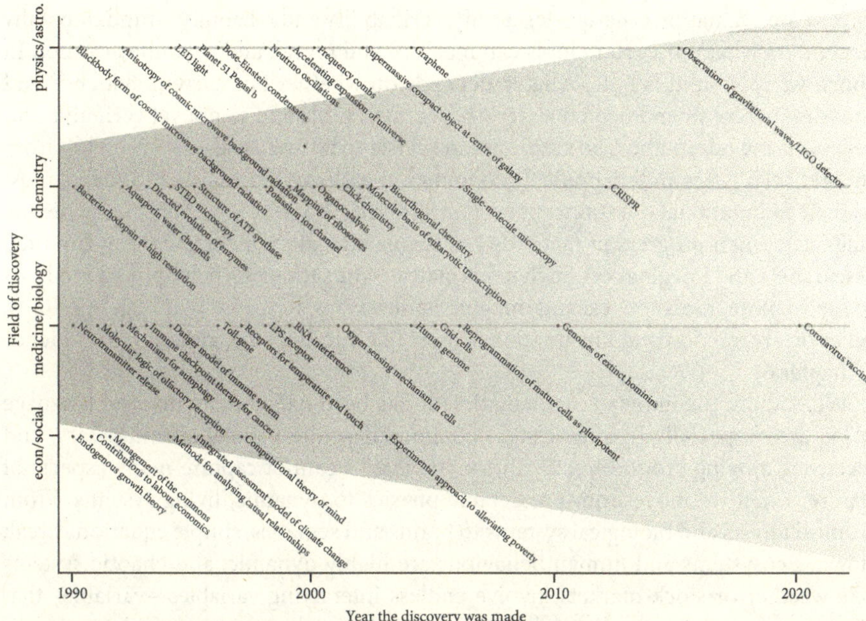

Figure 10.3 *The expanding boundaries of science: discoveries since 1990*

The data reflect major discoveries made since 1990—based on science's over 750 major discoveries.

How the current boundaries of our toolbox shape the current boundaries of science

So, where exactly do we reach the present borders of what we can observe, test and verify in the world? Our toolbox sets—at any given point—the current borders of what parts of the world we are able to get into contact with, and those still out of reach. Unlike any other species, we develop complex tools to uncover invisible chemical compounds, microbial life in our biosphere and infrared light. It is only by inventing tools that we—with our limited human senses—can reveal neurons firing in our brains and electrons inside atoms. Before the discovery of the human genome or elementary particles, these mysteries lay outside the periphery of science—invisible and inaccessible. But we redrew the borders of biology and physics with breakthrough methods like gene sequencing and particle accelerators—that detected and decoded them. Our scientific boundaries flexibly evolve as we extend our toolbox.

But our tools do not work equally well everywhere. *As we move into the realm of the tiniest particles, or the largest structures, or the most complex ecosystems, or the furthest distant stars, or the future of our species and human societies, our tools and methods start to strain and are not as powerful.* That is where our evidence and theories are not as reliable (Figure 10.2). Here we place our toolbox at the foundation of science and how we evaluate science: if we cannot test and verify evidence and theories using our toolbox, then they cannot be reliable—and remain uncertain. This is why science's core evaluation criteria—testability, verifiability and reliability—fundamentally depend on what our current tools can measure and reveal, and what they cannot. In short: what is—and is not—reliable depends on whether our current evidence and theories involve phenomena that are not yet observable and so not yet verifiable and reliable using our toolbox. So are we getting closer to hitting a wall in some fields, from theoretical physics and theoretical economics to philosophy of science? Indeed, most of their foundational questions remain unresolved. Progress in science has generally stagnated when progress in methods has stagnated—often independent of funding, researchers and imagination. Such fields plateau when they lack new applied methods to see, explore, measure, test and imagine in new ways. Research that largely, or only, relies on abstraction and imagination is most likely to stagnate without real evidence (Chapter 6).

What about the limits of mathematics? It has been hailed as a universal language of nature—especially by physicists. Equations describe planets orbiting stars, and electrons moving around nuclei. But mathematics cannot capture many aspects of nature, when we move from predictable physics to dynamic, living systems. From tropical forests and biological systems to brains and societies, simple equations break down. Ecosystems and human behaviour are highly dynamic, and chaotic systems like weather or stock markets involve endless interacting variables—variables that we cannot control or predict well. Mathematical models can come at a cost: simplified, ideal conditions of a complex world. Statistical and computational methods help manage and model complexity. Machine learning methods can bridge some gaps by detecting patterns hidden deep in complex systems that other methods cannot—from diagnosing cancer using medical images to predicting protein shapes.[90] But even these powerful methods cannot always provide exact answers.

Consider a thought experiment: imagine what science might look like if we were beings with optimal cognitive architecture—not limited by our specific human niche? We would have a remarkable capacity to quickly process millions of observations—without cutting-edge computers and statistical methods. We would have astonishing eyesight to see through atoms and across galaxies—without advanced microscopes, telescopes and spectrometers. We would have an incredible memory of everything we have observed—without immense datasets. We would understand and predict the world with remarkable accuracy—transcending our human lens. Yet because we are a product of specific evolutionary adaptations within our niche, we do not have these superhuman capacities. But the extraordinary tools we invent do have these capacities—by stretching our evolved mind and senses in ways unimaginable before them.

Conclusion

The tool inventions we have made so far largely determine what we have discovered, what we can discover and what we cannot yet discover without better tools. This powerful principle of tool innovation flips how we think about science's limits: it is not just about mysteries or unanswered questions 'out there' in the universe, it is about tackling the actual roadblocks built into our current tools to explore it. It is about shifting the focus from the big puzzles to overcoming our tools' blind spots, resolution limits and design challenges—that have uncovered science's biggest puzzles so far. Because each of science's major breakthroughs has meant breaking the bottlenecks of what our instruments could detect or measure—whether it is mapping DNA or spotting distant galaxies.

We can only extend our frontiers to the parts of reality that our tools open up and let us see. Yet our mind and senses but also our time in history, human lens and institutions also influence where we look and what we explore. But the key limit across all fields is what our methods and tools—from MRI devices to quantum sensors—can actually measure, test and reveal. Across fields, our toolbox is largely what redraws and redefines the outer borders of discovery—more so than other factors can. Our toolbox is the foundations and limits of science: setting what and how we can see, experiment and understand in the world.

A key insight emerges here: if we want to understand the bounds of science and tackle them, we have to understand what our tools enable us to observe, test and verify—and what they currently do not enable us to do. The history of science is a history about all the ways we transcended the limits imposed by evolution and by the current tools we designed up to now. It is a history of building instruments that reach farther, see further, think faster and imagine deeper than nature alone ever enabled us to. Just as the limits of an engineer's tools largely define what we can and cannot build, the limits of a scientist's tools largely define what we can and cannot discover. Ultimately, science thrives on its limits and the exciting challenge of finding ways to push through them.

Breaking the frontier means leveraging new methods—while everyday science means using conventional methods. So recognising the limits of science is not a weakness, it is the first step to breaking through the uncovered limits—the topic in the next and final chapter. Ultimately, every generation faces the same challenge: do we accept the current boundaries of what we can know and discover? Or do we create new tools and tool combinations—physical, digital, experimental—that continually expand the horizon of the unknown?

11
Pushing the limits of science

How we accelerate new discoveries by extending our powerful toolbox

Summary

Science is a quest to push the boundaries of what we know and reveal fundamental mysteries of the universe, matter, human life and the mind. But how we break and redefine the limits of science is not with conventional methods that produce conventional research—it is with new tools that enable new discoveries. Inventing tools with cutting-edge innovations—sharper resolving power, greater computational strength, more statistical power, greater acceleration, more precise measurement—continually break the boundaries at the frontier. From developing the gene-editing method CRISPR to rewrite genes, to creating quantum computers to solve problems we otherwise could never do, to harnessing machine learning methods to uncover hidden patterns in massive datasets, the power of our tools is enormous. Creating a new generation of toolmakers is one of the single best ways to invest our time in the future of science. So how do we actually push our limits and accelerate discoveries? To break new ground in any field, we generally first have to spot where our tools run into bottlenecks—and tackle them. It is about amplifying our ability to see, measure and theorise about the world with better lenses. Whether better fighting climate change with cutting-edge carbon-capture methods or better mapping the brain's complexity with more advanced neuroimaging and AI analysis, our progress depends on constantly pushing our tools' boundaries. Here we map out the steps we need to take to expand the top ten most influential discovery tools—from x-ray devices to statistical methods. Researchers best equipped at the edge of science are those who can best see the gaps in our tools and what we know—and test new ones to fill these gaps. This principle—that new tools are the engine of discovery—can be widely applied across science. We also explore deeper questions here: are there fundamental limits to what we can discover in some domains? And what can the future of science look like—and how can we power it with new tools, including AI tools?

Overview

Breaking the limits of science largely relies on a simple but powerful principle: breaking the current limits of our best tools and methods. Tackling real bottlenecks facing

The Engine of Scientific Discovery. Alexander Krauss, Oxford University Press. © Oxford University Press (2026).
DOI: 10.1093/oso/9780197829790.003.0012

our tools may not seem exciting at first glance. But it is exactly the cutting-edge research that unlocks new ways to see, measure and uncover the world that are not possible without them. It is about sharpening lasers' focus and tuning, overcoming electron microscopes' depth and clarity constraints, and resolving chromatography's resolution and flow bottlenecks.[370–372] It means pushing past electrophoresis' size-separation limits, addressing statistics' sample size, scale and power challenges, and exceeding centrifuges' separation and speed thresholds.[234,373,374] It is about penetrating x-ray imaging's depth and timing barriers, surpassing spectroscopes' resolution and sensitivity boundaries, and bypassing spectrometers' mass accuracy and range restrictions.[375,376] It involves boosting advanced thermometers' accuracy and sensitivity, solving space telescopes' sharpness and light-gathering obstacles, and expanding particle accelerators' energy and brightness ceilings.[377,378] It depends on tackling particle detectors' spatial resolution and noise limits, pushing beyond PCR's sensitivity and mixing barriers, and so much more.[127]

It is this cutting-edge research that opens extraordinary new lenses to the world that would catalyse more discoveries than other strategies we have. Each major upgrade has unlocked discoveries that were impossible before—from detecting thousands of exoplanets with more precise spectrometers to decoding entire genomes with more advanced DNA sequencing methods. Each major invention—a better sensor, a faster sequencer, a deeper imager—redraws the map of what science can explore, imagine and understand. This is the core of a new way of thinking about discovery: a build-to-discover strategy. With the *new methods-driven discovery theory*, we can make an important prediction: developing new methods and tools would be the key catalyser for new discoveries across fields. The best way to predict the pace of new discoveries is commonly looking at the pace at which we scale up, recombine and reinvent our tools—and where and when we do so. After all, waiting for serendipity or revising theories is generally more context-specific, and it is not about more funding but better targeting existing funding and researchers towards tool development (Chapter 6).

It is crucial we begin directing much more energy to spotting and fixing the blind spots in what our tools can do. When an experiment stalls or fails, we should turn our attention to why and where the method hits a wall—and tackle the bottleneck. Think of better AI-assisted microscopes that not only image but diagnose disease, make predictions or flag anomalies. Think of better sequencing methods that decode entire genomes in minutes.

While science's tools are currently still made in a surprisingly scattered and improvised way by individual researchers and teams, a number of remarkable innovations in methods have been made in recent years. Take biology and chemistry that keep pushing their borders. Click chemistry, invented in 2001 by Barry Sharpless and Morten Meldal and—along with Carolyn Bertozzi—awarded the Nobel prize in 2022, lets scientists click (snap) molecular building blocks together like Lego pieces. This vastly simplifies making complex molecules and has been called 'lego chemistry'. At Scripps Research in California, Sharpless discovered it by building on new methods including hydroxylation methods he invented. With this new technology, scientists can now design cancer drugs and advanced industrial materials more efficiently than ever.[379] Take the mapping of ribosomes that was completed by the Indian-born Venkatraman Ramakrishnan and his colleagues in 2000 using sharper electron-density maps and x-ray crystallography to reveal its structure. The discovery opened the door to producing entirely new antibiotics to fight bacterial infections we face. The gene-editing method CRISPR was discovered by Emmanuelle Charpentier and Jennifer Doudna in 2012 leveraging a powerful new differential RNA sequencing technique. Used today by labs around the world, this method revamped the life sciences, our ability to alter the DNA of plants and animals and drastically advance the fight of cancer.[114]

The Neandertal genome was sequenced by Svante Pääbo and his colleagues in 2010—and was awarded the Nobel in 2022. This fascinating breakthrough relied on applying new specialised sequencing methods they developed the previous year to deal with the degraded ancient DNA.[218,380] In the landmark article published in Science, *A draft sequence of the Neandertal genome*, Pääbo shined new light on our human origins: humans in Eurasia carry about 2–4% Neandertal DNA. Neandertals were more closely related to present-day Europeans and Asians than present-day Africans.[218] The finding changed how we understand human origins.

Take also physics that keeps pushing through the outer edges of science. A supermassive object at the centre of our galaxy—orbited by stars—was discovered in 2002. This extraordinary observation was sparked using the Very Large Telescope built on a mountain in Chile and the Keck Observatory on a mountain in Hawaii—coupled with a powerful new adaptive optics spectrometer. It redefined our understanding of our galaxy and our place in it.[381] And in 2015, gravitational waves were detected using a more sensitive laser interferometer.[115]

Each of these breakthrough discoveries pushed a field's boundaries—and won a Nobel prize. Yet researchers have not yet explained how we actually push the boundaries of science.[346,350–353,355,358,362] Throughout this book, we have been explaining how new discoveries are preceded by tool innovations that enable a leap in the way we see and measure—and are the factor we can most actively shape. Such tool-powered discovery pushes us to think of scientists not just as question-askers or experimenters, but as extraordinary toolmakers who build the very means to ask and answer bigger questions.

Yet as science makes remarkable progress, there are in each generation those who think we essentially already solved the big mysteries. Newton revealed the laws of motion and gravitational force and Maxwell uncovered the nature of electricity, magnetism and light. Mendeleev put the elementary pieces of the periodic table together that make up the basis of chemistry. Darwin and Wallace provided evidence of evolution as the reason why we see so many forms of life on earth—and as the driving principle of biology. Smith described the forces behind the Wealth of Nations and the invisible hand of markets: the division of labour, freedom of trade and pursuit of personal interest that can drive societal benefit. These breakthroughs laid the early foundation of physics, chemistry, biology and economics, and many believed that the big questions in these fields were basically settled.

At the end of the 1800s, the laws of motion, gravity, electromagnetism, energy conservation and thermodynamics seemed to explain the entire physical world. Physics looked like it had uncovered all the principles that hold physical reality together. In 1900, Lord Kelvin—one of the most renowned physicists at the time—allegedly said at a meeting at the British Association of Science: 'There is nothing new to be discovered in physics now. All that remains is more and more precise measurement'. But then came the vast breakthroughs of relativity and quantum mechanics at the beginning of the 1900s. They revealed that major pieces of the puzzle explaining physical reality on both the largest and smallest scales were missing. Special relativity introduced a different concept of time and space,[253] while quantum theory revealed a probabilistic, counterintuitive world where particles can act like waves, and outcomes are uncertain until measured.[53,160] These breakthroughs did not add details, they drastically transformed physics. They gave rise to one of the biggest unsolved puzzles: can quantum theory, the physics of the very small, be unified with relativity theory, the physics of the very large? We still do not know how to tie the two into one framework.[346] String theory is an attempt, but no answer has emerged yet.

Many scientists in each generation think we have run into a wall, with only fine-tuning left to do. But history, again and again, shows they are wrong. Every time we think we have hit the limits or are closer to 'completion', new surprising instruments and experimental findings unveil new mysteries—and hidden layers of reality we never imagined. From the James Webb Space Telescope that spots ancient galaxies to cutting-edge particle accelerators that probe the building blocks of matter.

Redrawing the edges of science by reinventing our toolbox

The future of science is not just about finding answers to questions we already know to ask—questions about life and the universe that current theories may raise. The history of science shows that it commonly depends even more on uncovering entirely new, fundamental questions and answers that largely new tools reveal—and we did not yet even imagine asking. New tools discover new questions: the microscope made it possible to discover cells and bacteria and ask fundamental questions about what they and life are actually made of.[149] The telescope raised transformative questions about the structure of the universe, the motion of planets and the existence of galaxies beyond our own.[140] X-ray imaging allowed us to look inside the body without surgery and led to entirely new questions about internal anatomy and disease diagnosis.[118] The cathode ray tube (particle detector) enabled uncovering the electron and triggered new questions about what matter and reality are made of at the smallest scales, paving the way for particle physics.[194] Chromatography enabled us to ask how complex mixtures can be separated and analysed, revolutionising chemistry, pharmacology and environmental science.[235] These inventions and their discoveries provided new pillars for modern biology, astronomy, physics and chemistry.

These major tools and methods reflect a large methodological leap and an entirely new type of invention—not just an incremental upgrade or combination of tools. Such major leaps range widely beyond those tools: from the first thermometer and Geiger counter, to the first electrophoresis method and the PCR method. These groundbreaking inventions—each sparking multiple discoveries—emerged primarily from practical experimentation rather than a guiding theory. They were born from iterative trial and error rather than theoretical prediction—as we explored in Chapter 6. They demonstrate how practical innovation can precede theoretical understanding, with explanations often developing after the tools reveal what we did not even know we needed to explain.

This brings us back to the four main forces driving tool innovation—and so driving discovery itself—that we uncovered examining science's over 750 major discoveries:

- *We invent entirely new tools*—from the first telescope to the PCR method.
- *We upgrade tools in new ways*—from higher-resolution telescopes to faster DNA sequencing tools.
- *We combine tools in new ways*—from MRI scanners coupled with advanced statistical techniques, to CRISPR gene editing with machine learning methods.
- *We adopt new tools from other fields*—using the RCT method from medicine in economics, or machine learning methods from computer science in everything from chemistry to climate modelling (Chapters 1 and 6).

We can think of these four *innovation pathways* like this: first come these *large-leap tools* (first-wave tools)—like chromatography itself. Once invented, they can be vastly improved into *upgraded tools*—like gas chromatography providing faster, higher-resolution separations. They can be combined into *hybrid tools*—like liquid chromatography-mass spectrometry that merges separation with precise molecular identification. And they can also be adopted into other fields as *transfer tools*—like chromatography moving from chemistry into biology and environmental science, tracking everything from pollutants to DNA fragments. Large-leap and transfer tools are first-generation innovations; upgraded and hybrid tools are second- and later-generation innovations. *This pattern—a big methodological leap followed by layering invention on invention—is largely how science digs deeper, layer by layer, into reality through entirely new ways of seeing, measuring and understanding that are impossible without them. Today's lower-hanging-fruit discoveries are easier to reach by borrowing methods from other fields and recombining existing methods, while upgrading and especially inventing methods can often demand more time and resources.*

Yet until now, we have not had a general understanding for how we systematically extend the frontiers of science.[346,350–353,355,358,362] We push back the goalposts and bounds of science by upgrading our toolbox through these innovative pathways. Other strategies—from scaling to automating tools—fall into one of these four categories. Ultimately, it is crucial to identify the blind spots, biases and assumptions built into our methods—and design ways to address them (as we mapped out in Chapter 6).

Taking a step back, a crucial lesson emerges here. Most scientists generally focus on experimental and theoretical research—often without thinking about how they could improve and refine their own methods to reveal new kinds of findings. But that is exactly the spark that pushes science into new terrain: finding ways to explore and experiment in completely new ways. Yet at first glance, it can seem overwhelming to imagine methods or tools better than the ones we rely on every day.

But the first step is simply to notice when our own tools fall short—or hit a wall. It is when our x-ray experiments yield blurry images, our spectroscopes generate results too noisy to interpret clearly, our statistical methods cannot decipher well patterns in vast datasets, our computational power is not strong enough to simulate complex systems. Instead of seeing limitations, here we see where new innovations are needed. When all researchers begin deliberately looking for gaps and limits in the very tools they themselves use, science will begin spotting vast opportunities to resolve them. It opens opportunities to spark transformative innovations across science. What we need is a shift in mindset about tools and discovery: the key is recognising that the gaps and limits we hit with our tools are far from failures or dead ends. They are the critical signposts showing us exactly where to look, test and push forward. The imperfections are commonly the best inspiration and roadmap we have for where to innovate next—where to repurpose, scale up and optimise what we use to study the world. After all, our tools hit all kinds of bottlenecks: from low precision in measurement and rare signals to slow processing and limited resolution. The powerful principle here: each shows how close we are to breaking new ground.

We looked at how science's top ten most influential toolmakers worked exactly this way (in Chapter 6). They put science's fixation on chasing answers aside, they instead pursued what existing tools could not handle and filled the gaps—and ended up sparking an enormous cascade of discoveries across fields, arguably more than any other researchers. We already unpacked the question of how we innovate new tools (Figure 6.6); but here we come back to this core question from a different angle: laying out concrete examples of where science's most used methods and tools hit walls that hold back progress. We first map out the method-driven process of how we expand the boundaries of scientific fields (see Figure 11.1).[82]

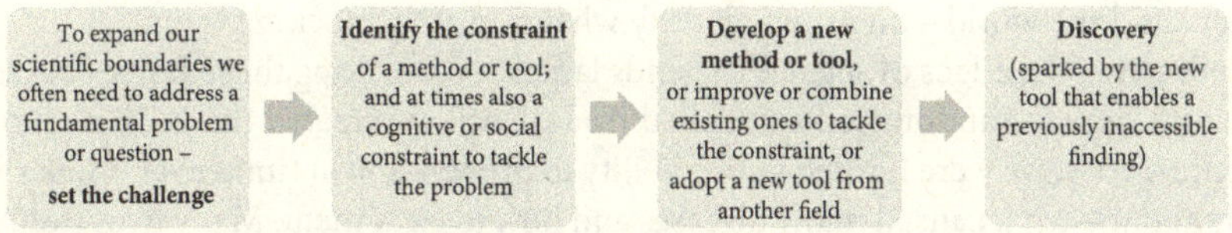

Figure 11.1 *General steps to push the boundaries of science*

So exploring science's top ten most used discovery tools, what do these instruments exactly let us do—and what barriers have they broken through? And crucially, what limits currently hold them back that we need to tackle to spur future discoveries? Each major tool extends what we can know, but also faces limits that define the edge of what we can know. Take the spectrograph—first built in 1859 by Gustav Kirchhoff and Robert Bunsen at the University of Heidelberg. Suddenly, we could see the hidden structures of atoms and molecules and the chemical makeup of distant planets and stars. This remarkable new instrument detected wavelengths of electromagnetic radiation we had no way to access before. It redefined the edge of astronomy and chemistry. Kirchhoff, a young professor from Königsberg Prussia, was even described 'as Bunsen's greatest discovery' because the tool transformed their entire field.[347] But today, our best spectrometers again hit limits—from sensitivity that cannot detect faint, distant signals, to computational limits and software that struggle to process large data streams from space telescopes. We need to design more sensitive detectors to capture those weaker signals, better algorithms to analyse the flood of data and faster ways to process them without missing details.[376,382]

This same pattern shows up everywhere; take the electron microscope. Ernst Ruska, the inventor, said: 'The light microscope opened the first gate to the microcosm. The electron microscope opened the second gate to the microcosm. What will we find opening the third gate?' Can quantum microscopy be the next key to opening up another layer of reality through even higher resolution? To get there, we have to overcome its current challenges—from electron damage to samples, to slow imaging times.[371] From supercomputers that struggle with processing power limits when modelling earth's complex climate, to MRI scanners that struggle with blurred motion in living patients, our tools all hit walls. Everywhere we look we find these bottlenecks waiting to be tackled, in each method we use—unlocking the next unexpected findings. Often the most important question we can ask ourselves to break new ground is: how can we better spot and tackle the bottlenecks facing our own tools? And the bottlenecks facing science's most influential discovery tools? That is the heart of much of how we push the current edges of science: not just running experiments and coming up with theories to explain what our existing tools reveal, but building the next generation of tools that let us see and imagine in entirely new ways. So while we have zoomed out and uncovered the broad patterns linking science's powerful tools to its biggest discoveries, here we zoom in further on science's top ten breakthrough tools (Table 11.1).

Spotting the bottlenecks in how we do science is crucial because we cannot design a better tool that aims to fix a bottleneck we are not yet fully aware of. That is why it is so important to continually scan for blind spots in how we observe, measure or understand the world—and then identify what is exactly blocking them.

Redrawing the edges of science depends largely on breaking those bottlenecks. Let us return here to randomised controlled trials. Until we created the RCT method in the mid-1900s, we were limited in our ability to prove—and at times even think rigorously about—what caused many diseases and how to treat them. Many lives were lost. Doctors often relied on observational comparisons that were more prone to bias. But with RCTs, we were finally able to systematically test how effective treatments actually are and better explain diseases. Like many methods, the RCT method is the outcome of continual reworking and many cumulative method refinements over time. It was built layer by layer—adding *randomisation* techniques to control for hidden variables, *blinding* techniques to reduce observer bias, *placebos* to handle patients' expectation effects.[234] The RCT's experimental design is a masterpiece, able to reduce uncertainty to provide life-saving evidence. It is an extraordinary example of how we create and upgrade methods to push past our limits. To understand phenomena from how our brains work to the origins of life and the universe, we rely on this kind of *method scaffolding*: experimenting with and reimagining our tools so they can take us further.

Yet it is not just about inventing new tools that can see and measure better—it is also about how reliable our evidence and theories are using our tools. That means we have to constantly test, validate and fine-tune our tools. They must be calibrated and their errors corrected. Medical imaging scanners for example need to be regularly calibrated to avoid misdiagnosis. Gravitational wave detectors demand astonishing precision to filter out faint ground vibrations and local noise.[115,138] AI models have to be checked for hidden biases or flaws in their training data.[263] Ultimately, the power of our tools comes not just from cutting-edge design, but from meticulously ensuring that they deliver the most reliable results.

Table 11.1 *Developing new methods and tools drives science by reducing our constraints: the ten central tools used most to trigger discoveries*

Method or tool	Year first developed	Human constraints (that the tool reduces)	What the tool enables us to study or explain (that would be impossible without it)	Current constraints of the tool (that we need to overcome to push our scientific frontiers further)
Laser	1960	Our inability to produce coherent light, make ultrafast measurements of distance and speed, and rapidly transmit large volumes of information	Study ultrafast interactions in matter, molecular structures, phenomena in deep space, and measure ultrashort time scales; develop fibre optics and laser surgery	Limits in power, efficiency and brightness of high-power diode lasers, and thermal constraints in solid-state lasers(370)
Electron microscope	1933	Our limited vision to magnify and perceive miniscule objects using visible light, and the constraints of optical microscopes and limited resolution	Observe nanoparticles, electrochemical reactions, microorganisms, crystals and molecules—by using the shorter wavelength of electrons	Resolution limitations, challenges in observing living cells, sample size restrictions, and damage to samples from the electron beam(371)
Chromatography	1931	Our constraints in separating and analysing chemical substances in a mixture	Analyse simple or complex mixtures containing vitamins, proteins and amino acids; isolate and purify natural compounds	No universal detector in high-performance liquid chromatography,(383) low detection limits at high resolution, difficulties with substances not volatile enough to separate(372)
Electrophoresis	1930	Our limited capacity to separate and identify molecules by size and electrical charge	Separate and analyse biological molecules like DNA, RNA and proteins—vital for biomedical research and forensic science	Challenges in analysing highly complex mixtures; heating blurs the resolution, and limited capacity for large-scale applications(373)
Statistics (modern)	1925	Our limited mental capacity to process large volumes of data, and complex relationships between variables	Study millions of data points across all fields, and more complex phenomena—from disease dynamics in populations and astronomical objects, to cells and the global economy	Constraints in sample size, scale and power; limited to studying phenomena we can represent statistically, constraints in the complexity of what we can model, and potential measurement error and sampling bias(234)

continued

Table 11.1 continued

Method or tool	Year first developed	Human constraints (that the tool reduces)	What the tool enables us to study or explain (that would be impossible without it)	Current constraints of the tool (that we need to overcome to push our scientific frontiers further)
Centrifuge	1924	Our inability to spin samples at ultrahigh speeds (greater than 20,000 rpm) to separate particles by density	Separate small biological particles like cells, viruses and nucleic acids from fluids	Limits on the volume of material we can process,[384] risks of cross-contamination between separated parts[385] and damage to samples from high-speed rotation[374]
X-ray diffraction/ crystallography	1912	Our limited capacity to visualise the atomic structures of matter because of short wavelengths	Analyse the atomic and molecular structure of matter like proteins and nucleic acids; and detect bone fractures, pneumonia and certain cancers	Low intensity of x-rays for light atoms, difficulties to analyse crystal growth in real-time;[386] and overlapping signals in some diffraction patterns (x-ray powder diffraction)[375]
Spectrograph/ spectrometer	1859	Our limited senses to detect and analyse electromagnetic radiation across wavelengths beyond the visible spectrum	Study structures of atoms and molecules, determine the chemical makeup of planets and stars, and electron arrangements of elements in different energy states	Low sensitivity such as in NMR;[376] often require coupling with methods like chromatography to probe complex samples (as in mass spectrometers); and face computational limits in interpreting spectral data[382]
Thermometer	1714	Our limitations in precisely perceiving and measuring variations in hot and cold temperature	Measure climate temperatures, surface temperatures in natural and industrial environments, and body temperature in health	Constraints in accurately measuring extreme temperatures, especially below −100°C and above +300°C, and issues of calibration and sensitivity[377]
Telescope	1608	Our limited vision to magnify and perceive distant objects, detect different regions of the electromagnetic spectrum	Observe exoplanets, stars, sunspots, galaxies, nebulae and other astronomical objects across multiple wavelengths	Blurred resolution from the Earth's surface caused by atmospheric disturbance; and space telescopes also face size and resolution limits[378]

Reducing such constraints will enable us to explore new phenomena and trigger new advances

These are the ten methods and tools most often used to make science's 761 major discoveries, including all nobel-prize and major non-nobel discoveries. The year marks when the method or tool was first developed—though all have been vastly improved since. The final column highlights some of their current limitations that need to be tackled.

It can seem odd to ask: how can we optimise not just our methods but also our own mind and senses? But it is precisely what we need to answer to invent sharper lenses, better sensors and more powerful processors—to stretch what we can understand. Which bottlenecks are currently holding us back? Which ones have we not noticed yet? These questions point us toward entirely unexplored frontiers. By spotting gaps in what our tools can do and relentlessly testing and improving them, we open the door to discoveries we cannot yet imagine. The speed at which we develop ever more sophisticated computational, statistical and experimental methods will be a key catalyst in accelerating science and discovery—in a cumulative loop. Yet many puzzles exist where we have not yet beat the current limits of our tools. Think about how neuroscientists try to map the tangled connections of the brain, or epidemiologists attempt to track vast global health trends over time with precision. Or geneticists struggle with enormous genomic datasets using big data methods and network analysis—they find much noise and at times can introduce as much noise as they aim to tackle. Pushing past our current barriers especially in computational power, machine learning, big data methods, etc. is likely to spur new frontier research—from medicine to astrophysics.

So what innovative, new methods can we design? Recent breakthroughs in quantum computing, machine learning, advanced sensors, neuroimaging, brain-computer interfaces, CRISPR, cryo-electron microscopes and so on offer a hint of what is possible. It is possible if the scientific community as a whole were to begin focussing on refining our tools to think and see in currently unimaginable ways—as those new tools enable us to.

We could create ever more powerful biosensors that detect more diseases even earlier and more accurately—before symptoms appear. We could design more advanced microscopes that reveal the inner workings of living cells in real time at higher resolution. We could run massive online studies with thousands of researchers collaborating in real time, continually improving a study without a preset outcome (like an evolving online encyclopaedia). In such studies, replication and peer review would be built into the research design.[274]

Up to this point, we have been describing where science can take us and where it hits a wall—and we now outline how we break those boundaries:

> Pushing the current limits of science is a cycle—the closer we get to the boundaries of discovery, the fewer tools and methods we often have available, such as electron microscopes and statistical simulations that are just not powerful enough. Here we commonly have to spot the bottlenecks holding our tools back and come up with ways to break them. Once we invent, upgrade, combine or adopt new tools of discovery—often by building on available ones—we unlock new ways to observe, measure and understand the world that are impossible without them. That is how we redraw the edges of science. And with each better tool or method in our hands, the cycle begins again: we explore further, hit new barriers and develop new ways to see and think past them. It is this continual loop of bottleneck and invention that drives science forward—with the pace of progress more in our hands than we realise.

Ultimately, the engine of science is building better tools that stretch the limits of what our current tools, minds and senses can do. Every time we sharpen how we see, measure and test the world, we expand the universe of what we can know.

About everyday knowledge, Ludwig Wittgenstein once famously said, 'the limits of my language are the limits of my world'. But when it comes to science and discovery, it is mainly the limits of our tools and methods that are the limits of our world. We hit our scientific boundaries when we run into the boundaries of what our best electron microscopes can resolve, when particle accelerators cannot reach higher energies, when machine learning tools cannot spot hidden relationships in noisy data. For our tools determine not just the experiments we have conducted so far but also largely the theories we have developed so far. Yet in fields like theoretical physics, some exceptional theoretical advances have involved larger conceptual leaps—leaps from existing experimental findings and tools to theory—than in most other fields (Chapter 6).

So here is the key insight: our understanding of the world only reaches as far as our current toolbox allows. Beyond our toolbox, there is often not much more we can say about the world. If we want to better solve mysteries like how consciousness arises, what dark matter is made of, how to tackle Alzheimer's and many others we cannot yet imagine, we need new tools. To expand the universe of possible questions we can even ask, we need to build better tools: better quantum sensors that can detect unimaginably tiny gravitational shifts, sharper cryo-electron microscopes that reveal proteins at nearly atomic resolution, faster AI systems that can accurately analyse immense genomic data in hours instead of years.

How artificial intelligence can be combined with our scientific tools to accelerate science

AI can be a powerful tool in a scientist's toolbox—amplifying how the world can be explored and predicted. AI systems can help scientists streamline the research process by automating repetitive, time-consuming tasks, making research faster and less prone to some human errors. AI-powered search tools can filter through vast scientific literature, extract information and spot trends across studies—from climate science to genetics. While we can delegate our vision to advanced microscopes to make better observations, we can delegate our mind to AI systems to better spot patterns, make better simulations and predictions and even can help design experiments—faster than scientists can alone.[90,249,268]

How can scientists combine AI with other tools to help tackle our current barriers? Coupling AI tools with high-resolution microscopes vastly improves imaging by using deep learning to sharpen blurry images and detect patterns. This can accelerate real-time disease diagnosis and provide guidance for surgeries. In astronomy, pairing machine learning algorithms with telescopes enables sifting through vast and noisy astronomical data to spot new exoplanets or classify supernovae by detecting subtle patterns in the data.[90]

Climate change is challenging to fully understand given its immense complexity. Machine learning can simulate climate systems—interactions between the ocean, atmosphere, ecosystems and human activity—more precisely by using enormous amounts of recorded data. It can integrate climate models with detailed local weather data to improve predictions about trends in rising sea levels, melting glaciers and extreme weather. This offers policymakers detailed analysis of the local effects of climate change.[90] In drug discovery, AI algorithms can help recognise hidden patterns in vast medical records and help speed up the search for new drugs by predicting how they may work and what side effects they may cause. Advanced AI now helps sharpen MRI and CT scans, making it easier to spot cancers early and detect signs of Alzheimer's. It helps in reducing some human diagnostic errors. By analysing patient data, machine learning models can help detect relationships between diseases, genes and drugs. This can help personalise and tailor medical treatments to individual needs. A remarkable breakthrough has been achieved in understanding proteins. To understand how proteins function in our body, we need to understand their structure. Predicting their structure enables us to identify proteins that affect diseases and so improve diagnosis and treatments. But establishing protein structures is difficult and time-intensive. The barriers of existing methods were broken with the new powerful AI programme AlphaFold developed by Google DeepMind: a programme that predicts protein structures with extremely high accuracy. It solved the decades-old 'protein folding problem'—and fuels drug discovery and advances disease research.[90]

So combining AI and self-improving systems with other tools can optimise research: increase speed, efficiency, scale, sensitivity and resolution. In short: AI can improve how our best tools perform. It is becoming more integral to accelerating innovation in different domains—with large language models also helping to mine vast research for hidden relationships. Building AI-enhanced scientific tools could help spur breakthroughs we cannot yet imagine. We could create ever more advanced lab systems that adapt and design their own experiments—learning from their own results. Yet machine learning systems often work like black boxes: making predictions without explaining how. They can also at times amplify biases built into the data they use—their training data. That means human judgement is essential. These systems and algorithms ultimately rely heavily on humans to be designed, understand how they generate results and detect and avoid errors.[90,249,341]

New approaches to designing methods and tools could emerge: a taxonomy of scientific methods—mapped out in Chapter 6—could be linked to AI to help screen for tools, suggest underused ones and help select new combinations of methods and tools for both general and specific problems. In other words, we could design AI systems in the future that do not just follow instructions but can learn and suggest new types of experimental methods and tools that push us beyond the borders of what we can currently imagine. In the future, we could also train machine learning programmes to analyse data on newly developed methods and tools (such as newly published methods papers) and explore the domains and problems the new methods and tools could best be applied to in spurring breakthroughs.

Are there set limits to human knowledge and what we can discover?

Since our mind's capacities evolved mainly to survive and reproduce, do we face ultimate limits to what we can know? Is there a ceiling on our understanding the biggest puzzles—like the nature of life, consciousness and the universe? As innovative toolmaking enables opening new layers of reality, the boundaries of discovery are not preset. Instead, we have—method by method—pushed them back. New methods of observing, measuring and experimenting—and new method combinations—have continually expanded what was previously beyond our reach. Our biology, our mind's constraints, our scientific institutions and the time and place we live influence the parts of the world we can access. But they do not dictate the types of tools we can come up with to tackle those very constraints.

The world has boundaries. Human life is short and finite. The speed of light is fixed. Our planet's natural resources are limited. But does that mean that science will eventually hit a wall too? Our methods and mind do not seem bounded by clearly defined limits—given the seemingly endless method combinations and our human imagination (Chapter 6).

Yet science has settled some of our fundamental puzzles. We will not re-discover the basis of biology (evolution and DNA's double helix), the backbone of chemistry (the elements that make up the periodic table) or the foundation of physics (relativity, quantum mechanics and the big bang). But we keep updating, expanding and better measuring them as we improve our methods. While it is true that the neutron, the DNA molecule, electron microscopy and CRISPR can only be discovered once, these landmark breakthroughs opened entire fields of research that we keep expanding today.

But some theoretical sciences may eventually slow down. Once major, successful theories become established—as in physics—establishing other grand theories would become less likely. In other words, continuously uncovering grand discoveries that lay a new foundation for a major field would be unlikely. So we are unlikely to continually developing fundamental theories that govern the world. The theory of evolution for example—underlying biology—is established with independent methods across diverse fields. It is unlikely that a new foundational discovery will make the theory of evolution obsolete. And that new methods and tools drive discoveries and push the frontier is also unlikely to be completely replaced by a fundamentally new, unknown driver of scientific progress—a new driver that does not rely on sharper microscopes, more sensitive detectors and more powerful computation methods.

Developing vast new fields that parallel in importance to biology, chemistry and physics is unlikely—just like we are not going to find an entirely new continent on our planet. But we will keep making major discoveries and opening new fields—from quantum biology to machine learning. We are learning to look better inward and outward at once unimaginable scales. At the frontier, breaking new ground unlocks more questions and mysteries, not fewer—in a seemingly endless loop.

What we know—at any moment—is largely determined by the tools we have built so far. Our future knowledge will remain on those parts of the world that our methods and technologies allow us to study. For some unknowns, we have not yet developed the right tools to uncover them—from dark matter to the nature of time. There are also imaginables—things we can imagine but do not yet know how to develop the needed tools and test them, like artificial consciousness or the possibility of finding life on other planets (or even entirely other forms of doing science).

We humans are fascinated by foundational questions about distant origins, vast complexity or fundamental structures. Yet are answers to such questions often beyond what we can currently observe and test? Indeed, it is here where science often hits walls and stops working—where we run into the boundaries of what we can discover. A methodological border of science emerges: science can study what is subject to observation, scientific methods and often quantification. This is why it is so difficult to get our grips on the nature of knowledge, consciousness, time, morality, subjective experience, freewill, politics, love and much more. Science also runs into barriers when asking 'why' questions: why do we exist? Why did the universe begin? Why are we conscious beings?

Even though our toolbox largely explains our current boundaries, it cannot tell us everything about them. Our frontier also partly depends on what research areas and instruments receive more policy attention, resources and public interest—from public health and cancer research to renewable energy (Chapter 10). Ethical issues and government decisions also partly influence some research questions: what research is legal? Should we edit human embryos? How should AI be used in informing or making medical diagnoses? Should we prohibit technologies and research that can foster climate change? Ultimately, the edge of science looks different in each field—and new fields keep emerging. Yet in the end, those aware of how new tools break open mysteries and those who get their hands on them will be the ones most likely to push the frontier forward.

Inventing the future of science—with new tools

Exploring science's future can help us better plan, innovate and anticipate it. But the further we look ahead, the blurrier and more uncertain things get—like our methods. The key is that the combinations of methods we can create seem almost limitless, and can open many new paths and fields to explore (Chapter 6). Science is an evolving, flexible system because we do science using our flexible methods and elastic mind. After all, our scientific methods—the ways we study the world—depend on creativity and generate even greater creativity in science that does not exist in the observable world itself.

As long as we live, we will keep investigating what affects our health, minds and societies. We will need new cures for evolving viruses, diseases, cancers and pandemics, and new solutions for our changing human and social challenges. So fields like medicine, biology, psychology and economics will keep growing. The same goes for fields like agriculture, engineering, environmental research and computer science—because they improve our lives and tackle the challenges we face. From new forms of agriculture that are better adapted to extreme weather and climate change, to green technology that replaces fossil fuels faster. In short, science that improves our lives—applied science—will likely always grow and not run out of problems to solve. In fact, it makes up most of science. So it does not seem that many parts of science will ever become finite and complete. Just as it was not possible to predict CRISPR, quantum computers, smartphones and the internet a hundred years ago, the same applies to scientific and technological developments in the next hundred years. And we will very likely look back at today's breakthroughs as just the beginning.

So will we near the limits of science? The answer depends on the feature of science we focus on: when it comes to method breakthroughs and experimental discoveries, we will very likely keep growing—especially across applied science—while uncovering entirely new fundamental theories and laws of nature will likely slow down. The pace will vary from field to field. Yet it is clear that new methods and problems open further methods and problems—in a cumulative loop. The future of science in any field will fundamentally depend on the tools we come up with.

With the past century marked by exponential scientific progress, can we keep growing as fast—or even faster? It seems very likely. But many earlier discoveries were low-hanging fruit because they were relatively easier to uncover. Think of looking through early telescopes to discover new planets and moons, to using the James Webb space telescope to estimate the age of the universe at about 14 billion years. Think of early microscopes uncovering bacteria and blood cells, to using cryo-electron microscopes to map protein structures at extraordinary atomic resolution. Or think of early x-rays methods revealing broken bones in the body, to using x-ray crystallography to determine the detailed structure of DNA and viruses. Many future breakthroughs will come from new method combinations—especially at the crossroads of different fields. Others will need entirely new methods that at times can be somewhat more challenging to invent.

At the research frontier, the future growth of science looks very promising for three reasons. First, making new method innovations is still surprisingly not yet coordinated in a strategic way across science—but these scattered breakthroughs consistently trigger new pathbreaking findings. There is an enormous untapped potential of new tool combinations that are almost seemingly endless and within our reach (Chapter 6). Second, the pace of major new discoveries and fields emerging and expanding does not seem to be slowing. From CRISPR gene editing, machine learning, quantum computing and environmental engineering, to nanotechnology, computational chemistry, quantum biology and even the science of science. Third, the past highlights that those in each generation who think we are nearing our limits have been incorrect. The real limit is largely how fast we can upgrade our toolbox.

Science is not something we may ever finish, with no more challenges and mysteries to resolve—from mutating viruses and changing ecosystems to evolving language and human systems. It is largely self-renewing. Only since the 20th century did we spark most of science's major discoveries. The future is much longer, holding many more advances through powerful new tool innovations. If we knew all the answers, it would mean we would already know all the questions and have already developed all the needed methods and tools we could ever need. But that is not how discovery works. This methods-powered view offers an alternative, more promising perspective of the future: if scientists commonly only chase the questions they already have, they are held back by what they know and can imagine today. But history shows that developing new methods and tools has unlocked most new questions we could not have predicted across fields, from biology, astronomy and geology to medicine, climate science and neuroscience. Each major tool we invent does that—from the PCR method to x-ray devices. With so many new method combinations and scaffolds easily within our reach today—small and big—we are in a better position than ever to keep pushing the boundaries of the unknown (Chapter 6). *That is the essence of frontier research: not just answering today's questions, but strategically extending our methods that opens and answers the questions of tomorrow we have not yet imagined. Because the real frontier of science is not just out there—it is mainly in building the tools that reveal what is out there—often unexpectedly.*

Conclusion: the science of discovery—and the tool revolution in science

Most people are familiar with big breakthroughs like Franklin, Crick and Watson's discovery of DNA's double-helix structure, Hubble's discovery that the universe is expanding and Einstein's theory of relativity. But who created the tools that made those surprising findings possible? Few people have heard of them: it was the Braggs' invention of x-ray crystallography that produced the extraordinary image revealing DNA's structure.[125] It was George Hale's massive telescope—the most advanced at the time—that made possible the earth-shaking observation of galaxies moving away, transforming how we understand the nature of our vast universe.[2] It was Albert Michelson's pioneering interferometer that allowed precisely measuring the speed of light and challenged the existing notions of absolute space and time.[50] Tools define what we can see, measure and experiment and largely shape the very theories we can even conceive to explain the surprising findings those tools reveal.

The invention of powerful new methods and tools has been the turning point behind fields across science. Better microscopes launched microbiology. Radio telescopes made radio astronomy possible. DNA sequencing methods sparked genomics. The PCR method unlocked modern genetics and virology. X-ray crystallography gave birth to structural biology. Carbon-14 dating transformed archaeology—from guesswork to precision science. Scanning tunneling microscopes laid the foundation for nanotechnology. Functional MRI powers cognitive neuroscience—giving us real-time images of the thinking brain. High-performance computing enables computational biology. Particle detectors and accelerators drive particle physics. Chromatography powers analytical chemistry. CRISPR drives genetic engineering—and so on. These tools open windows into worlds that are too small, too vast, too fast, too far beyond our mind's capacities to observe or imagine on our own. They have given birth to new fields and sparked more discoveries than a single theory—in their domains.

Again and again, new methods revealed parts of the world we could not see or conceive before. The observations often came first, with theories following to explain them. Without the right tools to detect or measure, our best theories cannot be meaningfully developed or tested, but also more funding and collaboration are not enough, and unexpected or serendipitous observations cannot be made. These other factors vary by discovery. *So what unites the diverse discoveries that electron microscopes have made possible—or that telescopes, or x-ray devices, or spectrometers have made possible? It is not a shared theory, research teams, more funding or serendipity, it is using the shared powerful tool. Each unlocked findings across different fields that changed how we understand the world and sparked the need for new theoretical explanations. This is what the new methods-driven discovery theory explains: new methods and tools are a necessary condition for major discoveries, and commonly the critical driver—they are the common thread driving our biggest breakthroughs across science and history.* The theory explains much, if not most, of scientific progress: from how new fields emerge, to why particular discoveries happen when they do, to why some problems resist solutions for decades, to why certain institutions become unexpected powerhouses of innovation.

This discovery—that new tools drive major breakthroughs—reveals a powerful principle in science: it not only explains how our past discoveries have emerged, but it also helps anticipate future discoveries. The theory is straightforward: when we develop a new tool, we can predict *where* the next breakthroughs can come from, and *when*. Tracking the speed and direction of tool innovations gives us powerful signals that discovery is soon to follow. The signal can come from a newly invented sensor, an upgraded gene-editing technique, a combined AI-powered microscope, or an experimental method newly adopted from another field. The best way to predict the pace of new discoveries is looking at *how fast*, *where* and *when* we scale up our tools.

But when people imagine science, they often think theories are at its heart, with methods just supporting them. But examining science's major discoveries, we reveal the opposite: methods are not just stepping stones, they enable sparking breakthrough findings that are not possible without them and often come first from inventing the method. Ultimately, the engine of discovery is inventing new ways to see, detect, measure, simulate, imagine and theorise about the world.

So why are tools commonly the central driving force of discoveries? The answer lies in six fundamental ways that they transform how we explore the unknown. One, tools do not just visualise, measure and test the world. They also enable us to conceive of reality in ways unthinkable without them. From electron microscopes that reveal viruses, to radio telescopes that uncover radio waves from distant galaxies. Two, tools make discovery less driven by chance and more driven by design—making it systematic and repeatable. Three, tools accelerate the pace and scale of discovery in extraordinary ways. Think of how computational, experimental and statistical methods have vastly sped up breakthroughs. Four, tools push us beyond the boundaries of existing theories by revealing unexpected phenomena current understanding cannot explain—from uncovering superconductivity to x-rays. Five, tools spark cascading breakthroughs across unrelated fields—with the laser, originally developed in physics, triggering discoveries in medicine (laser surgery) and communications (fibre optics). Six, tools make science scalable, precise and reproducible. Tools often transcend theories and what theoretical explanations can do (Chapters 1 and 6). In fact, new major tools are themselves the big discoveries. And discovering what drives discovery can mark a key advance for scientific progress.

The key to advancing science is inventing, upgrading, combining and adopting new methods and tools of discovery. It is how we break and redefine the current limits of science. If we do not place our tools at the heart of science, we cannot understand how we trigger major discoveries. But current explanations of science—whether from scientometricians, historians, psychologists, computer scientists or philosophers—do not recognise tools as the key driver. What we have uncovered in this book is the missing principle overlooked up to now: tool innovation is commonly the primary driver of discovery—and once recognised, we are much better able to spark new advances faster (Chapters 1–6).[346,350–353,355,358]

Expanding our toolbox is the thread connecting science—from the present, back to its origins, and forward to the cutting-edge frontier. It provides answers to the big questions about science: how we trigger new discoveries (Chapter 1), what sparks serendipity in science (Chapter 2), whether scientific progress is cumulative and even how we define the scientific method itself (Chapter 3), how our powerful tools interact with discoverers' demographics, institutions and resources (Chapter 4), how we actually create new scientific fields (Chapter 5), how we invent new methods and tools (Chapter 6), what the origins of science and our methods are (Chapters 7–8), what the common force driving science is (Chapter 9), what the current limits of science are and how we can break them (Chapters 10–11). At the centre of answers to these big puzzles is a common principle: method innovations. They provide a unified answer to many unsolved questions about science.

The efforts of millions of people using their eyes, minds and hands to explore the world are rarely as transformative as creating a new tool—like an advanced spectrometer or electron microscope—that can trigger multiple major breakthroughs by enabling us to see further and think deeper. In this book, we have adopted a new approach: examining science's over 750 major discoveries, the methods and tools they used and the traits and conditions of the discoverers to systematically assess what exactly sparks them. A striking finding emerged: it is new tools themselves that consistently trigger new breakthroughs—experimental, methodological and theoretical—by providing a new needed lens. The inspiration for this approach came from recognising a pattern across breakthroughs—that they rely on new methods, from randomised controlled trials to new statistical techniques. Testing the pattern here revealed that it surprisingly applies across scientific fields and history—leading to this new *method-to-discovery* principle. This key driver of discoveries has been overlooked because researchers had not yet reviewed science's over 750 discovery-making studies—one by one—to identify the underlying pattern and extract the data from each study and multiple other sources. It has been a meticulous and very time-consuming task that culminated in this book.

This *method-to-discovery* principle could also be applicable across other domains—from technology to business and even policy. Designing new methods could solve our large challenges across domains. The impact of new methods can be much greater than their initial intention as we can use them for different purposes in different fields. Just like the first microscopes and telescopes were not invented for scientific purposes but were then leveraged across science,[46] the internet was created for improving communication but we now use it worldwide for gaining knowledge—it is our global library.

Discovering this methods-driven principle has vast implications: scientists need to begin focusing on spotting the gaps and blind spots in their own tools and finding ways to tackle them. We mapped out the crucial pathways we take to invent new tools—the discovery engine. We also created a *scientific table of discovery methods*—a map of explored and unexplored method options that can help identify and predict untapped combinations of tools and where the next breakthroughs can emerge. This *method map* can coordinate efforts across scientists to create databases of methods and tools—cataloguing combinations, blind spots and unexplored regions. It can also guide resources and strategic investment. We also laid out seven major reforms that could drastically speed up scientific progress and develop the *Methodology of Science* (at the end of Chapter 6).

Here we summarise those critical reforms. One, we need far more researchers dedicated to tackling method bottlenecks across the scientific community. Two, we should make testing and expanding our best methods a regular part of what we all do. Three, we need to redefine research as not just question-driven but equally methods-driven. Four, we have to fix science's reward system, so journals, prizes and funding bodies equally incentivise method innovations. Five, we need to establish new *Methods Labs, Methods Hubs* and *Departments of Methodology of Science*—as incubators of innovation. Six, we have to shift our main measure of success from just outputs (especially article citations) to focus equally on inputs—method innovations. Seven, we need to train the next generation of tool innovators—and integrate deep methodological training into university degrees across science. These reforms would help us much better spot when even our best tools of discovery are holding us back—and how to overcome those barriers.

If we adopt these seven reforms, we could kick-start a tool revolution across science. It would break from the status quo: the era *before the tool revolution*, where we until now commonly made new methods and tools in a surprisingly improvised and scattered way. It would mark a shift towards *the tool revolution*, where we would design and develop new methods and tools in a deliberate, planned and strategic way. It would cut down the time it takes to identify new possible tools but also to create them. Until now, without yet a general theory of scientific progress and dedicated research programmes for tool development, each discoverer has had to improvise, test and work out on their own how to best make a breakthrough—and which new tools we need. The tool revolution would be driven by a shared understanding—across scientists, science policymakers and funders—of the incredible power of designing and engineering the very methods that generate new discoveries. Researchers lose much time and fields stall because we did not yet have a systematic roadmap for building better tools.

So what if we stopped waiting for the next discovery tools to arise by chance? How many breakthroughs do we overlook simply because we have not yet made tool development a clear priority? What if science labs start competing to invent discovery tools, not just to use them? Indeed, when most researchers stick with familiar tools just because that is what they know best, breakthroughs are slowed—or even missed (Chapter 1). That common comfort zone holds fields back. Yet science would then move from more unplanned insights to designing better ways to see, detect and think—to better *engineering discoveries. If researchers everywhere recognised that the tools and methods we use shape what we can discover with them, it will push us to extend them and invent better ones—far more systematically. It may be the most powerful, untapped frontier in science.* Ultimately, when we hit the limits of our tools, we know exactly where to innovate next. Each limitation we uncover is one step closer to unknown terrain and findings. Because the biggest leaps in science do not generally come from re-examining existing blurry data or re-measuring existing data samples. They come from building sharper lenses, more precise sensors and more accurate computer simulations.

A tool-centric science opens whole new lines of research and countless questions that hold large potential. How can we incentivise researchers to develop new methods that may not yield a paper, but could change a field in a few years? What are actually the best strategies to reward not just for solving scientific questions, but for inventing better ways to solve them? How could scientists best see their role not just as answering questions, but also as engineering better ways to answer them? How can we create a culture that values method design and toolmaking as the foundation of discovery? Can we develop models to help guide tool design and bootstrap innovation? Can we develop models to predict which new methods could enable bigger advances before they are developed? Could fields forecast their own growth by investing in the methods they will need in coming years (just like engineers plan major infrastructure in advance)? What ethical frameworks do we need to ensure responsible tool innovation, especially in fields like genetics and climate science?

Ultimately, new methods and tools are the engine of discovery—our keys to unlocking the mysteries of life, matter and the universe. As we invent more powerful tools, our capacity to reveal what is beyond our imagination today only grows. Each time we invent a better tool to see, measure or simulate the world, we expand what is possible—and what we can discover. The frontiers of science largely lie in uncovering underdeveloped and undeveloped tools. Designing better tools is set to become science's next frontier of competitive advantage. The logic of science and discovery is simple: conventional methods produce conventional research; new tools make new breakthroughs possible. If a science-wide awareness of this principle emerged and the scientific community made it a priority, it could begin a tool revolution in science. One that transforms how we understand, create and accelerate new scientific advances. One that changes how science is organised, practiced and funded. By recognising tool innovation as the key driver of science, we could make discoveries faster and more predictable over time. So there is a science of discovery: those who develop and use new methods and tools today are generally the ones who spark tomorrow's new breakthroughs.

Acknowledgements

I am thankful for comments from those I inflicted a draft of this book on, Uwe Peters, Alfonso García Lapeña and Corinna Peters (who I bothered more than should be legally permitted), and those who read one or a few chapters, Nikolas Schöll, Milan Quentel, Julia Hoefer Marti, Joseph Emmens, Alina Velias, J. P. Grodniewicz, Sarah Brierley, Jeremy Lewis, Jeffrey Tsao, Mar Bonet—and anonymous peer reviewers from Oxford University Press. I also want to thank Nebojsa Rodic, Laney Brager and Lorena Ortega. And I am grateful to my parents—not just for bringing me into the world but also as I could not have otherwise brought this project to life—and to my brother, for reminding me that spending years writing a book is not normal and to wrap it up. I received funding for this research from the European Commission (Marie Curie programme) and the Ministry of Science and Innovation of the Government of Spain (grant RYC2020-029424-I).

To get in contact and for collaborations and projects: a.krauss@lse.ac.uk or alexander.krauss@iae.csic.es

References

1. Watson, J. The Double Helix. New York: New American Library (1969).
2. Hubble, E. A relation between distance and radial velocity among extra-galactic nebulae. Proceedings of the National Academy of Sciences of the United States of America. 15 (3), 168–173 (1929).
3. Fortunato, S.; et al. Science of science. Science. 359 (2018).
4. Clauset, A; D. Larremore; R. Sinatra. Data-driven predictions in the science of science. Science. 355, 477–480 (2017).
5. Chen, C.; Y. Chen; M. Horowitz; H. Hou; Z. Liu; D. Pellegrino. Towards an explanatory and computational theory of scientific discovery. Journal of Informetrics. 3 (3), 191–209 (2009).
6. Stephan, P. How Economics Shapes Science. Cambridge, MA: Harvard University Press (2015).
7. Singh Chawla, D. U.S. agency launches experiments to find innovative ways to fund research. Nature. (2023).
8. U.S. National Science Foundation. Building Stronger Bridges Between Discovery, Innovation, and Prosperity. Alexandria VA: U.S. National Science Foundation. www.nsf.gov/science-matters/building-stronger-bridges-between-discovery (2022).
9. Xu, F.; L. Wu; J. Evans. Flat teams drive scientific innovation. Proceedings of the National Academy of Sciences of the United States of America. 119 (23), e2200927119 (2022).
10. Wu, L.; D. Wang; J. A. Evans. Large teams develop and small teams disrupt science and technology. Nature. 566, 378–382 (2019).
11. Wuchty, S.; B. Jones; B. Uzzi. The increasing dominance of teams in production of knowledge. Science. 316, 1036–1039 (2007).
12. Sinatra, R.; D. Wang; P. Deville; C. Song, A. Barabási. Quantifying the evolution of individual scientific impact. Science. 354, aaf5239 (2016).
13. Jones, B.; E. Reedy; B. Weinberg. Age and Scientific Genius. In D. K. Simonton (Ed.), The Wiley Handbook of Genius, pp. 422–450. John Wiley & Sons (2014).
14. Wang, D.; A. Barabási. The Science of Science. Cambridge: Cambridge University Press (2021).
15. Kuhn, T. The Structure of Scientific Revolutions. Chicago: University of Chicago Press (1st edition 1962; 4th edition 2012).
16. Kuhn, T. The Last Writings of Thomas S. Kuhn: Incommensurability in Science (Ed. B. Mladenovic). Chicago: University of Chicago Press (2022).
17. Popper, K. The Logic of Scientific Discovery. London: Routledge (2nd edition) (1959/2005).
18. Merton, R.; E. Barber. The Travels and Adventures of Serendipity: A Study in Sociological Semantics and the Sociology of Science. Princeton: Princeton University Press (2004).
19. Nature. The serendipity test. Nature. 554, 5 (2018).
20. De Rond, M.; I. Morley. Serendipity: Fortune and the Prepared Mind. Cambridge: Cambridge University Press (2010).
21. Uzzi, B.; S. Mukherjee; M. Stringer; B. Jones. Atypical combinations and scientific impact. Science. 342, 468–472 (2013).
22. Rzhetsky, A.; J. Foster; I. Foster; J. Evans. Choosing experiments to accelerate collective discovery. Proceedings of the National Academy of Sciences of the United States of America. 112, 14569–14574 (2015).
23. Chu, J; J. Evans. Slowed canonical progress in large fields of science. Proceedings of the National Academy of Sciences of the United States of America. 118 (41), e2021636118 (2021).
24. Azoulay, P.; et al. Toward a more scientific science. Science. 361 (6408), 1194–1197 (2018).
25. Reichenbach, H. Experience and Prediction: An Analysis of the Foundations and the Structure of Knowledge. Whitefish, MT: Literary Licensing, LLC (1949).
26. Ladyman, J. Understanding Philosophy of Science. London: Routledge (2002).

27. Chalmers, A. What is This Thing Called Science? Indianapolis, IN: Hackett Publishing Company (4th ed) (2013).
28. Schickore, J. Scientific Discovery. In The Stanford Encyclopedia of Philosophy, Edward Zalta & Uri Nodelman (eds.), https://plato.stanford.edu/archives/fall2025/entries/scientific-discovery/ (2018).
29. Copeland, S. On serendipity in science: discovery at the intersection of chance and wisdom. Synthese. 196, 2385–2406 (2019).
30. Murayama, K; M. Nirei; H. Shimizu. Management of science, serendipity, and research performance: Evidence from a survey of scientists in Japan and the U.S. Research Policy. 44, 4, 862–873, (2015).
31. Roberts, R. M. Serendipity: Accidental Discoveries in Science. Hoboken: John Wiley & Sons (1989).
32. Simonton, D. Scientific creativity: discovery and invention as combinatorial. Frontiers in Psychology. 12, 721104 (2021).
33. Normile, D. Global research chiefs seek ways to foster serendipity. Science. (2015).
34. Fink, T.; M. Reeves; R. Palma; R. Farr. Serendipity and strategy in rapid innovation. Nature Communications. 8, 2002 (2017).
35. Janosov, M.; F. Battiston; R. Sinatra. Success and luck in creative careers. EPJ Data Science. 9, 9 (2020).
36. Aslan, Y.; O. Yaqub; B. Sampat; D. Rotolo. Unexpectedness in medical research. Research Policy. 53, 8 (2024).
37. Liu, L. et al. Hot streaks in artistic, cultural, and scientific careers. Nature. 559, 396–399 (2018).
38. Sugimoto, C. Scientific success by numbers. Nature. 593 (2021).
39. Park, M.; E. Leahey; R. J. Funk. Papers and patents are becoming less disruptive over time. Nature. 613, 138–144 (2023).
40. Nature Human Behaviour. Broader scope is key to the future of 'science of science'. Nature Human Behaviour. 6, 899–900 (2022).
41. Ye, S.; R. Xing; J. Liu; F. Xing. Bibliometric analysis of Nobelists' awards and landmark papers in physiology or medicine during 1983–2012. Annals of Medicine. 45 (8), 532–538 (2013).
42. Thompson, G. Nobel Prizes That Changed Medicine. London: Imperial College Press (2012).
43. Breit, W.; B. Hirsch (Eds.). Lives of the Laureates: Eighteen Nobel Economists. Cambridge MA: MIT Press (2004).
44. Zuckerman, H. Scientific Elite: Nobel Laureates in the United States. New York: Free Press (1977).
45. Foster, J.; A. Rzhetsky; J. Evans. Tradition and innovation in scientists' research strategies. American Sociological Review. 80 (5), 875–908 (2015).
46. Bardell, D. The invention of the microscope. Bios. 75 (2), 78–84 (2004).
47. Nobel Prize. Press Release; Physics. Nobel Prize. https://www.nobelprize.org/prizes/physics/1986/press-release/ (1986).
48. Nobel Prize. Ernst Ruska – Biographical, and Nobel Lecture. Nobel Prize. https://www.nobelprize.org/prizes/physics/1986/ruska/ (1986a).
49. Nobel Prize. Archer J. P. Martin – Biographical, and Nobel Lecture. Nobel Prize. https://www.nobelprize.org/prizes/chemistry/1952/martin/ (1952).
50. Michelson, A; E. Morley. On the relative motion of the earth and the luminiferous ether. American Journal of Science. 34 (203), 333–345 (1887).
51. Einstein, A. Über das Relativitätsprinzip und die aus demselben gezogenen Folgerungen. Jahrbuchder Radioaktivität und Elektronik. 4, 411–462 (1907).
52. Van Dongen, J. On the role of the Michelson–Morley experiment: Einstein in Chicago. Archive for History of Exact Sciences. 63, 655–663 (2009).
53. Planck, M. Zur theorie des gesetzes der energieverteilung im normalspectrum. (On the theory of the energy distribution law of the normal spectrum). Verhandlungen der Deutschen Physikalischen Gesellschaft. 2, 237–245 (1900).
54. Einstein, A. Über einen die erzeugung und verwandlung des lichtes betreffenden heuristischen gesichtspunkt (On a heuristic point of view concerning the production and transformation of light). Annalen der Physik. 17, 132–148 (1905).

55. Bohr, N. On the constitution of atoms and molecules. Philosophical Magazine. 26, 1–25 (1913).
56. Jardine, B. Charles Darwin's microscopes. Whipple Museum of the History of Science. Cambridge: University of Cambridge (2016). www.whipplemuseum.cam.ac.uk/explore-whipple-collections/microscopes/charles-darwins-microscopes.
57. Darwin, C. On the Origin of Species by Means of Natural Selection, or the Preservation of Favoured Races in the Struggle for Life. London: John Murray (1859).
58. National Research Council. Progress in Science. In I. Feller; P. Stern (Eds.), A Strategy for Assessing Science: Behavioral and Social Research on Aging. Washington DC: National Academies Press, 67–94 (2007).
59. Galison, P. How Experiments End. Chicago: University of Chicago Press (1987).
60. Holmes, F. L. Lavoisier and the Chemistry of Life: An Exploration of Scientific Creativity. Madison: University of Wisconsin Press (1985).
61. Yu; Y.; D. Romero. Does the use of unusual combinations of datasets contribute to greater scientific impact? Proceedings of the National Academy of Sciences of the United States of America. 121 (41) (2024).
62. Dunbar, K. How scientists think in the real world: Implications for science education. Journal of Applied Developmental Psychology. 21, 49–58 (2000).
63. Dunbar, K. How Scientists Really Reason: Scientific Reasoning in Real-World Laboratories. In R. J. Sternberg; J. Davidson (Eds.), The Nature of Insight, pp. 365–395. Cambridge, MA: MIT Press (1994).
64. Nersessian, N. The Cognitive Basis of Model-Based Reasoning in Science. In P. Carruthers, S. Stich, M. Siegal, eds. The Cognitive Basis of Science. Cambridge: Cambridge University Press (2002).
65. Feist, G.; M. Gorman. Handbook of the Psychology of Science. Cham: Springer Press (2013).
66. Langley, P. Scientific discovery, causal explanation, and process model induction. Mind & Society. 18, 43–56 (2019).
67. Schmidt, M.; Lipson, H. Distilling free-form natural laws from experimental data. Science. 324, 81–85 (2009).
68. Džeroski, S.; Todorovski, L. Equation discovery for systems biology: Finding the structure and dynamics of biological networks from time course data. Current Opinion in Biotechnology. 19, 360–368 (2008).
69. Langley, P; H. A. Simon; G. L. Bradshaw; J. M. Zytkow. Scientific Discovery: Computational Explorations of the Creative Processes. Cambridge, MA: MIT Press (1987).
70. Raghu, M; E. Schmidt. A survey of deep learning for scientific discovery. ArXiv. https://doi.org/10.48550/arxiv.2003.11755 (2020).
71. Hesse, M. B. Models and Analogies in Science. Notre Dame, IN: University of Notre Dame Press (1966).
72. Thagard, P. Conceptual Revolutions. Princeton, NJ: Princeton University Press (1992).
73. Kuhn, T. Historical structure of scientific discovery. Science. 136 (3518), 760–764 (1962a).
74. Oxford English Dictionary. Science. Oxford: Oxford English Dictionary. www.lexico.com/definition/science (2022).
75. Krauss, A. Debunking revolutionary paradigm shifts: evidence of cumulative scientific progress across science. Proceedings of the Royal Society A. 480 (2024).
76. Krauss, A. Redefining the scientific method: as the use of sophisticated scientific methods that extend our mind. PNAS Nexus. 3 (4), 112 (2024).
77. Krauss, A. Science's greatest discoverers: a shift towards greater interdisciplinarity, top universities and older age. Nature, Humanities and Social Sciences Communications. 11, 272 (2024).
78. Krauss, A. Predictable serendipity: How new tools turn serendipity into systematic breakthroughs. Scientometrics (2025).
79. Krauss, A. New tools drive scientific discovery: A study of all nobel-prize and major non-nobel breakthroughs. Nature, Humanities and Social Sciences Communications. (2025).
80. Krauss, A. New scientific fields are triggered by powerful new methods. Nature, Humanities and Social Sciences Communications. 12, 1590 (2025).

81. Krauss, A. Homo methodologicus and the origin of science and civilisation. Heliyon, 9, 10 (2023).
82. Krauss, A. Science of Science: Understanding the Foundations and Limits of Science from an Interdisciplinary Perspective. Oxford: Oxford University Press (2024).
83. Szell, M.; Y. Ma; R. Sinatra. A Nobel opportunity for interdisciplinarity. Nature Physics. 14, 1075–1078 (2018).
84. Cai, X.; X. Lyu; P. Zhou. The relationship between interdisciplinarity and citation impact—a novel perspective on citation accumulation. Humanities & Social Sciences Communications. 10, 945 (2023).
85. Chan, H; B. Torgler. The implications of educational and methodological background for the career success of Nobel laureates: an investigation of major awards. Scientometrics. 102 (1):847–863 (2015).
86. Simonton, D. Creative productivity: A predictive and explanatory model of career trajectories and landmarks. Psychological Review. 104, 66–89 (1997).
87. Klahr, D.; H. Simon. Studies of scientific discovery: complementary approaches and convergent findings. Psychological Bulletin. 125 (5), 524–543 (1999).
88. Simon, H. A.; P. Langley; G. L. Bradshaw. Scientific discovery as problem solving. Synthese. 47, 1–27 (1981).
89. Bradshaw, G. F.; P. Langley; H. Simon. Studying scientific discovery by computer simulation. Science. 222 (4627), 971–975 (1983).
90. The Royal Society. The AI Revolution in Scientific Research. London: The Royal Society and The Alan Turing Institute (2019).
91. Gibney, E. Could machine learning fuel a reproducibility crisis in science? Nature. 608, 250–251 (2022).
92. Kitano, H. Nobel turing challenge: creating the engine for scientific discovery. Nature, Systems Biology and Applications. 7, 29 (2021).
93. Khurana, M. The Trajectory of Discovery: What Determines the Rate and Direction of Medical Progress? Introduction. Cambridge: Cambridge University Press (2023).
94. Holton, G.; H. Chang; E. Jurkowitz. How a scientific discovery is made: a case history. American Scientist. 84 (4), 364–375 (1996).
95. Nobel Prize. Nobel Media AB. Nobel Prize. www.nobelprize.org (2024).
96. Tiner, J. 100 Scientists Who Shaped World History. Naperville: Sourcebooks (2022).
97. Salter, C. 100 Science Discoveries That Changed the World. London: Pavilion (2021).
98. Gribbin, J. Britannica Guide to 100 Most Influential Scientists. Edinburgh: Encyclopedia Britannica, Robinson Publishing (2008).
99. Rogers, K. The 100 Most Influential Scientists of All Time. New York: Britannica Educational Publishing (2009).
100. Simmons, J. The Scientific 100: A Ranking of the Most Influential Scientists, Past and Present. New York: Citadel Press (2000).
101. Balchin, J. Quantum Leaps: 100 Scientists Who Changed the World. London: Arcturus Publishing Limited (2014).
102. Haven, K. 100 Greatest Science Discoveries of All Time. London: Libraries Unlimited (2007).
103. Encyclopaedia Britannica. Encyclopaedia Britannica. www.britannica.com (2024).
104. Daintith, J. Biographical Encyclopedia of Scientists - 3rd ed. Boca Raton: Taylor & Francis Group (2009).
105. Bunch, B.; A. Hellemans. The History of Science and Technology. Boston/New York: Houghton Mifflin Company (2004).
106. Oakes, E. Encyclopedia of World Scientists, Revised Edition. New York: Infobase Publishing (2007).
107. Simonis, D. (Ed.). Lives and Legacies: An Encyclopedia of People Who Changed the World Scientists, Mathematicians, and Inventors. Phoenix: The Oryx Press (1999).
108. Lerner, K. L.; B. Lerner (Eds). Gale Encyclopedia of Science, Third Edition. Michigan: Thomson/Gale (2004).
109. Nobel Prize. C.T.R. Wilson – Biographical, and Nobel Lecture. Nobel Prize. www.nobelprize.org/prizes/physics/1927/wilson/ (1927).

110. Nobel Prize. Donald A. Glaser - Facts. Nobel Prize. https://www.nobelprize.org/prizes/physics/1960/glaser/facts/ (1960).
111. Hooke, R. Micrographia: Or, Some Physiological Descriptions of Minute Bodies Made by Magnifying Glasses, with Observations and Inquiries Thereupon. Nashik: Science Heritage Limited (1665).
112. Knoll, M.; E. Ruska. Das elektronenmikroskop. Zeitschrift für Physik. 78, 318–339 (1932).
113. Human Genome Project. Human Genome Project. NIH, National Human Genome Research Institute. www.genome.gov/human-genome-project/ (2023).
114. Nobel Prize. Press Release: The Nobel Prize in Chemistry 2020. Nobel Prize. https://www.nobelprize.org/prizes/chemistry/2020/press-release/ (2020).
115. LIGO Laboratory. Laser Interferometer Gravitational-Wave Observatory - Facts. Pasadena: California Institute of Technology, LIGO Laboratory. https://www.ligo.caltech.edu/page/facts (2023).
116. Nobel Prize. Michel Mayor - Biographical, and Nobel Lecture. Nobel Prize. www.nobelprize.org/prizes/physics/2019/mayor/ (2019).
117. Nobel Prize. François Englert – Facts. Nobel Prize. https://www.nobelprize.org/prizes/physics/2013/englert/ (2013).
118. Röntgen, W. On a New Kind of Rays. Science. 3, (59), 227–231 (1896).
119. Nobel Prize. Max von Laue – Biographical, and Nobel Lecture. Nobel Prize. https://www.nobelprize.org/prizes/physics/1914/laue (1914).
120. von Laue, M. Eine quantitative Prüfung der Theorie für die Interferenzerscheinungen bei Röntgenstrahlen. Annals of Physics. 346, 989–1002 (1913).
121. Eckert, M. Max von Laue and the discovery of X-ray diffraction in 1912. Annals of Physics. 524, A83–A85 (2012).
122. Nobel Prize. Willard F. Libby – Biographical, and Nobel Lecture. Nobel Prize. www.nobelprize.org/prizes/chemistry/1960/libby/ (1960a).
123. University of California Academic Senate. 1980, University of California: In Memoriam. Berkeley: University of California; The Bancroft Library (1980).
124. Radebaugh, R. Historical Summary of Cryogenic Activity Prior to 1950. In K. D. Timmerhaus; P. R. Reed (Eds.), Cryogenic Engineering: Fifty Years of Progress. Reed. Cham: Springer, 3–27 (2007).
125. Nobel Prize. The Nobel Prize in Physics 1915. Nobel Prize. https://www.nobelprize.org/prizes/physics/1915/summary/ (1915).
126. Nobel Prize. John R. Vane – Nobel Lecture. Nobel Prize. https://www.nobelprize.org/prizes/medicine/1982/vane/lecture/ (1982).
127. Nobel Prize. Kary B. Mullis – Biographical, and Nobel Lecture. Nobel Prize. https://www.nobelprize.org/prizes/chemistry/1993/mullis/ (1993).
128. Nobel Prize. Press release; Chemistry. Nobel Prize. https://www.nobelprize.org/prizes/chemistry/1993/press-release/ (1993a).
129. Mullis, K. The Unusual Origin of the Polymerase Chain Reaction. Scientific American. 262 (4), 56–65 (1990).
130. Nobel Prize. Press release: The Nobel Prize in Physics 2020. Nobel Prize. https://www.nobelprize.org/prizes/physics/2020/press-release/ (2020a).
131. Nobel Prize. Arthur Ashkin – Facts, and Biographical. Nobel Prize. https://www.nobelprize.org/prizes/physics/2018/ashkin/ (2018).
132. Nobel Prize. Tim Hunt – Biographical. Nobel Prize. https://www.nobelprize.org/prizes/medicine/2001/hunt/biographical/ (2001b).
133. BMJ. Streptomycin treatment of pulmonary tuberculosis. British Medical Journal. 2 (4582), 769–782 (1948).
134. Cochrane Library. Cochrane Library. London. https://www.cochranelibrary.com/central (2024).
135. Kremer, M. Randomized evaluations of educational programs in developing countries: some lessons. American Economic Review. 93 (2), 102–106 (2003).
136. Dunn, P. James Lind (1716-94) of Edinburgh and the treatment of scurvy. Archives of disease in childhood. Fetal and Neonatal Edition. 76, F64–5 (1997).

137. Fisher, R. A. Statistical Methods for Research Workers. Edinburgh: Oliver & Boyd (1925).
138. Abbott, B. P.; et al. (LIGO Scientific Collaboration and VIRGO Collaboration). Observation of gravitational waves from a binary black hole merger. Physical Review Letters. 116, 061102. https://journals.aps.org/prl/abstract/10.1103/PhysRevLett.116.061102 (2016).
139. Nobel Prize. Scientific Background: The Laser Interferometer Gravitational-Wave Observatory and the First Direct Observation of Gravitational Waves. Advanced Information. Nobel Prize. https://www.nobelprize.org/uploads/2018/06/advanced-physicsprize2017.pdf (2017).
140. King, H. The History of the Telescope. Garden City, NY: Dover Publications (2011).
141. Townes, C. The First Laser. In G. Laura; L. Tim (Eds.), A Century of Nature: Twenty-One Discoveries that Changed Science and the World, pp. 107–112. Chicago IL: University of Chicago Press(2003).
142. Tiselius, A. The moving boundary method of studying the electrophoresis of proteins. Nova Acta Regiae Societatis Scientiarum Upsaliensis. 47 (4), 1–107 (1930).
143. The Svedberg and Herman Rinde. The ultra-centrifuge, a new instrument for the determination of size and distribution of size of particle in amicroscopic colloids. Journal of the American Chemical Society. 46 (12), 2677–2693 (1924).
144. Remer, T. Serendipity and the Three Princes. Norman, Oklahoma: University of Oklahoma Press (1965).
145. Ban, T. The role of serendipity in drug discovery. Dialogues in Clinical Neuroscience. 8 (3):335–344 (2006).
146. Dunbar, K.; J. Fugelsang. Causal Thinking in Science: How Scientists and Students Interpret the Unexpected. In M. Gorman; R. Tweney; D. Gooding; A. Kincannon (Eds.), Scientific and Technological Thinking, pp. 57–79. Taylor & Francis, Psychology Press (2005).
147. Anguera de Sojo, Á.; et al. Serendipity and the discovery of DNA. Foundations of Science. 19, 387–401 (2014).
148. Nobel Prize. Press Release. Nobel Prize. https://www.nobelprize.org/prizes/physics/1978/press-release/ (1978).
149. Hooke, R. Lectiones Cultlerianae. In R. T. Gunther (Ed.), Early Science in Oxford: The Life and Work of Robert Hooke. Oxford: Oxford University Press, Vol. 10 (1679/1935).
150. Bussmann-Holder, A; Keller, H. High-temperature superconductors: underlying physics and applications. Zeitschrift für Naturforschung B. 75, 1–2 (2020).
151. Carnap, R. Scheinprobleme in der Philosophie. (Transl. Pseudoproblems in Philosophy), pp. 301–343. Berlin: Bernary (1928).
152. Davisson, C.; L. Germer. Diffraction of electrons by a crystal of nickel. Physical Review. 30 (6), 705 (1927).
153. Nobel Prize. Award Ceremony Speech. Nobel Prize. https://www.nobelprize.org/prizes/physics/1946/ceremony-speech/ (1946).
154. Nobel Prize. Koichi Tanaka – Nobel Lecture. Nobel Prize. https://www.nobelprize.org/prizes/chemistry/2002/tanaka/lecture/ (2002).
155. Nobel Prize. Popular Information. Nobel Prize. https://www.nobelprize.org/prizes/chemistry/2000/popular-information/ (2000).
156. Nobel Prize. Award Ceremony Speech. Nobel Prize. https://www.nobelprize.org/prizes/chemistry/2003/ceremony-speech/ (2003).
157. Nobel Prize. Leland H. Hartwell – Nobel Lecture. Nobel Prize. https://www.nobelprize.org/prizes/medicine/2001/hartwell/lecture/ (2001).
158. Nobel Prize. Speed Read: Seeking Signs of Compatibility. Nobel Prize. https://www.nobelprize.org/prizes/medicine/1980/speedread/ (1980).
159. Nobel Prize. Baruch S. Blumberg – Facts. Nobel Prize. https://www.nobelprize.org/prizes/medicine/1976/blumberg/facts/ (1976).
160. Stuewer, R. H. "Max Planck". Encyclopedia Britannica. https://www.britannica.com/biography/Max-Planck (2024).
161. Merton, R. Singletons and multiples in scientific discovery: a chapter in the sociology of science. Proceedings of the American Philosophical Society. 105 (5), 470–486 (1961).
162. Ippoliti, E. Scientific discovery reloaded. Topoi. 39, 847–856 (2020).

163. Nobel Prize. Popular Information; Physics. Nobel Prize. https://www.nobelprize.org/prizes/physics/2007/popular-information/ (2007).
164. Nobel Prize. Gerald M. Edelman – Facts. Nobel Prize. https://www.nobelprize.org/prizes/medicine/1972/edelman/facts/ (1972).
165. Nobel Prize. Ernst Otto Fischer – Facts. Nobel Prize. https://www.nobelprize.org/prizes/chemistry/1973/fischer/facts/ (1973).
166. Pearce, R. M. Chance and the prepared mind. Science. 35 (912), 941–956 (1912).
167. UNESCO. UNESCO Science Report: the Race Against Time for Smarter Development. In S. Schneegans; T. Straza; J. Lewis (Eds.), Statistical Annex. Paris: UNESCO (2021).
168. de Solla Price, D. Science Since Babylon. Enlarged edition. New Haven and London: Yale University Press (1975).
169. Prus, B. [Aleksander Głowacki]. On Discoveries and Inventions. Public Lecture 23 March, Warsaw, Printed by F. Krokoszyńska (1873).
170. Richards, R.; L. Daston (Eds.). Kuhn's 'Structure of Scientific Revolutions' at Fifty: Reflections on a Science Classic. Chicago: University of Chicago Press (2016).
171. Marcum, J. Thomas Kuhn's Revolutions: A Historical and an Evolutionary Philosophy of Science? London: Bloomsbury Academic (2015).
172. Kindi, V.; T. Arabatzis (Eds.). Kuhn's The Structure of Scientific Revolutions Revisited. Chicago: University of Chicago Press (2012).
173. Wray, K. Kuhn's Evolutionary Social Epistemology. New York: Cambridge University Press (2011).
174. De Langhe, R. Towards the discovery of scientific revolutions in scientometric data. Scientometrics. 110, 505–519 (2017).
175. Hung, E. Beyond Kuhn: Scientific Explanation, Theory Structure, Incommensurability and Physical Necessity (1st ed.). London: Routledge (2005).
176. Ridgway E.; Baker P.; Woods J.; Lawrence M. Historical developments and paradigm shifts in public health nutrition science, guidance and policy actions: a narrative review. Nutrients. 11 (3), 531 (2019).
177. Anand, G.; Larson, E.; Mahoney, J. Thomas Kuhn on paradigms. Production and Operations Management Journal. 29, 1650–1657 (2020).
178. Merton, R. The Matthew effect in science. Science. 159, 56–63 (1968).
179. Clarivate. History of Citation Indexing. Clarivate. https://clarivate.com/webofsciencegroup/essays/history-of-citation-indexing/ (2022).
180. Popper, K. Conjectures and Refutations: The Growth of Scientific Knowledge. London, UK: Hassell Street Press (1962/2021).
181. Kitcher, P. The Advancement of Science: Science Without Legend, Objectivity Without Illusions. Oxford: Oxford University Press (1993).
182. Lakatos, I. Falsification and the Methodology of Research program. In I. Lakatos; A. Musgrave (Eds.), Criticism and the Growth of Knowledge, pp. 91–197. Cambridge: Cambridge University Press(1970).
183. Maxwell, N. Understanding Scientific Progress: Aim-Oriented Empiricism. St. Paul, USA: Paragon House (2017).
184. Sanger, F.; S. Nicklen; A. R. Coulson. DNA sequencing with chain-terminating inhibitors. Proceedings of the National Academy of Sciences. 74 (12), 5463–5467 (1977).
185. Narayanamurti, V.; J. Tsao. How technoscientific knowledge advances: A Bell-Labs-inspired architecture. Research Policy. 53, 4 (2024).
186. Copernicus, Nicolaus. On the Revolutions of the Heavenly Spheres. (Transl. Duncan et al.). New York: Barnes and Noble (1543/1976).
187. Baker, M. 1,500 scientists lift the lid on reproducibility. Nature. 533 (2016).
188. Goldacre, B. Make journals report clinical trials properly. Nature. 530, 7 (2016).
189. Allison, D.; et al. Reproducibility: A tragedy of errors. Nature. 530, 27–29 (2016).
190. Nobel Prize. Johannes Fibiger – Facts. Nobel Prize. https://www.nobelprize.org/prizes/medicine/1926/fibiger/facts/ (1926).
191. Nobel Prize. Egas Moniz – Facts. Nobel Prize. https://www.nobelprize.org/prizes/medicine/1949/moniz/facts/ (1949).

192. Markowitz, H. Harry Markowitz: Selected Works. World Scientific – Nobel Laureate Series, Vol. 1. pp. 32. London: World Scientific Publishing Company (2008).
193. Galilei, G. The Starry Messenger (Ed. A. van Helden). Chicago: University of Chicago Press (1989/1610).
194. Thomson, J. J. XL. Cathode rays. The London, Edinburgh, and Dublin Philosophical Magazine and Journal of Science. 44 (269), 293–316 (1897).
195. Needham, J. The Grand Titration: Science and Society in East and West. Toronto: University of Toronto Press (1969).
196. Merriam-Webster Dictionary. Scientific method. Merriam-Webster. www.merriam-webster.com/dictionary/scientific_method (2023).
197. Encyclopaedia Britannica. Scientific Method. Encyclopaedia Britannica. https://www.britannica.com/science/scientific-method (2023).
198. Collins English Dictionary. Science. New York City: HarperCollins Publishers. www.collinsdictionary.com/dictionary/english/science (2023).
199. Reiff, R; W. Harwood; T. Phillipson. A Scientific Method Based upon Research Scientists' Conceptions of Scientific Inquiry. In Proceedings of the Annual International Conference of the Association for the Education of Teachers in Science (2002).
200. McComas, W. Nature of Science in Science Instruction: Rationales and Strategies. Switzerland: Springer (2020).
201. McComas, W. F. Ten myths of science: Reexamining what we think we know about the nature of science. School Science and Mathematics. 96 (1), 10–16 (1996).
202. Emden, M. Reintroducing "the" scientific method to introduce scientific inquiry in schools? Science & Education. 30, 1037–1073 (2021).
203. O'Malley, M.; K. Elliot; C. Haufe; R. Burian. Philosophies of funding. Cell. 138, 611–615 (2009).
204. Bacon, F. Novum Organum (Ed. B. Fowler). Oxford: Clarendon Press (English version, 1878) (1620).
205. Znaniecki, F. The Subject Matter and Tasks of the Science of Knowledge (English Transl. 1982). In B. Walentynowicz (Ed.), Polish Contributions to the Science of Science. Dordrecht: Reidel Publishing Company (1923).
206. Galton, F. English Men of Science: Their Nature and Nurture. London: MacMillan (1874).
207. Leroy, F. (Ed.). A Century of Nobel Prize Recipients: Chemistry, Physics, and Medicine. New York: Marcel Dekker (2003).
208. Schlagberger E.; L. Bornmann; J. Bauer. At what institutions did Nobel laureates do their prize-winning work? An analysis of biographical information on Nobel laureates from 1994 to 2014. Scientometrics. 109 (2), 723–767 (2016).
209. Bjork, S.; A. Offer; G. Söderberg. Time series citation data: the Nobel Prize in economics. Scientometrics. 98, 185–196 (2014).
210. Zeng X. H. T.; et al. Differences in collaboration patterns across discipline, career stage, and gender. PLoS Biology. 14 (11), e1002573 (2016).
211. Lunnemann, P.; M. H. Jensen; L. Jauffred. Gender bias in Nobel prizes. Palgrave Communication. 5, 46 (2019).
212. Danús, L.; C. Muntaner; A. Krauss; M. Sales-Pardo; R. Guimerà. Differences in collaboration structures and impact among prominent researchers in Europe and North America. EPJ Data Science. (2023).
213. King, D. A. The scientific impact of nations. Nature. 430, 311–316 (2004).
214. Scellato, G.; C. Franzoni; P. Stephan. A mobility boost for research. Science. 356 (6339), 694 (2017).
215. Ioannidis, J. P.; et al. International ranking systems for universities and institutions: a critical appraisal. BMC Medicine. 5, 30 (2007).
216. Krauss, A.; L. Danús; M. Sales-Pardo. Early-career factors largely determine the future impact of prominent researchers: evidence across eight scientific fields. Nature Scientific Reports. 13, 18794 (2023).
217. Encyclopaedia Britannica. Svante Pääbo. Encyclopaedia Britannica. https://www.britannica.com/biography/Svante-Paabo (2023).
218. Richard E. G.; et al. A Draft Sequence of the Neandertal Genome. Science. 328, 710–722 (2010).

219. Encyclopaedia Britannica. Hermann von Helmholtz. Encyclopaedia Britannica. https://www.britannica.com/biography/Hermann-von-Helmholtz (2023).
220. Hellyer, M. Editor's Introduction: What was the Scientific Revolution? In Marcus Hellyer (ed). The Scientific Revolution: The Essential Readings. London, UK: Wiley-Blackwell (2003).
221. Rabesandratana, T. Age is an advantage. Science. (2014).
222. Encyclopedia Britannica. Alan Turing. Encyclopedia Britannica. https://www.britannica.com/biography/Alan-Turing (2025).
223. QS World University Rankings. QS World University Rankings. London: QS World University Rankings. www.topuniversities.com/university-rankings/world-university-rankings/2021 (2021).
224. Federal Reserve Economic Data. Investment in Government Fixed Assets. Federal Reserve Bank of St. Louis. https://fred.stlouisfed.org/release/tables?rid=356&eid=145951#snid=145952 (2020).
225. NSF. Survey of Earned Doctorates. National Science Foundation; National Center for Science and Engineering Statistics. https://ncses.nsf.gov/pubs/nsf21308/data-tables#group1 (2020).
226. Sun, X.; et al. Social dynamics of science. Scientific Reports. 3, 1069 (2013).
227. Frickel, S.; N. Gross. A general theory of scientific/intellectual movements. American Sociological Review. 70 (2), 204–232 (2005).
228. Schizas, D.; D.; Psillos; G. Stamou. Nature of science or nature of the sciences? Science Education. 100, 706–733 (2016).
229. Smith, V. L. An experimental study of competitive market behavior. Journal of Political Economy. 70 (2), 111–137 (1962).
230. Frisch, R.; F. V. Waugh. Partial time regressions as compared with individual trends. Econometrica. 1 (4), 387–401 (1933).
231. Gigerenzer, G.; D. Murray. The Inference Revolution. In Cognition as Intuitive Statistics. Hove: Psychology Press (1987).
232. Encyclopaedia Britannica. Sir Ronald Aylmer Fisher. Encyclopaedia Britannica. https://www.britannica.com/biography/Ronald-Aylmer-Fisher (2023).
233. Dawkins, R. River Out of Eden: A Darwinian View of Life. New York: Basic books (1995).
234. Krauss, A. Why all randomised controlled trials produce biased results. Annals of Medicine. 50 (4), 312–322 (2018).
235. Nobel Prize. Award Ceremony Speech in Chemistry. Nobel Prize. www.nobelprize.org/prizes/chemistry/1952/ceremony-speech/ (1952).
236. Nobel Prize. Richard L.M. Synge – Biographical, and Nobel Lecture. Nobel Prize. https://www.nobelprize.org/prizes/chemistry/1952/synge/ (1952).
237. Martin, A.; R. Synge. A new form of chromatogram employing two liquid phases. The Journal of Biochemistry. 35 (12), 1358–1368 (1941).
238. Mertz, J. Introduction to Optical Microscopy (2nd ed.). Cambridge: Cambridge University Press (2019).
239. Hepburn, Brian; Hanne Andersen. Scientific Method. In N. Z. Edward (Ed.). The Stanford Encyclopedia of Philosophy (Summer 2021 Edition) (2021).
240. Bloch, F.; W. Hansen; M. Packard. The nuclear induction experiment. Physical Review Journals. 70, 474 (1946).
241. Nobel Prize. Ernest Lawrence – Biographical, and Nobel Lecture. Nobel Prize. www.nobelprize.org/prizes/physics/1939/lawrence/ (1939).
242. Forman, P. Inventing the maser in postwar America. Osiris. 7, 105–134 (1992).
243. American Institute of Physics. Interview of Felix Bloch by Charles Weiner on 1968 August 15. College Park: American Institute of Physics. www.aip.org/history-programs/niels-bohr-library/oral-histories/4510 (1968).
244. Lawrence Livermore National Laboratory. 'Atom Smasher' Taught Science World to Think Big. Livermore: Lawrence Livermore National Laboratory. https://www.llnl.gov/news/'atom-smasher'-taught-science-world-think-big (2001).
245. Heilbron, J. L.; R. Seidel. Lawrence and His Laboratory: A History of the Lawrence Berkeley Laboratory. Berkeley: University of California Press, Vol. I, pp. 312 (1989).

246. Nobel Prize. Charles H. Townes – Biographical, and Nobel Lecture. Nobel Prize. https://www.nobelprize.org/prizes/physics/1964/townes/ (1964).
247. Willcox, W. The founder of statistics. Review of the International Statistical Institute. 5 (4), 321–328 (1938).
248. Dondelinger, R. Centrifuges: Take a Spin. Biomedical Instrumentation & Technology (2010).
249. Gil, Y.; M. Greaves; J. Hendler; H. Hirsh. Artificial intelligence. Amplify scientific discovery with artificial intelligence. Science. 346 (6206), 171–172 (2014).
250. Piwowar, H; et al. The state of OA: a large-scale analysis of the prevalence and impact of Open Access articles. Peer-Reviewed Journal. 6, e4375 (2018).
251. Center for Open Science. Center for Open Science. Washington DC. www.cos.io (2025).
252. Eisenberg, R. Academic freedom and academic values in sponsored research. Texas Law Review. 66; 1363 (1988).
253. Einstein, A. On the electrodynamics of moving bodies (Zur Elektrodynamik bewegter Körper). Annalen der Physik. 17 (10), 891–921 (1905).
254. Janssen, M; M. Mecklenburg. From Classical to Relativistic Mechanics: Electromagnetic Models of the Electron. In Interactions: Mathematics, Physics and Philosophy, 1860–1930. Dordrecht: Springer (2007).
255. Boughn, Stephen. Hasenöhrl and the equivalence of mass and energy. European Physical Journal H. (2013).
256. Bartocci, U. Albert Einstein e Olinto De Pretto. La vera storia della formula più famosa del mondo. (Transl 'Albert Einstein and Olinto De Pretto. The true history of the most famous formula in the world'). Bologna: Andromeda, (1999).
257. Einstein, A. Das Prinzip von der Erhaltung der Schwerpunktsbewegung und die Trägheit der Energie. Annalen der Physik. 20 (8), 627–633 (1906).
258. Einstein, A. On the inertia of energy required by the relativity principle (Original: Über die vom Relativitätsprinzip geforderte Trägheit der Energie). Annalen der Physik. 23, 371–384 (1907).
259. Romer, P. Endogenous technological change. Journal of Political Economy. 98 (5), S71–102 (1990).
260. Solow, R. Technical change and the aggregate production function. The Review of Economics and Statistics. 39 (3), 312–320 (1957).
261. Schumpeter, J. Capitalism, Socialism and Democracy. New York: Harper, 3rd ed. (1942/2008).
262. Merton, R. Science and technology in a democratic order. Journal of Legal and Political Sociology. 1, 115–126 (1942).
263. Newton, I. Philosophiæ Naturalis Principia Mathematica. London: Pepys (in English 1728) (1687).
264. Snow, J. On the Mode of Communication of Cholera. London: John Churchill (1855).
265. Krauss, A. How nobel-prize breakthroughs in economics emerge and the field's influential empirical methods. Journal of Economic Behavior and Organization. 221, 657–674, (2024).
266. Reeder, Blaine; Alexandria David. Health at hand: A systematic review of smart watch uses for health and wellness. Journal of Biomedical Informatics. 63, 269–276, (2016).
267. Glimcher, P; E. Fehr. Neuroeconomics: Decision Making and the Brain. New York: Academic Press (2014).
268. LeCun, Y.; Y. Bengio; G. Hinton. Deep learning. Nature. 521 (7553), 436–444 (2015).
269. Litjens, G.; et al. A survey on deep learning in medical image analysis. Medical Image Analysis. 42, 60–88 (2017).
270. Nobel Prize. The Svedberg – Biographical, and Nobel Lecture. Nobel Prize. www.nobelprize.org/prizes/chemistry/1926/svedberg/ (1926).
271. Claesson, S.; K. Pedersen. The Svedberg. 1884-1971. Biographical Memoirs of Fellows of the Royal Society. 18, 595–627 (1972).
272. Nobel Prize. Arne Tiselius – Biographical, and Nobel Lecture. Nobel Prize. https://www.nobelprize.org/prizes/chemistry/1948/tiselius/ (1948).
273. Hjertén, S. The history of the development of electrophoresis in Uppsala. Electrophoresis. 9, 3–15 (1988).

274. Krauss, A. Multi-studies: A Novel Approach to Addressing Irreplicability in RCTs. In MacDougall (Ed.), Gold Standard or Unhealthy Fixation? A Medical Educator's Guide to Current Thinking on the Status of Randomized Controlled Trials in Clinical Research. Cham, Switzerland: Springer Press, 163–180 (2023).
275. Melo, M.; J. Maasch; C. de la Fuente-Nunez. Accelerating antibiotic discovery through artificial intelligence. Communication Biology. 4, 1050 (2021).
276. Laland, K. Darwin's Unfinished Symphony: How Culture Made the Human Mind. Princeton: Princeton University Press (2017).
277. Henrich, J. The Secret of Our Success: How Culture is Driving Human Evolution, Domesticating our Species, and Making us Smarter. Princeton: Princeton University Press (2016).
278. Tomasello, M. Becoming Human: A Theory of Ontogeny. Cambridge: Harvard University Press (2019).
279. Diamond, J. Guns, Germs, and Steel: The Fates of Human Societies. New York: W.W. Norton & Company (1997).
280. Diamond, J. Collapse: How Societies Choose to Fail or Succeed. New York: Viking, Penguin group (2005).
281. Zilsel, E. The Social Origins of Modern Science (Eds. K. Raven; Cohen). Dordrecht: Kluwer Academic Publishers (2003).
282. Merton, R. Science, technology and society in seventeenth century England. Osiris. 4, 360–632 (1938).
283. Weber, M. The Protestant Ethic and the Spirit of Capitalism (Trans. 1930). London and Boston: Unwin Hyman (1905).
284. Eisenstein, E. The Printing Press as an Agent of Change. Cambridge: Cambridge University Press (1980).
285. Needham, J. Foreword. In K. Raven; Cohen (Eds.), Zilsel; The Social Origins of Modern Science. Dordrecht: Kluwer Academic Publishers, xi–xv, (2003).
286. Oxford English Dictionary. Civilisation. Oxford: Oxford English Dictionary. www.lexico.com/definition/civilization (2022a).
287. Sterelny, K. The Evolved Apprentice: How Evolution Made Humans Unique. Cambridge, MA: MIT Press (2012).
288. Mithen, S. Human Evolution and the Cognitive Basis of Science. In Carruthers P, Stich S, Siegal M, eds. The Cognitive Basis of Science. Cambridge: Cambridge University Press (2002).
289. Thorndike, E. Animal Intelligence. Darien: Hafner Publishing Company, 1911 (1970).
290. de Waal, F. Are We Smart Enough to Know How Smart Animals Are? New York: W. W. Norton & Company (2016).
291. McGrew, W. Chimpanzee Material Culture. Cambridge: Cambridge University Press (1992).
292. Boesch, C.; H. Boesch. Possible causes of sex differences in the use of natural hammers by wild chimpanzees. Journal of Human Evolution. 13, 415 (1984).
293. Morell, V. Watch this genius bird plan for its 'handyman' job. Science. (2023).
294. Harmand, S.; et al. 3.3-million-year-old stone tools from Lomekwi 3, West Turkana, Kenya. Nature. 521, 310–315 (2015).
295. Susman, R. Who made the Oldowan tools? Fossil evidence for tool behaviour in Plio-Pleistocene hominids. Journal of Anthropological Research. 47, 129–151, (1991).
296. Gowlett, J. Mental Abilities of Early Man: A Look at Some Hard Evidence. In R. Foley (Ed.), Hominid Evolution and Community Ecology. Cambridge, MA: Academic Press, 199–220, (1984).
297. Carruthers, P. The Roots of Scientific Reasoning: Infancy, Modularity and the Art of Tracking. In Carruthers P, Stich S, Siegal M, eds. The Cognitive Basis of Science. Cambridge, UK: Cambridge University Press, 73–96, (2002).
298. Proctor, R; E. Capaldi. Psychology of Science: Implicit and Explicit Processes. Oxford: Oxford University Press (2012).
299. Herrmann, E.; et al. Humans have evolved specialized skills of social cognition: the cultural intelligence hypothesis. Science. 317, 1360 (2007).
300. Giere, R. Explaining Science: A Cognitive Approach. Chicago: University of Chicago Press (1988).

301. Freeden, W.; C. Heine; Z. Nashed. An Invitation to Geomathematics. Cham: Springer Press (2019).
302. Wilkins, R. Neurosurgical Classics. Park Ridge: American Association of Neurological Surgeons (2nd ed.) (1992).
303. Chalmers, I.; E. Dukan; S. Podolsky; G. Davey Smith. The advent of fair treatment allocation schedules in clinical trials during the 19th and early 20th centuries. Journal of the Royal Society of Medicine. 105, 221–227 (2012).
304. Peirce, C. S. Illustrations of the logic of science. The Popular Science Monthly. (1877).
305. Nisbett, R.; K. Peng; I. Choi; A. Norenzayan. Culture and systems of thought: holistic vs. analytic cognition. Psychological Review. 108 (2), 291–310 (2001).
306. Logan, R. The Alphabet Effect. New York: Morrow (1986).
307. Williams, L. P. History of Science. Britannica. www.britannica.com/science/history-of-science (2017).
308. Needham, J. Science and civilisation in China, Vol: 3, pp. 154, 167–168. Cambridge: Cambridge University Press (1956).
309. Bridgman, Roger. 1000 Inventions and Discoveries. Noida: DK Publishing (Dorling Kindersley) (2014).
310. Barnes, J.; et al. (Eds.). Philosophy and Science in the Hellenistic World. Cambridge: Cambridge University Press (1982).
311. Eratosthenes. Eratosthenes' Geography. Princeton University Press (Transl. Duane Roller) (2010).
312. Al-Haytham, I. The Book of Optics. In A. Sabra (Ed.). The Optics of Ibn al-Haytham. Vol. 1 and 2). Kuwait: The National Council for Culture, Arts and Letters (1021/2002).
313. Al-Khalili, J. In retrospect: Book of optics. Nature. 518, 164–165 (2015).
314. Encyclopaedia Britannica. Encyclopedias and Dictionaries. Encyclopaedia Britannica. Vol. 18 (15th ed.), pp. 257–286 (2007).
315. Evans, G. The conclusiones of Robert Grosseteste's commentary on the posterior analytics. Studi medievali ser. 3 (24), 729–734, (1983).
316. Bacon, R. Opus maius. Oxford (Ed. J. H. Bridges 1897). Oxford: Clarendon press (1267).
317. Aquinas, T. Questiones Disputatae de Veritate. Questions 1-9, (transl. Mulligan). Chicago: Henry Regnery Company (1256-1259/1952).
318. Galilei, G. Il Saggiatore (in Italian); The Assayer (English Translation). In Stillman Drake and C. D. O'Malley. Rome (1623).
319. Cohen, H. F. The Scientific Revolution: A Historiographical Inquiry. Chicago: University of Chicago Press (1994).
320. Principe, L. The Scientific Revolution: A Very Short Introduction. Oxford: Oxford University Press (2011).
321. Chia, L.; H. De Weerdt. Knowledge and Text Production in an Age of Print: China, 900-1400. Leiden: Brill (2011).
322. Henry, J. The Scientific Revolution and the Origins of Modern Science. Houndsmill: Palgrave Macmillan (3rd ed.) (2008).
323. Aristotle. Physics. In The Works of Aristotle, Vol. 2. (Ed. W. D. Ross). Oxford: Clarendon Press (1930/300s BCE).
324. Montelpare, W.; E. Read; T. McComber; A. Mahar; K. Ritchie. John Snow and the Natural Experiment. In Applied Statistics in Healthcare Research. Montreal: Pressbooks (2021).
325. Gascoigne, R. The historical demography of the scientific community, 1450–1900. Social Studies of Science. 22 (3), 545–573 (1992).
326. Saliba, G. Islamic Science and the Making of the European Renaissance. Cambridge, MA: MIT Press (2007).
327. Duhem, P. Le système du monde: histoire des doctrines cosmologiques de Platon à Copernic (The System of World: A History Cosmological Doctrines from Plato to Copernicus). Paris: Hermann (1913).
328. Henn, B.; L. Cavalli-Sforza; M. Feldman. The great human expansion. Proceedings of the National Academy of Sciences of the United States of America. 109, 17758–17764 (2012).

329. Stigler, S. M. The History of Statistics: The Measurement of Uncertainty before 1900. Cambridge, MA: Belknap Press/Harvard University Press (1986).
330. Popper, K. Normal Science and its Dangers. In I. Lakatos; A. Musgrave (Eds.), Criticism and the Growth of Knowledge. Cambridge, UK: Cambridge University Press, 51–58, (1970).
331. Kuhn, T. Logic of Discovery or Psychology of Research? In P. A. Schilpp, The Philosophy of Karl Popper, The Library of Living Philosophers, Vol. xiv, Book ii, pp. 798–819. La Salle: Open Court.(1974).
332. Hansson, S. O. Science and Pseudo-Science. The Stanford Encyclopedia of Philosophy (Fall Edition). (Ed. E. N. Zalta) (2021).
333. Hume, D. A Treatise of Human Nature. London: John Noon (1739).
334. Jerison, H. Evolution of the Brain and Intelligence. New York: Academic Press (1973).
335. Tooby, J.; Cosmides, L. The Theoretical Foundations of Evolutionary Psychology. In D. M. Buss (Ed.), The Handbook of Evolutionary Psychology, Vol. 1, 2nd ed. New Jersey: Wiley, 3–87, (2016).
336. Shettleworth, S. Cognition, Evolution and Behaviour. Oxford University Press (2nd ed.) (2010).
337. Darwin, C. The Descent of Man, and Selection in Relation to Sex. London: John Murray (1871).
338. Wallace, A. R. Contributions to the Theory of Natural Selection: A Series of Essays. Macmillan and Company (1870).
339. Boyd, R.; P. Richerson. Why culture is common, but cultural evolution is rare. Proceedings of the British Academy. 88, 77–93 (1996).
340. Pinker, S. How the Mind Works. New York: W. W. Norton & Company, 1997 first ed. (2009).
341. Krenn, M.; et al. On scientific understanding with artificial intelligence. Nature Reviews Physics. 4, 761–769 (2022).
342. Carruthers, P.; S. Stich; M. Siegal. Introduction: What makes science possible. In Carruthers P, Stich S, Siegal M, eds. The Cognitive Basis of Science. Cambridge: Cambridge University Press, 1–20, (2002).
343. Gigerenzer, G.; H. Brighton. Homo heuristicus: why biased minds make better inferences. Topics in Cognitive Science. 1, 107–143 (2009).
344. Hutmacher, F. Why is there so much more research on vision than on any other sensory modality? Frontiers in Psychology. (2019).
345. Mercier, H.; D. Sperber. The Enigma of Reason. Cambridge: Harvard University Press (2017).
346. Science. Special Issue: What Don't We Know? In D. Kennedy; C. Norman (Ed.), Science, 1 July, Vol. 78, pp. 309 (2005).
347. Weeks, M. E. The discovery of the elements. XIII. Some spectroscopic discoveries. Journal of Chemical Education. 9 (8) (1932).
348. Mumford, S.; M. Tugby (Eds). Metaphysics and Science. Oxford: Oxford University Press (2013).
349. Schrenk, M. Metaphysics of Science: A Systematic and Historical Introduction. London and New York: Routledge (2017).
350. Gleiser, M. The Island of Knowledge: The Limits of Science and the Search for Meaning. New York: Basic Books (2014).
351. du Sautoy, M. What We Cannot Know: Explorations at the Edge of Knowledge. London: 4th Estate (2016).
352. Barrow, J. Impossibility: The Limits of Science and the Science of Limits. Oxford: Oxford University Press, (1999).
353. Yanofsky, N. The Outer Limits of Reason: What Science, Mathematics, and Logic Cannot Tell Us. Cambridge, MA: The MIT Press (2013).
354. Fodor, J. The Modularity of Mind: An Essay on Faculty Psychology. Cambridge: MIT Press (1983).
355. Chomsky, N. New Horizons in the Study of language and Mind. Cambridge: Cambridge University Press (2000).
356. Chomsky, N. Science, Mind, and Limits of Understanding. In A. J. Gallego; R. Martin (Eds.), Language, Syntax, and the Natural Sciences. Cambridge: Cambridge University Press (2018).
357. McGinn, C. Problems in Philosophy: The Limits of Inquiry. Oxford: Blackwell (1993).

358. Cartwright, N. The Dappled World: A Study of the Boundaries of Science. New York: Cambridge University Press (1999).
359. Williamson, T. Knowledge and its Limits. Oxford: Oxford University Press (2002).
360. Rescher, N. The Limits of Science. Pittsburgh: University of Pittsburgh Press (1999).
361. Pigliucci, M.; M. Boudry (Eds.). Philosophy of Pseudoscience: Reconsidering the Demarcation Problem. Chicago, IL: University Chicago Press (2013).
362. Horgan, J. The End of Science: Facing the Limits Of Knowledge in the Twilight of the Scientific Age. New York: Basic books (2015).
363. Nobel Prize. Press Release; Physics. Nobel Prize. https://www.nobelprize.org/prizes/physics/1986/press-release/ (1986).
364. Tweney, R.; M. Doherty; C. Mynatt. On Scientific Thinking, pp. 325–326. New York: Columbia University Press (1981).
365. National Science Foundation (NSF). Federal Funds for R&D Series. American Association for the Advancement of Science (AAAS). https://www.aaas.org/programs/r-d-budget-and-policy/research-science-and-engineering-discipline (2019).
366. Stephan, P. The economics of science. Journal of Economic Literature. 34 (3), 1199–1235 (1996).
367. Partha, D.; P. David. Toward a new economics of science. Research Policy. 23 (5), 487–521 (1994).
368. Merton, R. Behavior Patterns of Scientists, pp. 213–220. Leonardo: MIT Press (1970).
369. Harwood, J. Styles of Scientific Thought: the German Genetics Community 1900-1933. Chicago: University of Chicago Press (1993).
370. Tränkle, G. High Power Laser Diodes: Improvements in Power, Efficiency, and Brilliance. In Proc. SPIE 10900, High-Power Diode Laser Technology XVII (2019).
371. de Jonge, N.; D. Peckys. Live Cell Electron Microscopy Is Probably Impossible. ACS Nano 2016 10 (10), 9061–9063, (2016).
372. Schomburg, G. Practical limitations of capillary gas chromatography. Journal of High Resolution Chromatography. 2, 461–474 (1979).
373. Rabilloud, T; A. R. Vaezzadeh; N. Potier; C. Lelong; E. Leize-Wagner; M. Chevallet. Power and limitations of electrophoretic separations in proteomics strategies. Mass Spectrometry Reviews. 28 (5), 816–843 (2009).
374. Rosado, M.; et al. Chapter Four - Advances in Biomarker Detection: Alternative Approaches for Blood-Based Biomarker Detection. In G. Makowski (Ed.), Advances in Clinical Chemistry. London: Elsevier, Vol. 92, pp 141–199 (2019).
375. Bunaciu, A.; E. Udriştioiu; H. Aboul-Enein. X-ray diffraction: instrumentation and applications. Critical Reviews in Analytical Chemistry. 45 (4), 289–299 (2015).
376. Emwas, A. H. M.; et al. NMR-based metabolomics in human disease diagnosis: applications, limitations, and recommendations. Metabolomics. 9, 1048–1072 (2013).
377. Hagart-Alexander, C. Temperature Measurement. In B. Noltingk (Ed.), Measurement of Temperature and Chemical Composition (Fourth Edition), pp. 1–66. Oxford: Butterworth-Heinemann (1992).
378. The Gale Encyclopedia of Science. 'Telescope' (Eds. H. K. Katherine; J. L. Longe). Farmington Hills, MI: Gale, a Cengage Company (2021).
379. Clery, D. 'Lego' chemistry that joins molecules wins this year's Nobel Prize. Science. (2022).
380. Krause, J.; et al. The complete mitochondrial DNA genome of an unknown hominin from southern Siberia. Nature. 464, 894–897 (2010).
381. Nobel Prize. Press Release: The Nobel Prize in Physics 2020. Nobel Prize. https://www.nobelprize.org/prizes/physics/2020/press-release/ (2020a).
382. Del Sole, R.; P. Chiaradia. Surfaces, Optical Properties of. (Eds. F. Bassani; G. L. Liedl; P. Wyder), Encyclopedia of Condensed Matter Physics, pp. 144–150. London: Elsevier (2005).
383. Dong, M. Modern HPLC for Practicing Scientists, pp. 5. Hoboken, NJ: John Wiley & Sons (2006).
384. Percival, S.; P. Wyn-Jones. Chapter Twenty-Two - Methods for the Detection of Waterborne Viruses. In S. L. Percival;et al. (Eds.), Microbiology of Waterborne Diseases (Second Edition), pp. 443–470. Cambridge, MA: Academic Press (2014).

385. Takeuchi, N.; S. Saheki. Evaluation and problems of ultracentrifugal technique for separation and analysis of serum lipoproteins: comparison with other analytical methods. The Japanese journal of clinical pathology. 41 (7), 750–758 (1993).
386. Raval, N.; et al. Importance of Physicochemical Characterization of Nanoparticles in Pharmaceutical Product Development. In R. Tekade (Ed.), Advances in Pharmaceutical Product Development and Research, Basic Fundamentals of Drug Delivery. Cambridge, MA: Academic Press, pp 369–400 (2019).
387. Nobel Prize. The Nobel Prize in Physiology or Medicine, 1901-2000. Nobel Prize. (Lindsten, J.; N. Ringertz). www.nobelprize.org/prizes/themes/the-nobel-prize-in-physiology-or-medicine-1901-2000/ (2001).
388. Maddison Project Database. Maddison Project Database. (developed by: Bolt, J.; R. Inklaar; H. de Jong; J. L. van Zanden). www.ggdc.net/maddison (2018).
389. Sherby, L. The Who's Who of Nobel Prize Winners 1901–2000. London: Oryx Press (2002).
390. Krauss, A. Discovery tools: How powerful new scientific methods and instruments emerge and catalyse innovation. Forthcoming (2026).

Index

Online Appendix

An online appendix with supplementary material is provided in the online edition of this book. The online appendix includes a methods section that outlines greater details on the data. It also contains conceptual frameworks and over a dozen figures that provide additional analyses.

Please visit Oxford Academic https://academic.oup.com/ and search this book's ISBN: 9780197829790